Optimization of Integer/Fractional Order Chaotic Systems by Metaheuristics and their Electronic Realization

Esteban Tlelo-Cuautle
Instituto Nacional de Astrofísica, Óptica y Electrónica
Tonantzintla, Puebla, Mexico

Luis Gerardo de la Fraga
Centro de Investigación y de Estudios Avanzados
Mexico City, Mexico

Omar Guillén-Fernández
Instituto Nacional de Astrofísica, Óptica y Electrónica
Tonantzintla, Puebla, Mexico

Alejandro Silva-Juárez
Instituto Nacional de Astrofísica, Óptica y Electrónica
Tonantzintla, Puebla, Mexico

CRC Press
Taylor & Francis Group
Boca Raton London New York

CRC Press is an imprint of the
Taylor & Francis Group, an **informa** business
A SCIENCE PUBLISHERS BOOK

First edition published 2021
by CRC Press
6000 Broken Sound Parkway NW, Suite 300, Boca Raton, FL 33487-2742

and by CRC Press
4 Park Square, Milton Park, Abingdon, Oxon OX14 4RN

ISBN-13: 978-0-367-48668-6 (hbk)

Typeset in Times New Roman
by Radiant Productions

Esteban wants to dedicate this book to his beloved family,

Luis Gerardo wants to dedicate this book to his son Alexei: "Try to understand this world, my son",

Omar wants to dedicate this book to his beloved family. I am especially thankful to my parents and siblings, whose have been a living example in life; I extend my gratitude to all my partners and friends,

Alejandro wants to dedicate this book to his wife Claudia and sons: Alejandro, Abraham and Aarón.

Preface

Recently, researchers around the world have introduced different chaotic systems that are modeled by integer or fractional-order differential equations, and whose mathematical models can generate chaos or hyper-chaos. The numerical methods to simulate those integer and fractional-order chaotic systems are quite different and their exactness is responsible in the evaluation of characteristics such as Lyapunov exponents, Kaplan-Yorke dimension, and entropy. One challenge is estimating the step-size to run a numerical method. It can be done analyzing the eigenvalues of self-excited attractors, while for hidden attractors it becomes difficult to evaluate the equilibrium points that are required to formulate the Jacobian matrices. Time simulation of fractional-order chaotic oscillators also requires estimating a memory length to achieve exact results, and it is associated with memories in hardware design. In this manner, simulating chaotic/hyper-chaotic oscillators of integer/fractional-order and with self-excited/hidden attractors is quite important to evaluate their Lyapunov exponents, Kaplan-Yorke dimension and entropy. Further, to improve the dynamics of the oscillators, their main characteristics can be optimized applying metaheuristics, which basically consists of varying the values of the coefficients of a mathematical model. The optimized models can then be implemented using commercially available amplifiers, field-programmable analog arrays (FPAA), field-programmable gate arrays (FPGAs), microcontrollers, graphic processing units, and even using nanometer technology of integrated circuits.

This book details the application of different numerical methods to simulate integer/fractional-order chaotic systems. Those methods are used within optimization loops to maximize positive Lyapunov exponents, Kaplan-Yorke dimension, and entropy. Single and multi-objective optimization approaches applying metaheuristics are described, as well as their tuning techniques to generate feasible solutions that are suitable for electronic implementation. The book details several applications of chaotic oscillators such as in random bit/number generators, cryptography, and secure communications that are quite useful in robotics and the Internet of Things.

Contents

Chapter 1

Numerical Methods

1.1 Introduction

Historically, the mathematical models associated with continuous-time dynamical systems have originated from several fields of science, such as: physics, chemistry, biology, geology and other applied engineering areas. It is well-known that in the real-world the majority of those mathematical models have a non-linear nature and therefore their behavior can be modeled by the formulation of systems of non-linear ordinary differential equations (ODEs) whose analytical solution is complex and takes a long time or sometimes is impossible to find. An alternative for the solution of these dynamical systems is the approximation of the mathematical models that generate non-linear phenomena through the discretization of the ODEs by applying numerical methods.

In the last years, a kind of continuous-time dynamical systems that have been used to improve the secure characteristics of communication systems are the chaotic dynamical systems. In particular, the field of chaos control and synchronization employs chaotic oscillators to develop chaotic secure communication systems, which allows the implementation of cryptographic applications to transmit information by preserving the privacy and security against attacks during the interchange of data that can be text, images, audio and video [1, 2, 3]. These applications depend on the dynamical characteristics of the chaotic oscillators. For instance, the main definitions associated with the characteristics of the chaotic dynamical systems agree that they posses a long-term aperiodic behavior, they are deterministic systems that exhibit high dependence and become extremely sensitive to initial conditions. The dynamical model of a chaotic oscillator is deterministic because its parameters are known; besides, as the time increases ($t \to \infty$) its behavior is difficult to distinguish, and the trajectories of the state variables do not tend to a fixed point, periodic, or quasi-periodic orbits. A chaotic oscillator has the same number of Lyapunov exponents as the number of ODEs, and the chaotic behavior exits when it has at least one positive Lyapunov exponent. Chaotic oscillators generate strange attractors that have a fractal dimension that can be estimated from the evaluation of the Lyapunov exponents. The majority of researchers agree that the most important characteristic of a chaotic oscillator is the high sensitive dependence to initial conditions, which means that observing one state from two initial conditions with infinitesimal difference, then the resulting state trajectories will diverge exponentially as time increases, but the range of values is bounded [4].

The advantage of a chaotic oscillator to have high sensitivity to initial conditions, makes it a good candidate for the development of applications, such as: the design of random number generators (RNG) [5], which can generate pseudo random sequences (PRNG) [6], or true random sequences (TRNG) [7]; the development of image encryption systems [8], cryptographic encryption algorithms [9], security of biometric models [10], among others [11, 12, 13].

Nowadays, the chaotic oscillators can be modeled by ODEs of integer and fractional order [4]. The most well known chaotic oscillators are Lorenz, Rössler, Chua, Chen, Lü, among others. In their integer-order version, they are solved by applying one-step, multi-step or special numerical methods. The challenge is the selection of the best numerical method for each type of chaotic oscillator, and the time-step (h) to guarantee the convergence and stability of the solution. In the case of solving fractional-order chaotic oscillators, one should apply different definitions to solve the fractional derivative, and among them the most well known are the Caputo, Grünwald Letnikov and Adams Bashforth-Moulton (predictor-corrector) approximations. These methods require an adequate h and in addition, since they base the solution on memory, the challenge is the estimation of the memory size to develop electronic implementations. Other issues are oriented to ensure the exactness of the numerical method to diminish computational chaos or superstability [14].

1.2 Numerical methods for ODEs of integer order

The chaotic oscillators that are case study herein are autonomous, so that their solution depends on initial conditions. The mathematical models of the chaotic systems are described by ODEs that include nonlinear functions or multiplications among their state variables, and they are formulated as initial value problems. The number of ODEs determine the dimension of the chaotic dynamical system and they can also be modeled in higher orders performing transformations as described in [15]. An initial value problem must accomplish the continuity of the non-linear functions in some interval $a \leq x \leq b$, and it must guarantee the principle of existence and uniqueness, as detailed in [16].

The analytical solution of ODEs that are modeling an autonomous chaotic oscillator is impossible to find in almost all the cases because of the complexity of the system. In this manner, the most simple way to get the solution is by performing an approximation of their dynamics by applying numerical methods. The ODEs are then discretized according to the numerical method, which remains an open problem to find the most suitable method that guarantees exactness and convergence of the solution. In addition, one must verify the stability of the method by an appropriate estimation of the step-size h. Higher order systems can be solved by choosing among one-step, multi-step or special numerical methods [17]. It is well-known that for linear dynamical systems, h can be determined from the evaluation of the eigenvalues, which are related to the natural frequencies of the system. On the contrary, in non-linear dynamical systems, the problem is quite complex so that one needs to analyze the initial value problem of the form $\dot{x} = f(x,t)$, which solution depends on the appropriate selection of the initial condition given by $x(0) = \eta$, and the estimation of the step-size $h = t_{n+1} - t_n$. In addition, in some non-linear problems, h can vary and can be adapted to reduce numerical errors.

The main goal of a numerical method is devoted to provide a good approximation of the behavior of a continuous-time dynamical system, modeled by ODEs, through a computational method in a digital machine. During the solution of the ODEs, the errors with respect to the ideal or correct solution of the initial value problem can be associated with the truncation of the computer arithmetic, the truncation of higher-order terms in the numerical method, the round-off by floating point arithmetic and the type of method, e.g., one-step and multi-step. The numerical solution by applying one-step and multi-step methods is found at discrete points related to h and therefore the fitting of these points provides a different numerical error when comparing to the theoretical solution in the continuous interval $a \leq x \leq b$, where the bounds a and b are finite [16].

Some one-step and multi-step methods can be derived from polynomials. For instance, the general linear multi-step methods can be described by (1.1), where the coefficients α_j and β_j are constants. Assuming that $\alpha_k = 1$ and not both α_0 and β_0 are zero, then if $\beta_k = 0$ the derived method is said to be explicit, otherwise, if $\beta_k \neq 0$ the derived method is said to be of implicit type. In both cases, the derived methods iterate from initial conditions and the evaluation of the next iteration depends on past discrete values.

$$\sum_{j=0}^{k} \alpha_j x_{n+j} = h \sum_{j=0}^{k} \beta_j f(x_{n+j}, t_{n+j}) \tag{1.1}$$

When an explicit numerical method computes the next value x_{n+k}, one must use just the past values x_{n+j}, $f(x_{n+j}, t_{n+j})$, with $j = 0, 1, ..., k-1$. Those past values must have already been computed previously. By contrast, the implicit methods compute x_{n+k} by updating values at the same iteration index (using an explicit method) and past values to approach the solution by $x_{n+k} = h\beta_k f(x_{n+k}, t_{n+k}) + g$, where g is a function of values previously computed at x_{n+j}, $f(x_{n+j}, t_{n+j})$. The explicit and implicit methods have a different accuracy and convergence regions that basically depend on the step-size h.

1.2.1 One-step methods

The simplest explicit and one-step method that can be derived from (1.1), is by setting $k = 1$, $\alpha_1 = \alpha_0 = \beta_0 = 1$ and $\beta_1 = 0$ to obtain the well-known Forward-Euler method whose iterative formulae is given in (1.2). Basically, the approximation to the solution is based in the evaluation of the slope that is estimated in the function being extrapolated from an actual value (at iteration n) to a next value (at iteration $n+1$). In a more detailed description, this one-step explicit method is derived from Taylor series in which the higher order terms, from the second power, are truncated. For this reason, this is the method with the lowest exactness due to the fact that the error when comparing the approximated solution with the analytical one can be very high if the step-size h is not chosen appropriately, e.g., sufficiently small.

The simplest implicit and one-step method can be derived from (1.1) by setting $k = 1$, $\alpha_1 = \alpha_0 = \beta_1 = 1$ and $\beta_0 = 0$ to obtain the well-known Backward Euler method whose formulae is given in (1.3). As one sees, it requires a preliminary computation to approximate the solution of the problem at iteration $n+1$ to evaluate $f(x_{n+1}, t_{n+1})$, and afterwards one can evaluate x_{n+1}. This computation can be performed by applying an explicit method such as Forward Euler. So that more computation operations are required, first to compute the predictive values by applying an explicit method to evaluate $f(x_{n+1}, t_{n+1})$, and afterwards one can evaluate the corresponding value x_{n+1}.

$$x_{n+1} = x_n + h \cdot f(x_n, t_n) \tag{1.2}$$

$$x_{n+1} = x_n + h \cdot f(x_{n+1}, t_{n+1}) \tag{1.3}$$

Among the different types of implicit and explicit numerical methods, the solution to the majority of problems formulated as initial value ones, can be obtained by applying Runge-Kutta algorithms, which are derived by taking more terms in the Taylor's series expansion. The most used methods go from order one to order four, and they require the same number of evaluations of the functions $f()$, as the order of the method to estimate the approximation at the same time-step. In this manner, the methods proposed by Runge and developed by Kutta and Heun, are called long pass methods and they are of one-step type [18, 19]. The fourth-order Runge-Kutta method is the most used due to its high exactness in approximating the solution of ODEs. It has the disadvantages of losing linearity, error analysis and the extra evaluations of the function to approximate mid-points. For these reasons it is considered more difficult than by applying a lineal multi-step method. There exist implicit Runge-Kutta methods whose main advantage is the improvement of the stability characteristics. For instance, Table 1.1 shows the most used explicit Runge-Kutta methods having orders from one to four.

Table 1.1: Runge-Kutta methods.

Order	Algorithm
1_{st}	$y_{j+1} = y_j + hf(y_j, t_j)$
2_{nd}	$y_{j+1} = y_j + hf(y_j + \frac{h}{2} f(y_j, t_j), t_k + \frac{h}{2})$
3_{rd}	$y_{j+1} = y_j + \frac{h}{4}(k_1 + 3k_3)$ $k_1 = f(y_j, t_j)$ $k_2 = f(y_j + \frac{hk_1}{3}, t_j + \frac{h}{3})$ $k_3 = f(y_j + \frac{2hk_2}{3}, t_j + \frac{2h}{3})$
4_{th}	$y_{j+1} = y_j + \frac{h}{6}(k_1 + 2k_2 + 2k_3 + k_4)$ $k_1 = f(y_j, t_j)$ $k_2 = f(y_j + \frac{hk_1}{2}, t_j + \frac{h}{2})$ $k_3 = f(y_j + \frac{hk_2}{2}, t_j + \frac{h}{2})$ $k_4 = f(y_j + hk_3, t_j + h)$

1.2.2 Multi-step methods

These kind of numerical methods re-use past information and then more than one discrete points are averaged to estimate the next step. Compared to the fourth-order Runge-Kutta method, which requires the evaluation of four nested functions at the same time-step to estimate the value of y_{k+1}, the general linear multi-step method evaluates different number of functions depending on the order of the method, but the computed functions are re-used to estimate the next iteration. In this manner, from (1.1) one can derive explicit and implicit multi-step methods. In the side of explicit-type methods one can find the ones called Nth-order Adams-Bashforth, which are derived by setting $\alpha_k = 1$ and $\beta_k = 0$ in (1.1). The resulting iterative equations are given in Table 1.2, listing from order one to order six. It is worth mentioning that the first-order Adams-Bashforth method is equivalent to the Forward Euler one.

In the side of polynomial and implicit-type methods one can find the ones called Nth-order Adams-Moulton, which are generated from (1.1) by setting $\alpha_k = 1$ and $\beta_k \neq 0$. This type of multi-step numerical methods compute the solution by requiring the estimation of $f(y_{j+1}, t_{j+1})$, which can be evaluated by applying an explicit method. Some authors recommend the use of an explicit method of the same order as the implicit method. For example: If one uses the 6th-order

Table 1.2: Adams-Bashforth methods.

Order	Iterative Formulae
1_{st}	$y_{j+1} = y_j + hf(y_j, t_j)$
2_{nd}	$y_{j+1} = y_j + \frac{h}{2}\{3f(y_j, t_j) - f(y_{j-1}, t_{j-1})\}$
3_{rd}	$y_{j+1} = y_j + \frac{h}{12}\{23f(y_j, t_j) - 16f(y_{j-1}, t_{j-1}) + 5f(y_{j-2}, t_{j-2})\}$
4_{th}	$y_{j+1} = y_j + \frac{h}{24}\{55f(y_j, t_j) - 59f(y_{j-1}, t_{j-1}) + 37f(y_{j-2}, t_{j-2})$ $-9f(y_{j-3}, t_{j-3})\}$
5_{th}	$y_{j+1} = y_j + \frac{h}{720}\{1901f(y_j, t_j) - 2774f(y_{j-1}, t_{j-1}) + 2616f(y_{j-2}, t_{j-2})$ $-1274f(y_{j-3}, t_{j-3}) + 251f(y_{j-4}, t_{j-4})\}$
6_{th}	$y_{j+1} = y_j + \frac{h}{1440}\{4277f(y_j, t_j) - 7923f(y_{j-1}, t_{j-1}) + 9982f(y_{j-2}, t_{j-2})$ $-7298f(y_{j-3}, t_{j-3}) + 2877f(y_{j-4}, t_{j-4}) - 475f(y_{j-5}, t_{j-5})\}$

Adams-Moulton method given in Table 1.3, the prediction of $f(y_{j+1}, t_{j+1})$ can be done by using the 6th-order Adams-Bashforth method given in Table 1.2. However, for engineering applications, one may use any order for the explicit methods, and as one can infer, for sure the less expensive computationally speaking is the use of the first-order or Forward Euler method. The selection of the order of the implicit and explicit methods depends on the problem at hand, and also one can find methods that allow for the adaptation of the order and the step-size.

Table 1.3: Adams-Moulton methods.

Order	Algorithm
1_{st}	$y_{j+1} = y_j + hf(y_{j+1}, t_{j+1})$
2_{nd}	$y_{j+1} = y_j + \frac{h}{2}\{f(y_{j+1}, t_{j+1}) + f(y_j, t_j)\}$
3_{rd}	$y_{j+1} = y_j + \frac{h}{12}\{5f(y_{j+1}, t_{j+1}) + 8f(y_j, t_j)) - f(y_{j-1}, t_{j-1})\}$
4_{th}	$y_{j+1} = y_j + \frac{h}{24}\{9f(y_{j+1}, t_{j+1}) + 19f(y_j, t_j)) - 5f(y_{j-1}, t_{j-1})$ $+f(y_{j-2}, t_{j-2})\}$
5_{th}	$y_{j+1} = y_j + \frac{h}{720}\{251f(y_{j+1}, t_{j+1}) + 646f(y_j, t_j) - 264f(y_{j-1}, t_{j-1})$ $+106f(y_{j-2}, t_{j-2}) - 19f(y_{j-3}, t_{j-3})\}$
6_{th}	$y_{j+1} = y_j + \frac{h}{1440}\{475f(y_{j+1}, t_{j+1}) + 1427f(y_j, t_j) - 798f(y_{j-1}, t_{j-1})$ $+482f(y_{j-2}, t_{j-2}) - 173f(y_{j-3}, t_{j-3}) + 27f(y_{j-4}, t_{j-4})\}$

1.3 Stability of the numerical methods

The one-step and multi-step numerical methods described above perform the approximation of a solution through evaluating discrete points, and therefore a question arises: What precision can be reached? Or in other words one must answer: How many errors generates the method? In general, two types of errors can be evaluated: the local error that is introduced by a single step during the execution of the integration process and the global error, which is the overall error caused by repeated application of the integration formulae [15]. Both the local and global errors are divided into rounded-off and truncation errors. The rounded-off error is inevitable and results from performing arithmetic with real numbers on a digital computer. The truncation error is the local error that would be the result if the numerical algorithm is implemented on an infinite-precision computer. In general any numerical method can diminish these kind of errors, but some methods possesses more errors than the others. However, even if the error is low, this will not be useful at all if the method is numerically unstable. In this case, when the method is unstable the sum of all the small errors can become unlimited and the trajectory of the discrete points can have an inevitable divergency.

The computation of the stability region of a numerical method can be obtained as follows: One should apply the numerical method to a first-order lineal equation of the type $y' = f(x,y)$, with $y(0) = y_0$, and where $f(x,y) = \lambda y$, and for which λ is a eigenvalue that can be a complex number. For example: when applying the Forward-Euler method given in (1.2) to the linear equation, then one defines $f(x) = \lambda x$, and the discretization gives the iterative formulae $y_{n+1} = y_n + h\lambda y_n = (1 + h\lambda)y_n$. In this case, the Forward-Euler method is stable if it accomplishes the inequality: $|1 + h\lambda| < 1$, which is called stability region and it can be illustrated as shown in Fig. 1.1(a).

The evaluation of the stability region of the Backward Euler method given in (1.3), is performed in the same manner as for the Forward Euler method. Therefore, using the same linear equation, one gets the iterative equation: $y_{n+1} = y_n + h\lambda y_{n+1} = (\frac{1}{1-h\lambda})y_n$, so that the stability region is defined by $|1 - h\lambda| > 1$, and it is shown in Fig. 1.1(b).

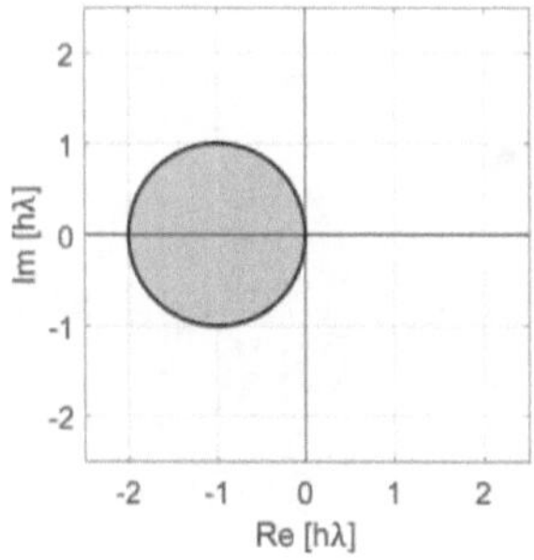

(a)

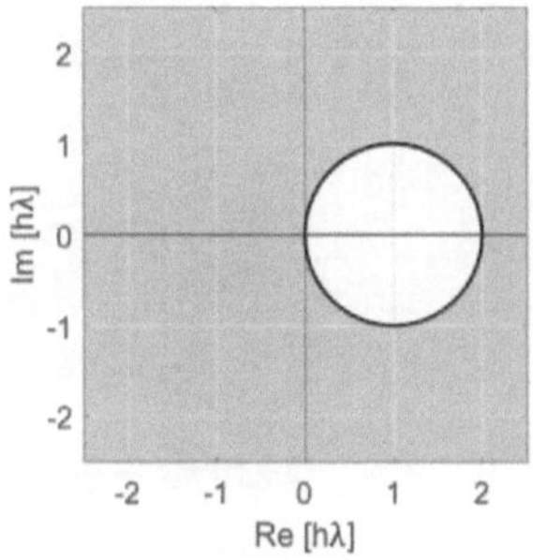

(b)

Figure 1.1: Stability region of the: (a) Forward-Euler and (b) Backward-Euler method.

As one sees, the stability region of the numerical methods is plotted in the complex plane because the eigenvalue λ can be real or complex. From Fig. 1.1 one can appreciate that the stability region relates inversely the eigenvalue with the step-size h value. In this case, if the eigenvalues of a function increases, it means that the step-size must decrease and viceversa. In such a case, before the application of a numerical method, one must analyze the eigenvalues of the dynamical system to estimate an appropriate step-size to accomplish the stability of the method. Although, this stability criteria can be taken as guarantee when solving linear problems, fortunately it applies in the majority of non-linear problems, and it is useful in solving ODEs modeling chaotic oscillators.

Figure 1.2 shows the stability regions of the Runge-Kutta method from its first to its fourth-order. It can be appreciated that as the order augments, the stability region also augments, so that this property makes the fourth-order Runge-Kutta method the most suitable for solving non-linear problems and therefore nowadays it is the most used one.

The stability regions of the Adams-Bashforth methods are shown in Fig. 1.3, and the ones for the Adams-Moulton methods are given in Fig. 1.4. These stabil-

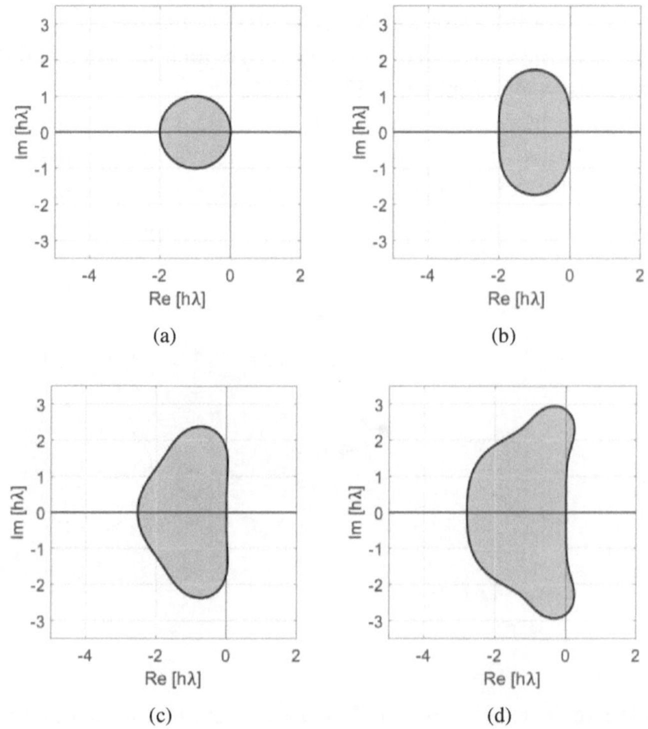

Figure 1.2: Stability regions of Runge-Kutta methods: (a) 1st, (b) 2nd, (c) 3rd, and (d) 4th-order.

ity regions are computed in a similar way as done for the Euler methods. In fact, as one can see, the equivalent methods of the one-step type for these families are the Forward and Backward Euler. Figures 1.3 and 1.4 show the stability regions of the multi-step methods from the 2nd to the 5th-order.

The solution of initial value problems by applying numerical methods give rise to a trade-off between the expected result and the payment to get the most exact one. For this reason this section shows details of the most classical numerical methods, which have the lowest consumption of computer resources to provide the best approximation to the exact solution. The analysis of the stability regions of the one-step and multi-step methods help to choose an adequate step-size h, to guarantee exactness and stability of the method. In [4], one can find the comparison of different numerical methods (Forward Euler, fourth-order Runge-Kutta, third-order Adams Bashforth and second-order Adams Moulton), to approximate the initial value problem: $\frac{dy}{dt} = -y^2$, with $y(0) = 1$, and which analytical solution is $y(t) = 1/t$, and its evaluation at $t = 4$ provides the exact solution $y(4) = 0.2$. In that reference one can appreciate that the fourth-order Runge-Kutta method generates the lowest numerical error, generating errors around 10^{-6} from $h = 2^{-1}$.

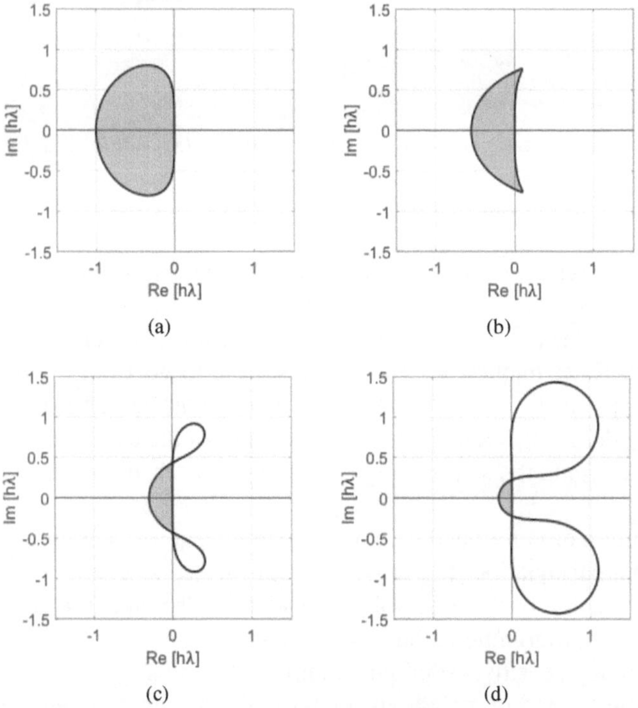

Figure 1.3: Stability regions of the Adams-Basforth methods: (a) 2nd, (b) 3rd, (c) 4th, and (d) 5th-order.

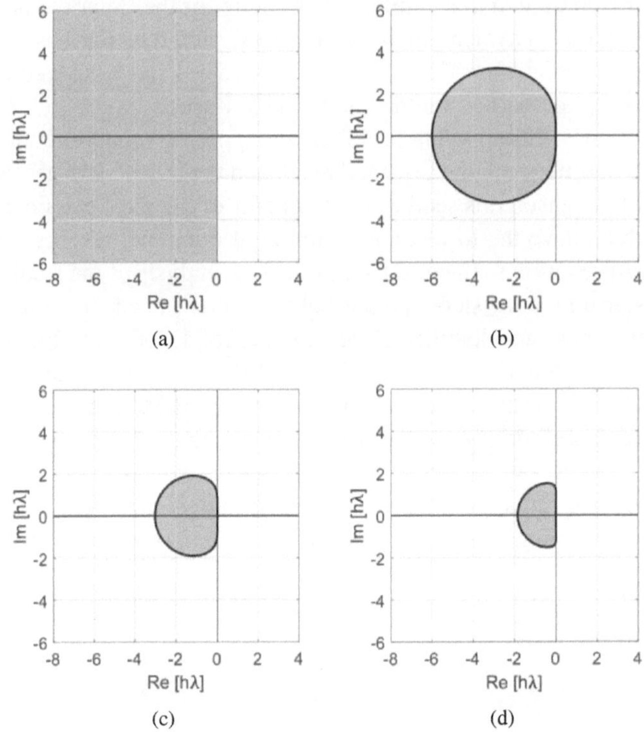

Figure 1.4: Stability regions of Adams-Moulton methods: (a) 2nd, (b) 3rd, (c) 4th, and (d) 5th-order.

This is a very high time-step value compared to the one when applying the Forward Euler, which reaches that error value but with a very small time-step as $h = 2^{-14}$. This means that compared to the fourth-order Runge-Kutta method, the Forward Euler method will take longer time to go from the discrete time 0 to 4 in steps of 2^{-14}. This is the main reason why researchers prefer using the fourth-order Runge-Kutta method, which allows the use of higher step-sizes without abandoning the stability region, and provides better exactness of the results.

The solution of non-linear dynamical systems requires a previous analysis of the mathematical equations to choose an adequate step-size that can be estimated from the evaluation of the eigenvalues. One must also take into account practical issues when programming the numerical methods to solving chaotic systems, which are highly sensitive to initial conditions. For this type of oscillatory systems, one must perform a rigorous analysis, good programming and selection of the step-size to avoid practical problems as computational chaos and superstability. The computational chaos is a kind of false chaotic behavior that arises when choosing a bad step-size so that chaotic behavior arises even if the problem is not of chaotic nature. The superstability arises when the bad selection of the step-size makes that chaotic system lose its behavior and converge towards one point [14].

1.4 Fractional-order chaotic systems

In engineering, all integer-order chaotic oscillators can be transformed to its fractional-order version. In this manner, as nowadays one can find many applications of autonomous integer-order chaotic oscillators, their fractional-order version may enhance those engineering applications. All reviews on the history of fractional calculus agree that this topic being more than 300 years ago, and the first reference may probably have been associated with Leibniz and L'Hospital in 1695, where the half-order derivative was mentioned. The majority of researchers agree that the main reason for the delay on dealing with fractional-order chaotic oscillators was the absence of solution methods for fractional-order differential equations. Therefore, and as already mentioned in [20], at the present time there are some methods that have demonstrated its suitability to find, with a good approximation, the solution of fractional-order derivatives and integrals. In this manner, fractional calculus can easily be used in wide areas of applications such as in physics, electrical engineering, control systems, robotics, signal processing, chemical mixing, bioengineering, and so on.

In all the history of chaos, various mathematical definitions have been proposed, and all of them have the common characteristic related to the super-sensitivity or sensitive dependence to the initial conditions. This issue can be characterized by Lyapunov instability as a main property of the chaotic oscillation. Detailed information on the fundamentals of the theory of nonlinear oscillations, which were laid in 1960's and 1970's by A. Poincaré, B. Van der Pol, A. A. Andronov, N. M. Krylov, A. N. Kolmogorov, D. V. Anosov, Ya. G. Sinai, V. K. Melnikov, Yu. I. Neimark, L. P. Shilnikov, G. M. Zaslavsky, and their collaborators, can be found in [20], where the authors provide a study of fractional-order chaotic systems accompanied by MATLAB® programs for simulating their state space trajectories.

As already summarized in [4], one of the seminal papers dealing with fractional-order circuits is [21], which discusses the major disadvantages of integer order sinusoidal oscillators, and then the authors proposed to apply fractional calculus to develop noninteger-order filters used in oscillator systems of order greater than 2. More recently, related references on nonlinear non-integer order circuits and systems are [22, 23, 24, 25, 26]. A major contribution of [26] is the novel classification of nonlinear dynamical systems by highlighting the existence of two kinds of attractors: self-excited and hidden attractors. In fact, revising recent literature one can see that the localization of self-excited attractors can be performed by applying standard computational procedures, but hidden attractors are difficult to locate and therefore it requires more specific computational procedures because the equilibrium points do not help in their localization. Some examples of chaotic oscillators embedding hidden attractors are dynamical systems with no equilibrium points, with only stable equilibria, curves of equilibria, surfaces of equilibria, and with non-hyperbolic equilibria.

Recent and relevant applications of fractional-order chaotic oscillators can be found in [27, 28, 29, 30, 31], all of them reporting outstanding research on applications of chaotic systems described by fractional-order dynamic models. Those applications must be simulated before their implementation, and this is an open challenge because the numerical method must be suitable for electronic implementation, and it must require low computational time in order to be used within an optimization loop to enhance the dynamical characteristics of a chaotic oscillator. On the one hand, the authors in [20] discuss some numerical simulation approaches for the solution of fractional-order systems. On the other hand, some time-domain numerical methods have been studied and implemented on embedded systems like field-programmable gate arrays (FPGAs), namely: Grünwald-Letnikov method [32], and Adam-Bashforth-Moulton method [33]. The fractional-order chaotic oscillators can also be approximated and simulated in the frequency domain, and in that case it can be implemented by using discrete amplifiers or field-programmable analog arrays (FPAA), as recently shown in [34].

1.5 Definitions of fractional order derivatives and integrals

Fractional calculus is a generalization of integration and differentiation to the non-integer-order fundamental operator ${}_{\alpha}D^{t}_{a}$, where a and t are the bounds of the operation and $\alpha \in R$. The continuous integro-differential operator is defined as

$$ {}_{\alpha}D^{t}_{a} = \begin{cases} \frac{d^{\alpha}}{dt^{\alpha}}, & \alpha > 0 \\ 1, & \alpha = 0 \\ \int_{a}^{t} (d\tau)^{\alpha}, & \alpha < 0 \end{cases} \tag{1.4} $$

The three most frequently used definitions for the general fractional differintegral are: Grünwald -Letnikov, Riemann-Liouville and Caputo definitions [35, 36, 37]. Other definitions are connected with well-known names as, for instance, Weyl, Fourier, Cauchy, Abel, Nishimoto, etc. In this book, the fractional-order chaotic oscillators are simulated by applying Grünwald -Letnikov definition. This consideration is based on the fact that for a wide class of fractional functions, Grünwald -Letnikov, Riemann-Liouville and Caputo definitions are equivalent under some conditions [37]. In addition, we also apply the time-domain simulation method based on the predictor-corrector Adams-Bashforth-

Moulton method. These methods are then suitable to be used within an optimization loop to optimize the dynamical characteristics of the fractional-order chaotic oscillators, such as Lyapunov exponents, Kaplan-Yorke dimension and entropy. This is shown in the following chapters by applying mono-objective and multi-objective optimization algorithms based on metaheuristics.

1.5.1 Grünwald-Letnikov fractional integrals and derivatives

Lets us consider the continuous function $f(t)$. Its first derivative can be expressed as (1.5), where when this is used twice, one obtains a second derivative of the function $f(t)$ in the form of (1.6). Afterwards, using both (1.5) and (1.6) one can get a third derivative of the function $f(t)$ in the form of (1.7).

$$\frac{d}{dt}f(t) \equiv f'(t) = \lim_{h\to 0}\frac{f(t)-f(t-h)}{h} \tag{1.5}$$

$$\begin{aligned}\frac{d^2}{dt^2}f(t) \equiv f''(t) &= \lim_{h\to 0}\frac{f'(t)-f'(t-h)}{h}\\ &= \lim_{h\to 0}\frac{1}{h}\left\{\frac{f(t)-f(t-h)}{h}-\frac{f(t-h)-f(t-2h)}{h}\right\}\\ &= \lim_{h\to 0}\frac{f(t)-2f(t-h)+f(t-2h)}{h^2}.\end{aligned} \tag{1.6}$$

$$\frac{d^3}{dt^3} \equiv f'''(t) = \lim_{h\to 0}\frac{f(t)-3f(t-h)+3f(t-2h)-f(t-3h)}{h^3} \tag{1.7}$$

According to this rule, one can write a general formulae for the $n-$th derivative of a function $f(t)$, by t for $n \in N$, $j > n$ in the form of (1.8), which expresses a linear combination of function values $f(t)$ in the variable t. Henceforth, Binomial coefficients with alternating signs for positive values of n are defined by (1.9), while in the case of negative values of n, one gets (1.10), where $\begin{bmatrix} n \\ j \end{bmatrix}$ is defined by (1.11).

$$\frac{d^n}{dt^n}f(t) \equiv f^{(n)}(t) = \lim_{h\to 0}\frac{1}{h^n}\sum_{j=0}^{n}(-1)^j\binom{n}{j}f(t-jh). \tag{1.8}$$

$$\binom{n}{j} = \frac{n(n-1)(n-2)\cdots(n-j+1)}{j!} = \frac{n!}{j!(n-j)!}. \tag{1.9}$$

$$\binom{-n}{j} = \frac{-n(-n-1)(-n-2)\cdots(-n-j+1)}{j!} = (-1)^j\begin{bmatrix} n \\ j \end{bmatrix}, \tag{1.10}$$

$$\begin{bmatrix} n \\ j \end{bmatrix} = \frac{n(n+1)\cdots(n+j-1)}{j!} \tag{1.11}$$

Substituting n in (1.8) by $-n$, one can write (1.12), where n is a positive integer number.

$$\frac{d^{-n}}{dt^{-n}} f(t) \equiv f^{(-n)}(t) = \lim_{h \to 0} \frac{1}{h^n} \sum_{j=0}^{n} \begin{bmatrix} n \\ j \end{bmatrix} f(t - jh), \tag{1.12}$$

According to (1.5)-(1.8), one can write the fractional-order derivative definition of order α, $(\alpha \in R)$ by t, which has the form of (1.13). In this case, the binomial coefficients can be calculated by using the relation between Euler's Gamma function and factorial, which becomes defined as in (1.14), for $\binom{a}{0} = 1$.

$$D_t^{\alpha} f(t) = \lim_{h \to 0} \frac{1}{h^{\alpha}} \sum_{j=0}^{\infty} (-1)^j \binom{a}{j} f(t - jh). \tag{1.13}$$

$$\binom{a}{j} = \frac{\alpha!}{j!(\alpha - j)!} = \frac{\Gamma(\alpha + 1)}{\Gamma(j+1)\Gamma(\alpha - j + 1)} \tag{1.14}$$

The definition given in (1.14) requires the sum of infinity data, which is not reachable using power computing resources. In this manner, one can consider $n = \frac{t-a}{h}$, and supposing that α is a real constant, which expresses a limit value, then (1.14) can be updated to the equation given in (1.15), where $\lfloor x \rfloor$ is associated to the integer part of x, and α and t are the bounds of the operation in ${}_aD_t^{\alpha} f(t)$. These equations will be discussed later for the FPGA-based implementation of fractional-order chaotic oscillators.

$${}_aD_t^{\alpha} f(t) = \lim_{h \to 0} \frac{1}{h^{\alpha}} \sum_{j=0}^{\left[\frac{t-a}{h}\right]} (-1)^j \binom{a}{j} f(t - jh). \tag{1.15}$$

1.5.2 Riemann-Liouville fractional integrals and derivatives

The Riemann-Liouville definition can be described under some assumptions, for example one can consider the Riemann-Liouville n-fold integral, which can be defined by (1.16) for $n \in N$, $n > 0$.

$$\underbrace{\int_a^t \int_a^{t_n} \int_a^{t_{n-1}} \cdots \int_a^{t_3} \int_a^{t_2}}_{n-fold} f(t_1) dt_1 dt_2 \cdots dt_{n-1} dt_n = \frac{1}{\Gamma(n)} \int_a^t \frac{f(\tau)}{(t-\tau)^{1-n}} d\tau \tag{1.16}$$

The fractional-order integral α for the function $f(t)$ can then be expressed from (1.16) as given in (1.17), for α, $a \in R$, $\alpha < 0$.

$$ {}_aI_t^{\alpha} f(t) \equiv {}_aD_t^{-\alpha} f(t) = \frac{1}{\Gamma(-\alpha)} \int_a^t \frac{f(\tau)}{(t-\tau)^{\alpha+1}} d\tau \tag{1.17} $$

From the relation expressed in (1.17), one can write the formulae for the Riemann-Liouville definition of the fractional derivative of order α in the form of (1.18), for $(n-1 < \alpha < n)$, where α and t are the limits of the operator ${}_aD_t^{\alpha} f(t)$.

$$ {}_aD_t^{\alpha} f(t) = \frac{1}{\Gamma(n-\alpha)} \frac{d^n}{dt^n} \int_a^t \frac{f(\tau)}{(t-\tau)^{\alpha-n+1}} d\tau \tag{1.18} $$

For the particular case of considering that $0 < \alpha < 1$ and $f(t)$ being a causal function of t, that is, $f(t) = 0$ for $t < 0$, then the fractional-order integral is defined by (1.19), and the expression for the fractional order derivative is defined by (1.20), where $\Gamma(\cdot)$ is the Euler's Gamma function [35].

$$ {}_0D_t^{-\alpha} f(t) = \frac{1}{\Gamma(\alpha)} \int_0^t \frac{f(\tau)}{(t-\tau)^{1-\alpha}} d\tau, \text{ for } 0 < \alpha < 1, t > 0 \tag{1.19} $$

$$ {}_0D_t^{\alpha} f(t) = \frac{1}{\Gamma(n-\alpha)} \frac{d^n}{dt^n} \int_0^t \frac{f(\tau)}{(t-\tau)^{\alpha-n+1}} d\tau \tag{1.20} $$

1.5.3 Caputo fractional derivatives

The Caputo definition of fractional order derivatives can be written as [37, 38] given in (1.21). As considered in several works, under the homogenous initial conditions the Riemann-Liouville and the Caputo derivatives are equivalent. For instance, describing the Riemann-Liouville fractional derivative by ${}_a^{RL}\mathrm{D}_t^{a} t(t)$, and the Caputo definition by ${}_a^{C}\mathrm{D}_t^{a} f(t)$, then the equivalent relationship between them is given in (1.22), for $f^{(k)}(a) = 0$, and with $(k = 0, 1, ..., n-1)$.

$$ {}_aD_t^{\alpha} f(t) = \frac{1}{\Gamma(n-\alpha)} \int_a^t \frac{f^{(n)}(\tau)}{(t-\tau)^{\alpha-n+1}} d\tau, \text{ for } n-1 < \alpha < n. \tag{1.21} $$

$$ {}_a^{RL}\mathrm{D}_t^{a} t(t) = {}_a^{C}\mathrm{D}_t^{a} f(t) + \sum_{k=0}^{n-1} \frac{(t-a)^{k-a}}{\Gamma(k-\alpha+1)} f^{(k)}(a) \tag{1.22} $$

The initial conditions for the fractional-order differential equations with the Caputo derivatives are in the same form as for the integer-order differential equations. This is an advantage because the majority of problems require definitions of fractional derivatives, where there are clear interpretations of the initial conditions, which contain $f(a)$, $f'(a)$, $f''(a)$, etc.

1.6 Time domain methods for fractional-order chaotic oscillators

The fractional-order derivatives and integrals are generalizations of the integer-order ones that are particular cases, so that any integer-order dynamical system can be adapted to become a system of arbitrary order, including fractional-order. However, yet there is not a generalized consensus on the definition of the fractional-order derivatives and integrals, so that one can find a variety of proposals, but for sure the most known and applied in different fields for research are: fractional derivative and integral of Riemmann-Liouville, Grünwald-Letnikov and Caputo [39].

The approximation of the solution of the fractional-order derivatives can be performed by applying the Grünwald-Letnikov algorithm, which approximation expression is given in (1.15) in page 14, and also one can apply the predictor-corrector Adams-Bashforth-Moulton method given in (1.23) and (1.24), respectively.

$$y_h^p(t_n+1) = \sum_{k=0}^{m-1} \frac{t_{n+1}^k}{k} y_0^{(k!)} + \frac{1}{\Gamma(q)} \sum_{k=0}^{n} b_{j,n+1} f(t_j, y_n(t_j)), \tag{1.23}$$

$$\begin{aligned} y_h(t_{n+1}) = \sum_{k=0}^{m-1} \frac{t_{n+1}^k}{k!} y_0^{(k)} + \frac{h^q}{\Gamma(\alpha+2)} f(t_{n+1}, y_h^p(t_{n+1})) + \\ + \frac{h^q}{\Gamma(\alpha+2)} \sum_{j=0}^{n} a_{j,n+1} f(t_j, y_n(t_j)), \end{aligned} \tag{1.24}$$

Both algorithms are applied herein to approximate the numerical solution of the fractional-order derivatives associated to different types of fractional-order chaotic oscillators. In several articles, one can find that for a wide class of functions, both definitions are equivalent. For instance, the relation of Grünwald-Letnikov with the explicit approximation given in (1.23) for the $q-th$ derivative at the points $kh, (k = 1, 2, \ldots)$ has the form of (1.25) [37, 39, 40], where L_m denotes the "memory length", $t_k = kh$, where h is related to the time-step, and $(-1)^j \binom{q}{j}$ are the binomial coefficients $C_j^q (j = 0, 1, \ldots)$. In general, the binomial coefficients can be approached by an infinite number of values, however it is a challenge to implement on digital hardware as the resources are limited. In this manner, one can take advantage of the concept on "short memory" [39], to reduce hardware resources and then to have the possibility of implementing fractional-order chaotic oscillators on FPGAs [4], which is commonly applied in the Grünwald-Letnikov method through (1.26).

$${}_{(k-L_m/h)}D_{tk}^q f(t) \approx h^{-q} \sum_{j=0}^{k} (-1)^j \binom{q}{j} f(t_{k-j}) \tag{1.25}$$

$$c_0^q = 1, \qquad c_j^q = \left(1 - \frac{1+q}{j}\right) c_{j-1}^q \tag{1.26}$$

Applying the Grünwald-Letnikov approximation of the fractional-order derivative of the form: ${}_aD_t^q y(t) = f(y(t),t)$, leads us to deal with the discrete equation given by (1.27), which takes advantage of the short memory concept and then the binomial coefficients are truncated by a desired length of memory.

$$y(t_k) = f(y(t_k),t_k)h^q - \sum_{j=v}^{k} c_j^{(q)} y(t_{k-j}) \tag{1.27}$$

The sum in (1.27) is associated to the memory of the algorithm, and it can be truncated according to the short memory principle [39]. That way, the lowest index in the sum operation should be $v = 1$ for the case when $k < (L_m/h)$, and $v = k - (L_m/h)$ for the case $k > (L_m/h)$. Besides, without applying the short memory principle one must set $v = 1$ for all values of k. It is then quite obvious that when performing this truncation the error should increase and then the solution may not converge. Besides, one can estimate the length of memory, for example: if $f(t) \leqslant M$, L_m can be estimated using (1.28), which barely includes a required precision ε.

$$L \geqslant \left(\frac{M}{\varepsilon|\Gamma(1-q)|}\right)^{1/q} \tag{1.28}$$

The time domain Grünwald-Letnikov method is an explicit one, and as mentioned above, one can apply an implicit one, for instance the well-known predictor-corrector Adams-Bashforth-Moulton method [41]. This approximation for the fractional-order derivatives of the chaotic oscillators is more exact than Grünwald-Letnikov, but it requires higher number of operations and hardware resources. The Adams-Bashforth-Moulton method is based on the fact that the fractional order derivative of the form given in (1.29), is equivalent to Volterra's integral equation given in (1.30).

$$D_t^q y(t) = f(y(t),t), y^{(k)}(0) = y_0^{(k)}, k = 0,1,\ldots,m-1 \tag{1.29}$$

$$y(t) = \sum_{k=0}^{[q]-1} y_0^{(k)} \frac{t^k}{k!} + \frac{1}{\Gamma(q)} \int_0^t (t-\tau)^{q-1} f(\tau, y(\tau)) d\tau \tag{1.30}$$

Discretizing (1.30) for a uniform array $t_n = nh$ $(n = 0,1,\ldots,N)$, $h = T_{sim}/N$ and using the short memory principle (fixed or logarithmic [42]) one gets a numerical approximation very close to the truth one $y(t_n)$, of the fractional differential equation while conserving the exactness of the fractional order. Therefore, supposing that one has the approximations in $y_h(t_j), j = 1,2,\ldots,n$, and one wants to get the solution $y_h(t_{n+1})$ using (1.31), where $a_{j,n+1}$ is evaluated by (1.32).

$$y_h(t_{n+1}) = \sum_{k=}^{m-1} \frac{t_{n+1}^k}{k!} y_0^{(k)} + \frac{h^q}{\Gamma(\alpha+2)} f(t_{n+1}, y_h^p(t_{n+1})) +$$

$$\frac{h^q}{\Gamma(\alpha+2)} \sum_{j=0}^{n} a_{j,n+1} f(t_j, y_n(t_j)) \tag{1.31}$$

$$a_{j,n+1} = \begin{cases} n^{q+1} - (n-q)(n+1)^q & if\ j=0 \\ (n-j+2)^{q+1} + (n-j)^{q+1} + 2(n-j+1)^{q+1} & if\ 1 \leqslant j \leqslant n \\ 1 & if\ j = n+1 \end{cases} \tag{1.32}$$

In (1.31) the preliminar approximation $y_h^p(t_{n+1})$ is named predictor and is given by (1.33), where $b_{j,n+1}$ is evaluated by (1.34).

$$y_h^p(t_{n+1}) = \sum_{k=0}^{m-1} \frac{t_{n+1}^k}{k!} y_0^{(k)} + \frac{1}{\Gamma(q)} \sum_{j=0}^{n} b_{j,n+1} f(t_j, y_n(t_j)) \tag{1.33}$$

$$b_{j,n+1} = \frac{h^q}{q}((n+1-j)^q - (n-j)^q) \tag{1.34}$$

As already shown in [43], both time domain numerical methods (Grünwald-Letnikov (1.8) and Adams-Bashforth-Moulton ((1.23)-(1.24)) have approximately the same exactness and a good approximation of the solution of a fractional-order chaotic oscillators. However, fractional-order chaotic oscillators require to accomplish the following theorems:

Definition: Lets us consider the general fractional-order system of n dimensions given by (1.35), in which the roots of evaluating $f(X) = 0$ are the equilibrium points. In this case, $D^q(X) = (D^q x_1, D^q x_2, \ldots, D^q x_n)^T, X = (x_1, x_2, \ldots, x_n)^T \in R^n$.

$$D^q(X) = f(X), \tag{1.35}$$

Theorem 1. A fractional-order system modeled by three state variables $n = 3$ is asymptotically stable at the equilibrium point equal to 0, if and only if $|\arg(\lambda_i(J))| > q\pi/2$, $i = 1, 2, 3$. In this case J denotes the Jacobian matrix of $f(X)$, and λ_i are the eigenvalues of J [44].

Theorem 2. The equilibrium point O of a dynamical system described by (1.35) is unstable if and only if the order of q satisfies the condition imposed by (1.36), for at least one eigenvalue, where $\text{Re}(\lambda)$ and $\text{Im}((\lambda)$ denote the real and imaginary parts of λ [45].

$$q > \frac{2}{\pi} \arctan \frac{|\text{Im}((\lambda)|}{|\text{Re}(\lambda)|}. \tag{1.36}$$

Theorem 3. For $n = 3$, if one of the eigenvalues $\lambda_1 < 0$ and the other two complex conjugated $|\arg(\lambda_2)| = |\arg(\lambda_2)| < q\pi/2$, then the equilibrium point O is called saddle point of index 2. If one of the eigenvalues $\lambda_1 > 0$ and the other two complex conjugated $|\arg(\lambda_2)| = |\arg(\lambda_3)| > q\pi/2$, then the equilibrium point O is called saddle point of index 1 [41].

The algorithms that are used in this book to approach the solution of fractional-order chaotic oscillators are given in Algorithm 1 for the Grünwald-Letnikov method, and Algorithm 2 for the predictor-corrector Adams-Bashforth-Moulton method. These Algorithms are applied in the next chapters to simulate different families of fractional-order chaotic oscillators and to include them in an optimization loop to maximize their dynamical characteristics, such as the Kaplan-Yorke dimension and entropy by applying metaheuristics. Algorithm 1 is adapted to simulate the fractional-order Lorenz chaotic oscillator.

1.7 Simulation of the fractional-order derivative ${}_0D_t^q y(t) = x(t)$

This section shows the evaluation of the fractional-order derivative ${}_0D_t^q y(t) = x(t)$ by applying Grünwald-Letnikov, FDE12 predictor-corrector and Adams-Bashforth-Moulton methods.

1.7.1 Approximation by applying Grünwald-Letnikov method

The fractional-order derivative can be approached by applying the Grünwald-Letnikov definition, where the solution is computed taking into account (1.37), in order to obtain the iterative formulae given in (1.38), which includes the short memory principle.

$$y(t_k) = f(y(t_k), t_k)h^q - \sum_{j=v}^{k} c_j^{(q)} y(t_{k-j}) \tag{1.37}$$

$$x(t_k) = (x(t_{k-1})h^q - \sum_{j=v}^{k} c_j^{(q)} x(t_{k-j}), \tag{1.38}$$

1.7.2 Approximation by applying FDE12 predictor-corrector method

The fractional-order derivative ${}_0D_t^q y(t) = x(t)$ can directly be simulated applying FDE12 available into MATLAB™ [46, 47]. This method is taken as reference to compare the results with respect to Grünwald-Letnikov and Adams Bashforth-Moulton methods. In the case of fractional-order chaotic oscillators,

Algorithm 1 Grünwald-Letnikov to solve a fractional-order system.

Require: f, q, y_0, T, T_0 and h.

Output Variables
Y matrix of $n+1 \times m$ real numbers that contain the approximate solutions
t array of $n+1$ real numbers with the time samples from T_0 to T
Internal Variables
m the system dimension
n the number of time-steps that the algorithm considers
i, j variables used as indexes
o the memory length
c arrays of $o+1$ real numbers that contain the coefficient values
$n = \text{floor}(\,(T - T_0)/h\,);\; m = \text{length}(y_0)$
$Y = \text{zeros}(n+1, m)$
$c(1) = 1$
for $j = 2$ to $o+1$ **do** ▷ Calculate binomial coefficients
    $c(j) = (\,1 - (1+q)/j\,)\, c(j-1)$
end for
$t(0) = 0.0;\; Y(1,:) = y_0$
for $i = 2$ to $o-1$ **do**
    $t(i) = i \cdot h$
    $s = 0.0$
    for $j = i$ to 1 **do** ▷ In decreasing order
        $s \leftarrow s + c(j) \cdot Y(i-j+1,:)$
    end for
    $Y(i,:) = h^q \cdot f(\,t(i-1),\, Y((i-1),:)\,) - s$
end for
for $i = o$ to $n+1$ **do**
    $t(i) = i \cdot h$
    $s = 0.0$
    for $j = o$ to 1 **do** ▷ In decreasing order
        $s \leftarrow s + c(j) \cdot Y(i-j+1,:)$
    end for
    $Y(i,:) = h^q \cdot f(\,t(i-1),\, Y((i-1),:)\,) - s$
end for

the Grünwald-Letnikov method may require a lower time-step to reach similar accuracy than FDE12. This will be detailed in the following chapters, and will be considered as a challenge to improve time computation for the implementation of fractional-order chaotic oscillators using FPGAs.

Algorithm 2 Predictor-corrector Adams-Bashforth-Moulton method.

Require: f, q, y_0, T, T_0 y h.

Output Variables:

Y matrix of size $n+1 \times m$ of real numbers that contain the approximate solutions

t vector of $n+1$ real numbers with the time samples

Internal Variables:

m the system dimension

n the number of time steps that the algorithm is to consider

i, j variables used as indexes

a, b arrays of $N+1$ real numbers that contain the values of the corrector and predictor, respectively

p the predicted value

$n = \text{floor}((T - T_0)/h)$

$Y = \text{zeros}(n+1, m)$

for $j = 1$ to n **do**

  $a(j) = j^q - (j-1)^\alpha$

  $b(j) = (j+1)^{(q+1)} - 2j^{(q+1)} + (j-1)^{(q+1)}$

end for

$t(0) = 0.0; Y(1,:) = y_0$

for $i = 1$ to n **do**

  $t(i) = ih$

  $p = Y(i,:) + \frac{h^q}{\Gamma(q+1)} \sum_{j=1}^{i} b(j)\, f(jh, Y(j,:))$

  $Y(i+1\ :) = Y(i,:) + \frac{h^q}{\Gamma(q+2)}(f(ih,p) + ((i-1-q)i^{j^q})f(0,Y(1)) + \sum_{j=1}^{i} a(j)\, f(jh, Y(j)))$

end for

1.7.3 Approximation by applying Adams-Bashforth-Moulton method

The predictor-corrector approximation for solving integer-order chaotic oscillators is similar for fractional-order ones. In this case, the Adams-Bashforth-Moulton method detailed in [48] is applied herein to approach the solution of the fractional-order derivative [49]. The fractional-order differential equation given in (1.39) is considered.

$$D_t^q y(t) = f(y(t), t), \;\; y^{(k)}(0) = y_0^{(k)}, \;\; k = 0, 1, \ldots, m-1 \tag{1.39}$$

The discrete form of the method to obtain the corrector and iterative equation is expressed by (1.40), where $a_{j,n+1}$ is evaluated by (1.41), and using as predictor the discrete equation given in (1.42), in which $b_{j,n+1}$ is computed by using (1.43).

$$y_h(t_{n+1}) = \sum_{k=}^{m-1} \frac{t_{n+1}^k}{k!} y_0^{(k)} + \frac{h^q}{\Gamma(\alpha+2)} f(t_{n+1}, y_h^p(t_{n+1})) + \frac{h^q}{\Gamma(\alpha+2)} \sum_{j=0}^{n} a_{j,n+1} f(t_j, y_n(t_j)) \tag{1.40}$$

$$a_{j,n+1} = \begin{cases} n^{q+1} - (n-q)(n+1)^q & \text{if } j = 0 \\ (n-j+2)^{q+1} + (n-j)^{q+1} + 2(n-j+1)^{q+1} & \text{if } 1 \leqslant j \leqslant n \\ 1 & \text{if } j = n+1 \end{cases} \tag{1.41}$$

$$y_h^p(t_{n+1}) = \sum_{k=0}^{m-1} \frac{t_{n+1}^k}{k!} y_0^{(k)} + \frac{1}{\Gamma(q)} \sum_{j=0}^{n} b_{j,n+1} f(t_j, y_n(t_j)) \tag{1.42}$$

$$b_{j,n+1} = \frac{h^q}{q}((n+1-j)^q - (n-j)^q) \tag{1.43}$$

1.7.4 Numerical simulation of ${}_0D_t^{0.5}y(t) = x(t)$ and ${}_0D_t^{0.9}y(t) = x(t)$

Lets us consider the fractional-order derivative ${}_0D_t^{0.5}y(t) = x(t)$. Its numerical simulation by applying Grünwald-Letnikov (GL), FDE12 and Adams-Bashforth-Moulton (ABM) methods is shown in Fig. 1.5. One can perform numerical simulations by using different lengths of memory by applying GL, for example 10, 20, 50 and 200 and the results will be quite similar. Taking as reference the results provided by FDE12 [50], one can appreciate that ABM is more close to FDE12 than GL. FDE12 combines some integration rules and its stability properties can be found in [51]. Table 1.4 shows the values for the 600 iterations in steps of 60 and the corresponding standard deviation of GL and ABM with respect to FDE12.

Figure 1.6 shows the numerical simulation of ${}_0D_t^{0.9}y(t) = x(t)$ by applying FDE12, GL and ABM, and the standard deviation of GL and ABM with respect to FDE12 is given in Table 1.5. As one sees, the error is lower compared to a fractional-order of $q = 0.5$. It means that the approximations of GL and ABM are much better when solving fractional-order derivatives near to unity.

As one can infer, the Adams-Bashforth-Moulton method approaches the solution of the fractional-order derivative in a similar exactness than it is done when applying the FDE12 method available into MATLAB. The error generated by applying the Grünwald-Letnikov method can be diminished by varying the conditions of simulation. This means that for exactness issues, it will be much

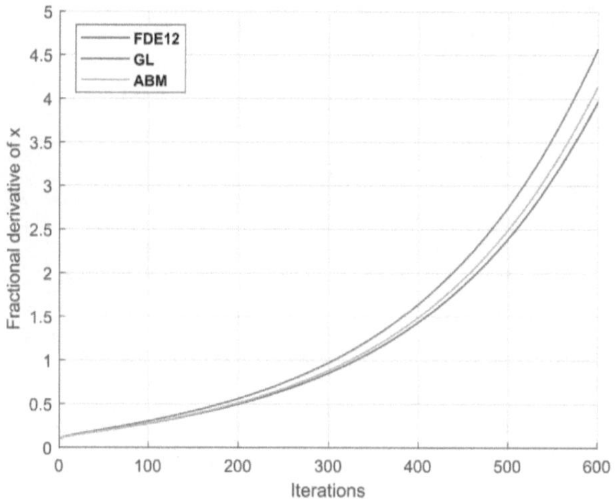

Figure 1.5: Numerical simulation of ${}_0D_t^{0.5}y(t) = x(t)$ by applying FDE12, GL and ABM methods with $h = 0.005$, $x(0) = 0.1$, and 600 iterations.

Table 1.4: Standard deviation of GL and ABM with respect to FDE12 for the numerical simulation of ${}_0D_t^{0.5}y(t) = x(t)$.

Iteration	FDE12	GL error	ABM error	σ (GL)	σ (ABM)
60	0.20912	0.01661	0.00019	0.01175	0.00013
120	0.31255	0.03283	0.00445	0.02321	0.00314
180	0.44497	0.05354	0.01033	0.03786	0.00730
240	0.62002	0.08089	0.01833	0.05720	0.01296
300	0.85397	0.11744	0.02919	0.08304	0.02064
360	1.16809	0.16651	0.04388	0.11774	0.03103
420	1.59083	0.23253	0.06374	0.16442	0.04507
480	2.16043	0.32149	0.09059	0.22733	0.06406
540	2.92844	0.44144	0.12687	0.31215	0.08971
600	3.96440	0.60323	0.17588	0.42655	0.12436

better to apply Adams-Bashforth-Moulton method than the Grünwald-Letnikov one, but as shown in the following chapters, it will require a huge number of hardware resources, which may not be available in an FPGA for some special cases of fractional-order chaotic oscillators, and therefore, finding the appropriate parameters for the Grünwald-Letnikov method, to generate low error, may be the best choice when trading exactness vs. hardware resources. This is discussed and detailed in the rest of this book as a trade-off between accelerating time simulation by using high time-step h values and guaranteeing exactness of

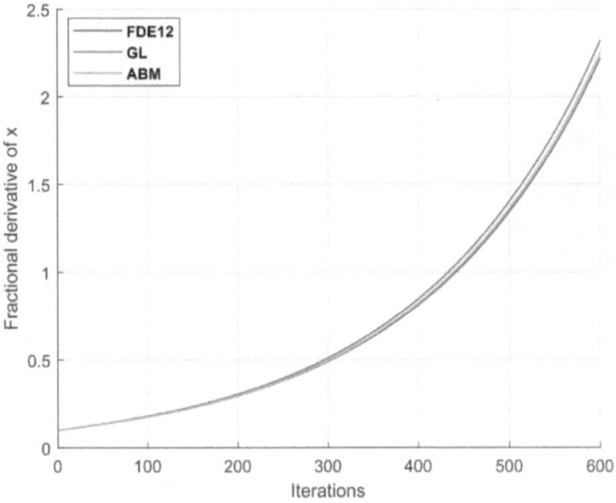

Figure 1.6: Numerical simulation of ${}_0D_t^{0.9}y(t) = x(t)$ by applying FDE12, GL and ABM methods with $h = 0.005$, $x(0) = 0.1$, and 600 iterations.

Table 1.5: Standard deviation of GL and ABM with respect to FDE12 for the numerical simulation of ${}_0D_t^{0.9}y(t) = x(t)$.

Iteration	FDE12	GL error	ABM error	σ (GL)	σ (ABM)
60	0.14227	0.00191	0.00025	0.00135	0.00018
120	0.19586	0.00441	0.00029	0.00312	0.00020
180	0.26716	0.00774	0.00121	0.00547	0.00086
240	0.36287	0.01221	0.00256	0.00863	0.00181
300	0.49172	0.01822	0.00444	0.01289	0.00314
360	0.66540	0.02633	0.00702	0.01862	0.00497
420	0.89967	0.03726	0.01054	0.02634	0.00745
480	1.21575	0.05200	0.01531	0.03677	0.01082
540	1.64229	0.07190	0.02176	0.05084	0.01539
600	2.21797	0.09876	0.03050	0.06984	0.02156

the solution when the optimized fractional-order chaotic oscillators are implemented on reconfigurable hardware like FPAA and FPGA.

Chapter 2

Integer-Order Chaotic/Hyper-chaotic Oscillators

2.1 Introduction

Chaotic behavior was exploited by Edward Lorenz to introduce weather models in 1963, which are quite sensitive to small changes and produce the butterfly effect. Lorenz discovered deterministic chaos in systems that vary in irregular form and even at very long time they do not repeat their previous history [52]. All dynamical systems can be modeled by ordinary differential equations (ODEs) including nonlinear functions and the derivatives can have integer or fractional orders [53]. A chaotic system can be modeled by three ODEs [52, 54, 55, 56, 57, 58, 59], and can include piecewise-linear (PWL) functions to generate multi-scroll and multi-directional attractors [60, 61]. If the model includes more than three ODEs, they can generate hyper-chaotic behavior that is associated to have more than one positive Lyapunov exponent [62, 63, 64, 65, 66].

Nowadays, due to the properties of the ODEs, chaotic systems can be classified as self-excited and hidden chaotic attractors. If one can evaluate the equilibrium or stable points, they are said to be self-excited, otherwise they are hidden chaotic attractors. Self-excited attractors are formed from inestable sets of initial conditions that are near the vicinity of the equilibrium points, and they can be simulated by applying numerical methods, as the ones listed in Chapter 1. Hidden attractors do not possesses equilibrium points or it is quite difficult to find them and therefore one cannot guarantee that a traditional numerical method can simulate it [67]. In both cases, one can estimate their dynamical characteristics as Lyapunov exponents, Kaplan-Yorke dimension (D_{KY}), and entropy. These characteristics depend on the values of the coeficients of the ODEs and they can be set by exploring their associated bifurcation parameters.

2.2 Self-excited chaotic oscillators

Some of the most known self-excited chaotic attractors are Lorenz [52], Rössler [54], Chen [55], and Sprott [58]. The Lorenz system can be modeled by (2.1),

$$\begin{aligned}\dot{x} &= \sigma(y-x),\\ \dot{y} &= x(\rho-z)-y,\\ \dot{z} &= xy-\beta z,\end{aligned} \tag{2.1}$$

where the chaotic behavior is generated by setting $\sigma = 10, \rho = 28$, and $\beta = 8/3$. The associated equilibrium points can be evaluated from (2.2). From the first equation one gets $x = y$, and by substituting this equality into the second equation $x(28-z)-x = 0$ one gets $z = 27$. The same is done for the third equation to have $x^2 - 8z/3 = 0$, and finally $x = \pm\sqrt{8z/3}$. From this analysis one can identify three equilibrium points: one located at $x = 8.4852$, $y = 8.4852$ and $z = 27$; and the second located at $x = -8.4852$, $y = -8.4852$ and $z = 27$. A third equilibrium point is located at the origin: $x = 0$, $y = 0$ and $z = 0$ because with these values the equation (2.2) is also solved.

$$\begin{aligned}10(y-x) &= 0,\\ x(28-z)-y &= 0,\\ xy-8z/3 &= 0,\end{aligned} \tag{2.2}$$

The eigenvalues are obtained from the evaluation of the Jacobian of the ODEs, and they are evaluated by using the values of the equilibrium points. For the Lorenz system one must evaluate three Jacobians to find the corresponding eigenvalues associated to the three equilibrium points, and therefore one can determine the stability region of a numerical method that helps to estimate the step-size to perform the numerical solution, as shown in Chapter 1.

The Jacobian matrix associated with Lorenz is given in (2.3), from which one must replace the state variables x, y, and z with the sets of equilibrium points to derive the characteristic equation from the evaluation of the relationship $|\lambda I - J| = 0$. By using the equilibrium point located at the origin one gets (2.4). From this polynomial one can compute the three eigenvalues associated with the equilibrium point located at the origin. One can do the same to find the other eigenvalues by using the other two equilibrium points, and at the end one can list nine eigenvalues.

$$J = \begin{bmatrix} -10 & 10 & 0 \\ 28-z & -1 & -x \\ y & x & -8/3 \end{bmatrix} \tag{2.3}$$

$$\lambda^3 + \frac{41\lambda^2}{3} - \frac{722\lambda}{3} - 720 = 0 \tag{2.4}$$

The evaluation of the eigenvalues of the Lorenz system by using the three equilibrium points leads us to the first set of eigenvalues given as: $\lambda_{1,2} = 0.0939 \pm j10.1945$ and $\lambda_3 = -13.8545$. The second set of eigenvalues becomes: $\lambda_{1,2} = 0.0939 \pm j10.1945$ and $\lambda_3 = -13.8545$. The third set of eigenvalues is given as: $\lambda_1 = 11.8277$, $\lambda_2 = -22.8277$ and $\lambda_3 = -2.6667$. These eigenvalues can help to determine if the dynamical system based on Lorenz equations given in (2.1) can generate chaotic behavior. Each set of eigenvalues must be in equilibrium, it means that if an eigenvalue is purely real and positive, the real part of the complex eigenvalues must be negative and viceversa. This is one of the criteria to evaluate the chaotic behavior of a dynamical system [15].

The simulation of the Lorenz system requires the analysis of the numerical method. For example, Forward Euler has a stability defined by ($|1 - h\lambda| > 1$), and by substituting the nine eigenvalues, one finds that the step-size that accomplishes those eigenvalues is located between $0 < h < 0.08$. If the step-size accomplishes the inequality, one can ensure that the numerical method will be stable and can be applied to solve the initial value problem around the neighborhood of the equilibrium points. There could be chaotic oscillators that have critical points where the relative stability analysis is not satisfied, in such a case one must reduce the value of step-size to guarantee stability. For instance, Fig. 2.1 shows the chaotic time series simulation and Fig. 2.2 shows attractors views of (2.1) by applying Forward Euler with $h = 0.005$.

The coefficients of the ODEs can be varied to enrich the chaotic behavior, but one needs to explore the sets of values and in such a case one can take advantage of the bifurcation diagrams. Figure 2.3 shows the bifurcation diagrams of Lorenz by using β as bifurcation parameter. As one sees, the three state variables show that the chaotic behavior can be guaranteed when beta (β) takes values to generate many points in the vertical axes. This can be exploited to search for the best values of the coefficients to improve the dynamics of the positive Lyapunov exponent, maximize D_{KY} and entropy, as shown in the following chapters. The Lyapunov spectrum of the Lorenz chaotic oscillator computed by TISEAN is given in Fig. 2.4.

Other chaotic oscillators commonly used in the literature and modeled by 3 ODEs are given in Table 2.1. The associated parameter values to generate chaotic behavior is also included to determine the equilibrium points and eigenvalues. The Lyapunov exponents values are also given as well as their D_{KY}. Figure 2.5

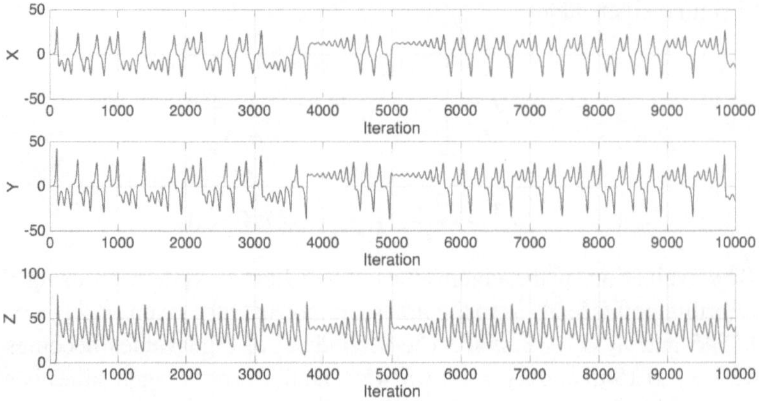

Figure 2.1: Chaotic time series of the state variables of Lorenz system given in (2.1) and solved by Forward-Euler with h=0.005.

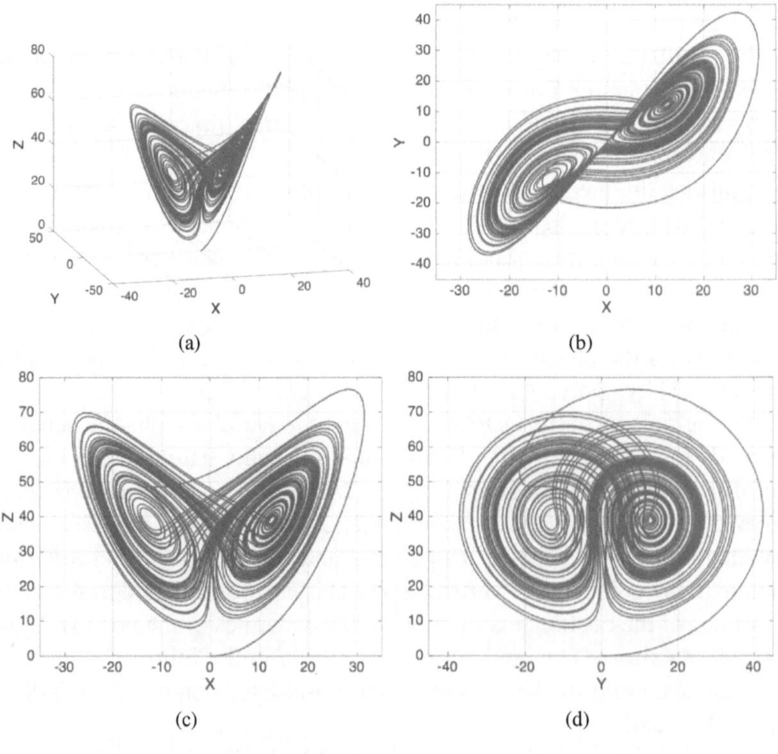

Figure 2.2: Views of the Lorenz chaotic attractor given in (2.1).

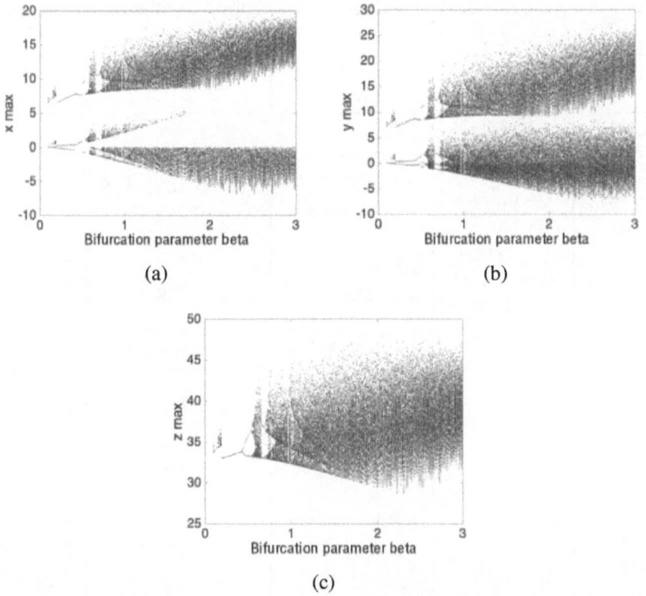

Figure 2.3: Bifurcation diagrams of Lorenz by using β as bifurcation parameter, for the state variables: (a) x, (b) y, and (c) z.

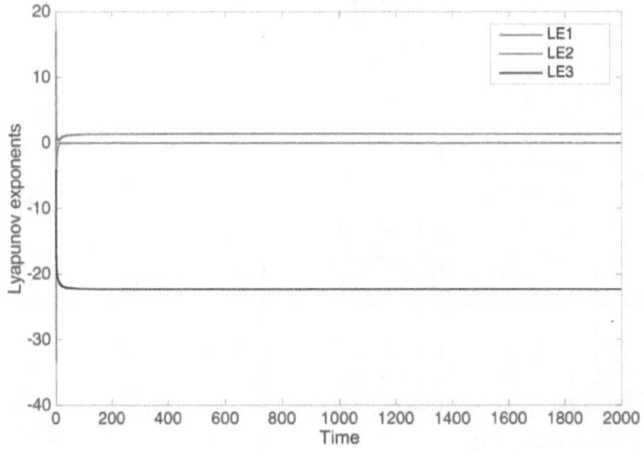

Figure 2.4: Lyapunov spectrum of the Lorenz chaotic oscillator.

shows views of the attractors of these chaotic systems, and some bifurcation diagrams are shown in Figs. 2.6–2.9.

The common characteristic of the ODEs modeling the 3D and integer order chaotic oscillators shown in Table 2.1, is that the nonlinear behavior is generated from the multiplication among two state variables. Other chaotic systems are

Table 2.1: 3D integer order chaotic oscillators.

System	Parameters	Equilibrium Points	Eigenvalues	Lyapunov Exp. and D_{KY}
Rössler $\dot{x}=-y-z,$ $\dot{y}=x+ay,$ $\dot{z}=b+z(x-c)$	$a=0.2,$ $b=0.2,$ $c=5.6,$ $x_0=0.007,$ $z_0=-0.03,$ $y_0=0.03$	$EP_1:$ $x=5.6929$ $y=-28.4648$ $z=28.4648$ $EP_2:$ $x=0.0070$ $y=-0.0351$ $z=0.0351$	$\lambda_{1,2}=-4\cdot10^{-7}$ $\pm j5.428$ $\lambda_3=0.1929,$ $\lambda_{1,2}=0.097$ $\pm j0.995$ $\lambda_3=-5.6869,$	$LE_1=0.07635$ $LE_2=0.00357$ $LE_3=-5.40493$ $D_{KY}=2.0148$
Chen $\dot{x}=a(y-x)$ $\dot{y}=(c-a)x$ $-xz+cy$ $\dot{z}=xy-bz$	$a=35,$ $b=3,$ $c=28$ $x_0=0.01,$ $y_0=0.01,$ $z_0=0.01$	$EP_1:$ $x=7.9372$ $y=7.9372$ $z=21.0$ $EP_2:$ $x=-7.9372$ $y=-7.9372$ $z=21.0$ $EP_3:$ $x=0.0$ $y=0.0$ $z=0.0$	$\lambda_{1,2}=4.2139$ $\pm j14.884$ $\lambda_3=-18.4279$ $\lambda_{1,2}=4.2139$ $\pm j14.884$ $\lambda_3=-18.4279$ $\lambda_1=-30.8359$ $\lambda_2=23.8359$ $\lambda_3=-3$	$LE_1=2.05974$ $LE_2=0.00455$ $LE_3=-12.06430$ $D_{KY}=2.1711$
Sprott case h $\dot{x}=-y+z^2$ $\dot{y}=x+ay$ $\dot{z}=x-bz$	$a=0.5,$ $b=1,$ $x_0=0.01,$ $y_0=0.01,$ $z_0=0.01$	$EP_1:$ $x=0.0$ $y=0.0$ $z=0.0$ $EP_2:$ $x=-2.0$ $y=4.0$ $z=-2.0$	$\lambda_{1,2}=0.25$ $\pm j0.968$ $\lambda_3=-1$ $\lambda_{1,2}=-0.3574$ $\pm j2.127$ $\lambda_3=0.2148$	$LE_1=0.11424$ $LE_2=-0.00042$ $LE_3=-0.61383$ $D_{KY}=2.1854$
Sprott case s $\dot{x}=-x-ay$ $\dot{y}=x+z^2$ $\dot{z}=b+x$	$a=4,$ $b=1,$ $x_0=-1.0,$ $y_0=0.2,$ $z_0=-0.8$	$EP_1:$ $x=-1.0$ $y=0.25$ $z=-1.0$ $EP_2:$ $x=-1.0$ $y=0.25$ $z=1.0$	$\lambda_{1,2}=-1.1014$ $\pm j2.331$ $\lambda_3=1.2029$ $\lambda_{1,2}=0.3037$ $\pm j2.21$ $\lambda_3=-1.6075$	$LE_1=0.18606$ $LE_2=0.00369$ $LE_3=-1.18975$ $D_{KY}=2.1595$

based on nonlinear functions and some of them can be linearized to approximate the behavior by PWL functions. These type of functions can be exploited to generate multi-scrolls and in multiple directions, and therefore one can derive a

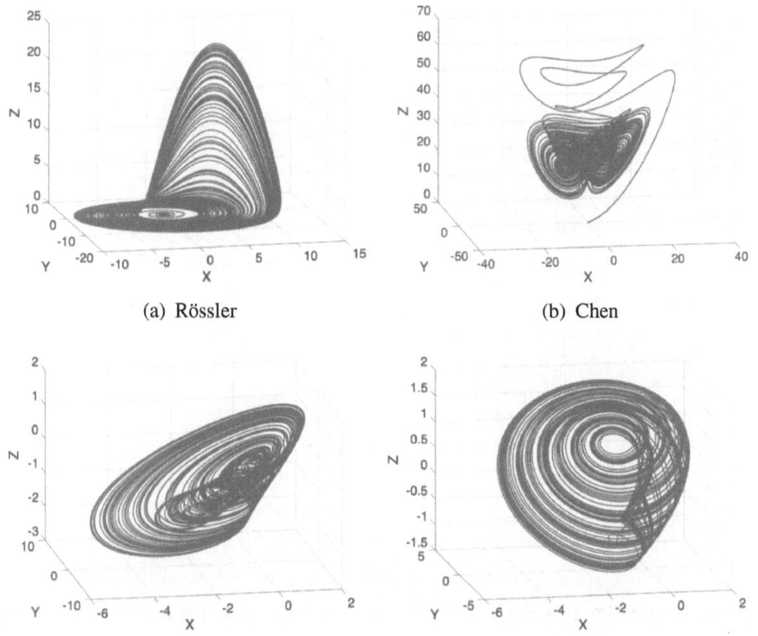

(a) Rössler (b) Chen

(c) Sprott case h (d) Sprott case s

Figure 2.5: Integer order chaotic attractors simulated by applying Forward-Euler with h=0.005 and the parameters given in Table 2.2 for: (a) Rössler, (b) Chen, (c) Sprott case h, and (d) Sprott case s.

systematic approach to synthesize the PWL functions according to the number of scrolls and directions being generated. This property is also associated with an increase in the value of the positive Lyapunov exponent, if the number of scrolls increase, the positive Lyapunov exponent increases [68]. For instance, Table 2.2 shows the 3D dynamical system to generate multi-scroll chaotic attractors, where $f(x)$ is the PWL function well-known as saturated nonlinear function (SNLF). One can synthesize an odd number of saturated levels in the PWL function to generate an odd-number of scrolls and one can synthesize an even number of saturated levels to generate an even-number of scrolls. The common values of the coefficients of the 3D ODEs are given as $a = b = c = d_1 = 0.7$, from which one can estimate the equilibrium points and eigenvalues.

Figure 2.10, shows the simulation of 2 and 4 scroll chaotic attractors. An extension of the PWL function and of the ODEs with more PWL functions leads us to generate 4×4 and $4 \times 4 \times 4$ attractors as shown in Fig. 2.11. Figure 2.12 shows the bifurcation diagram of the chaotic oscillator based on SNLF by choosing the parameter a as control. Recall that one can test all the coefficient values to identify the most suitable for a better control of the chaotic behavior.

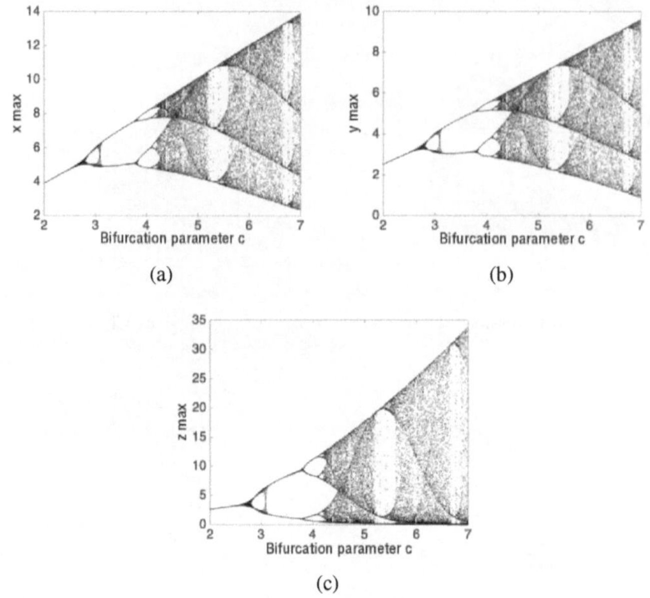

Figure 2.6: Bifurcation diagrams of Rössler system by varying parameter c vs. the state variables: (a) x, (b) y, and (c) z.

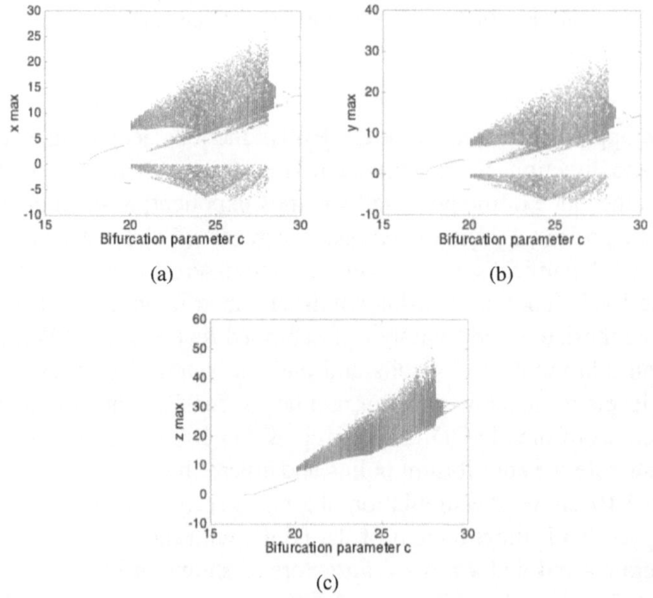

Figure 2.7: Bifurcation diagrams of Chen system by varying parameter c vs. the state variables: (a) x, (b) y, and (c) z.

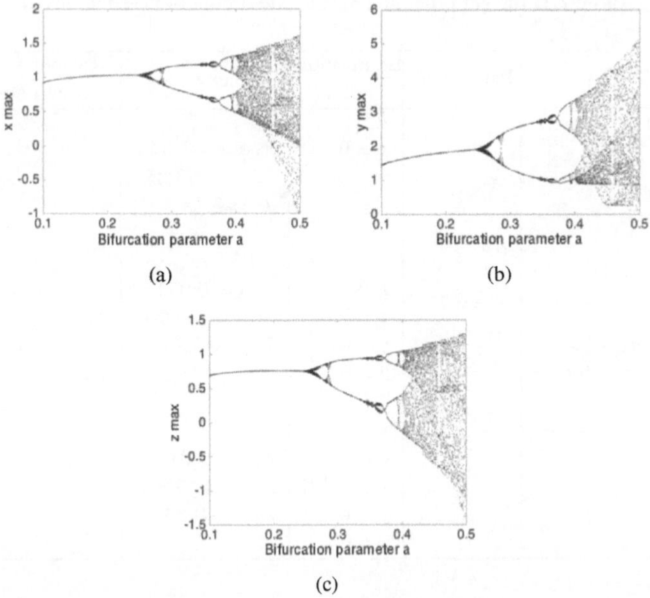

Figure 2.8: Bifurcation diagrams of Sprott case h system by varying parameter *a* vs. the state variables: (a) x, (b) y, and (c) z.

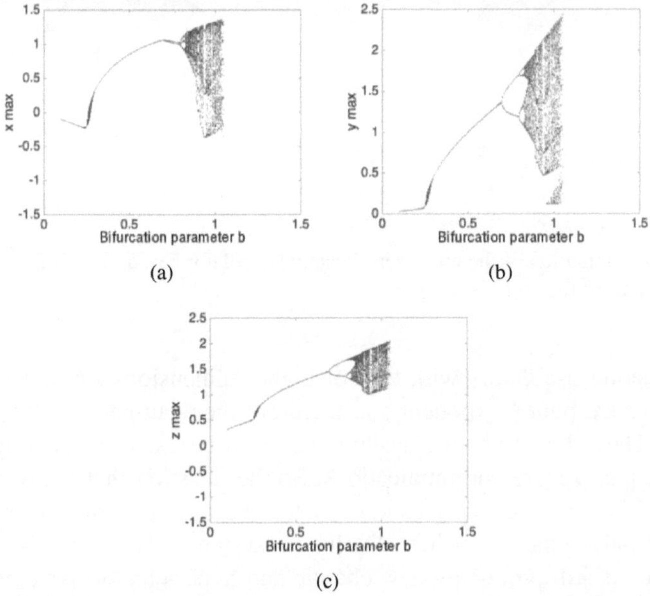

Figure 2.9: Bifurcation diagrams of Sprott case s system by varying parameter *b* versus the state variables: (a) x, (b) y, and (c) z.

Table 2.2: Integer order and 3D chaotic oscillator based on SNLF.

System	Parameters	Equilibrium Points	Eigenvalues	Lyapunov Exp. and D_{KY}
$\dot{x}=y$ $\dot{y}=z$ $\dot{z}=-ax-by$ $-cz+d_1f(x)$	$a=0.7$ $b=0.7$ $c=0.7$ $d_1=0.7$ $x_0=0.01$ $y_0=0.01$ $z_0=0.01$	EP_1: $x=0$ $y=0$ $z=0$ EP_2: $x=1.0$ $y=0$ $z=0$ EP_3: $x=-1.0$ $y=0$ $z=0$	$\lambda_{1,2}=-1.943$ $\pm j3.05$ $\lambda_3=3.18$ $\lambda_{1,2}=0.074$ $\pm j0.905$ $\lambda_3=-0.84$ $\lambda_{1,2}=0.074$ $\pm j0.905$ $\lambda_3=-0.84$	$LE_1=0.09816$ $LE_2=0.00153$ $LE_3=-0.79969$ $D_{KY}=2.1247$

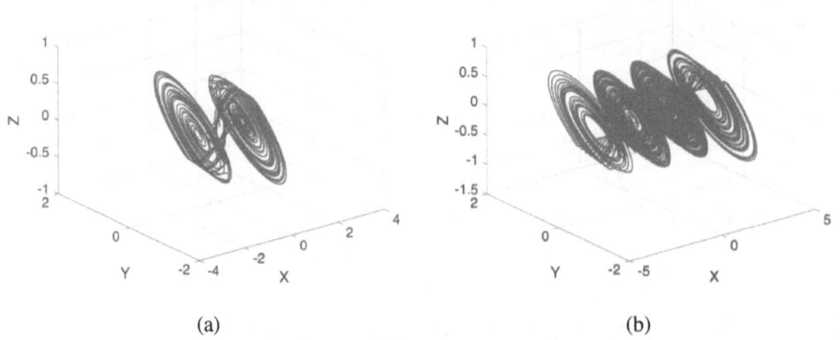

Figure 2.10: Simulation of the multi-scroll chaotic oscillator based on SNLF to generate: (a) two-scrolls and (b) four-scrolls.

The chaotic oscillators with four or higher-dimensions can have more than one positive Lyapunov exponent and therefore they can generate hyper-chaotic behavior. These hyper-chaotic systems can also be used in many applications such as chaotic secure communications. So the question that many researchers have asked is: Can the hyper-chaotic system be used to generate more unpredictability rather than by using 3D chaotic systems? Is it for this reason that the research is still producing new chaotic and hyper-chaotic systems, and they are compared with respect to their dynamical characteristics, such as Lyapunov exponent values, D_{KY} and entropy. In the side of applications, the research focuses on the hardware resources required for their implementation by using ana-

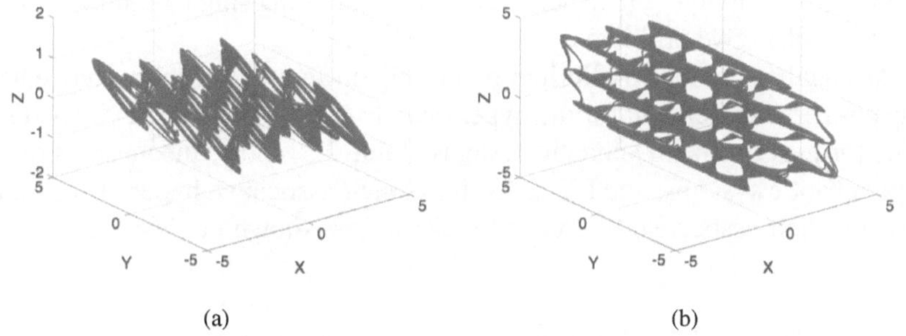

Figure 2.11: Simulation of the multi-scroll chaotic oscillator based on SNLF to generate: (a) 4-scrolls, and (b) $4 \times 4 \times 4$-scrolls.

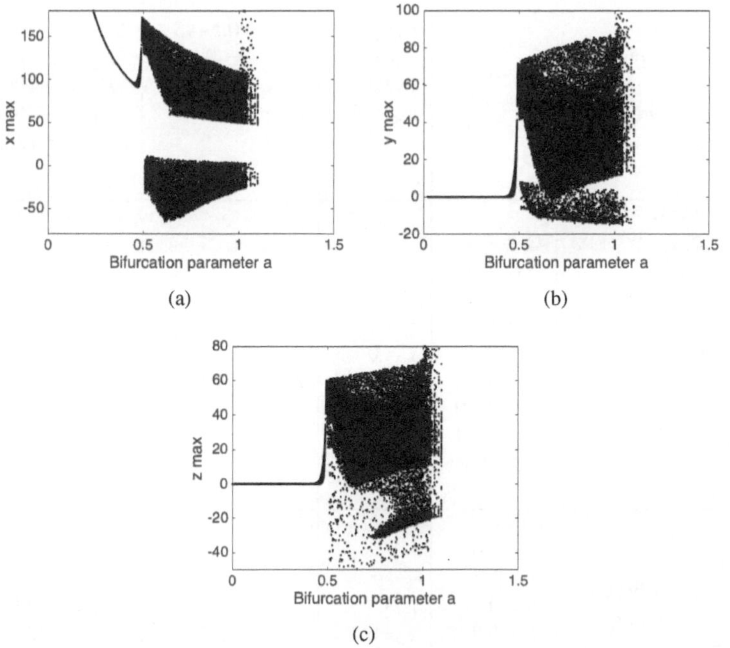

Figure 2.12: Bifurcation diagrams of the multi-scroll chaotic oscillator based on SNLF by varying parameter a and observing the state variables: (a) x, (b) y, and (c) z.

log or digital electronics. In this manner, one can infer that the best chaotic or hyper-chaotic system must provide the best dynamical behavior, allow electronic implementation and improve engineering applications. Henceforth, in the next chapters some of these questions are answered from the point of view of electronic implementations using analog and digital electronics. The selection of the best chaotic system is performed by optimizing the dynamical characteristics of the systems by applying metaheuristics and considering single or mono-objective and multi-objective problems.

The analysis of the equilibrium points, eigenvalues, Lyapunov exponents and Kaplan-Yorke dimension for the hyper-chaotic systems, can be performed in the same form as for the 3D chaotic systems. Table 2.3, shows the analysis of some hyper-chaotic systems, and Fig. 2.13 shows their associated hyper-chaotic attractors, and their associated bifurcation diagrams are shown in Figs. 2.14–2.16.

Table 2.3: Integer order hyper-chaotic oscillators.

System	Parameters	Equilibrium Points	Eigenvalues	Lyapunov Exp. and D_{KY}
$Volos[63]$ $\dot{x}=ax+y+$ $+yz-cw$ $\dot{y}=-xz+yz$ $\dot{z}=-z-mxy$ $+b$ $\dot{w}=x$	$a=1.0$ $b=0.25$ $c=0.7$ $m=1.0$ $x_0=1.0$ $y_0=-1.0$ $z_0=1.0$ $w_0=1.0$	$EP_1:$ $x=0$ $y=0$ $z=0.25$ $w=0$	$\lambda_{1,2}=0.5445$ $\pm j0.889$ $\lambda_3=0.1609$ $\lambda_4=-1.0$	$LE_1=0.19537$ $LE_2=0.06159$ $LE_3=-0.00019$ $LE_4=-2.09682$ $D_{KY}=3.1225$
$Karakaya[64]$ $\dot{x}=\alpha(y-W(w)x)$ $\dot{y}=-z-x$ $\dot{z}=-\beta y-\gamma z$ $\dot{w}=x$ $where:$ $W(w)=d$ $\text{if}\lvert w\rvert>1$ or $W(w)=c$ $\text{if}\lvert w\rvert<1$	$\alpha=4.0$ $\beta=1.0$ $\gamma=0.65$ $c=0.2$ $d=10.0$ $x_0=0.01$ $y_0=0.01$ $z_0=0.01$ $w_0=0.01$	$EP_1:$ $x=0$ $y=0$ $z=1.0$ $w=0$	$\lambda_{1,2}=-0.2671$ $\pm j2.148$ $\lambda_3=0.3842$ $\lambda_4=0.0$	$LE_1=0.00012$ $LE_2=-0.41573$ $LE_3=-0.42670$ $LE_4=-0.57618$ $D_{KY}=3.0002$
$Yu[66]$ $\dot{x}=ax+dW(\varphi)y$ $+yz-cw$ $\dot{y}=yz-xz$ $\dot{z}=-z-mxy+b$ $\dot{w}=x$ $\dot{\varphi}=y$ $where:$ $W(\varphi)=e+3n\varphi^2$	$a=1.0$ $b=1.0$ $c=0.7$ $m=1.0$ $d=0.2$ $e=0.01$ $n=0.01$ $x_0=0.01$ $y_0=0.01$ $z_0=0.01$ $w_0=0.01$ $\varphi_0=0.01$	$EP_1:$ $x=0$ $y=0$ $z=1.0$ $w=0$ $\varphi=0$	$\lambda_{1,2}=0.8378$ $\pm j1.206$ $\lambda_3=0.3242$ $\lambda_4=0$ $\lambda_5=-1.0$	$LE_1=0.26044$ $LE_2=0.02307$ $LE_3=-0.00475$ $LE_4=-0.07893$ $LE_5=-4.91462$ $D_{KY}=4.0407$

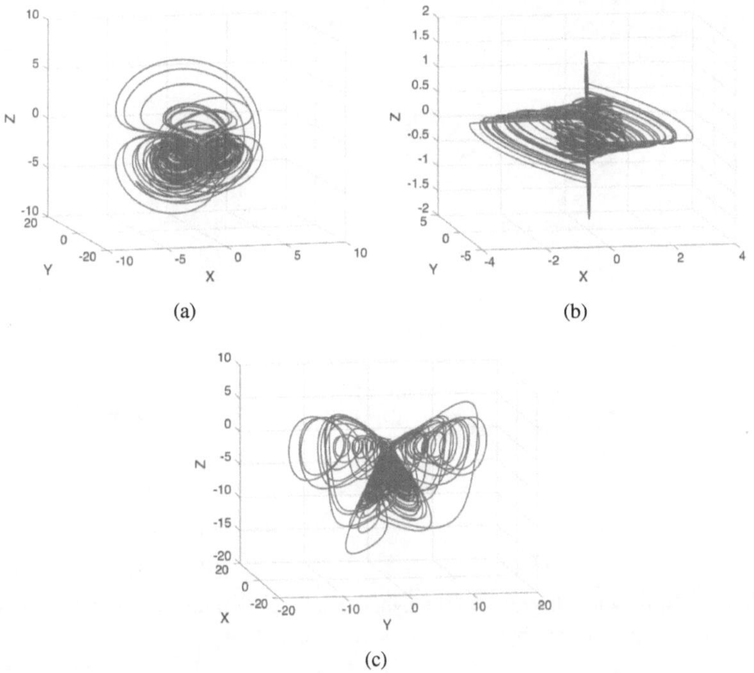

Figure 2.13: Simulation of integer order hyper-chaotic attractors using the parameters given in Table 2.2. Oscillators introduced in (a) [63], (b) [64], and (c) [66].

2.3 Amplitude scaling for the analog implementation of autonomous chaotic oscillators

The first electronic implementations of autonomous chaotic oscillators were developed using operational amplifiers and diodes. Basically, those implementations were performed from a state-variables synthesis approach in which the derivatives were solved by designing integer-order integrators. Nowadays, the integrators can be implemented using different kinds of amplifiers working in voltage-mode, current-mode or mixed-mode and they can be designed using complementary-metal-oxide-semiconductor (CMOS) technology of integrated circuits, as already summarized in [69]. Three decades ago some authors used the operational transconductance amplifier (OTA) to design CMOS chaotic oscillators and to synthesize PWL functions, as recently shown in [70]. Having integer-order integrators implemented with operational amplifier topologies, they can be transformed to OTA-based circuits, as shown in [71]. This design process can be extended to implement fractional-order chaotic oscillators, as detailed in the following chapters. The topologies of the chaotic oscillators can also be synthesized into a field-programmable analog array (FPAA), but in such a case the

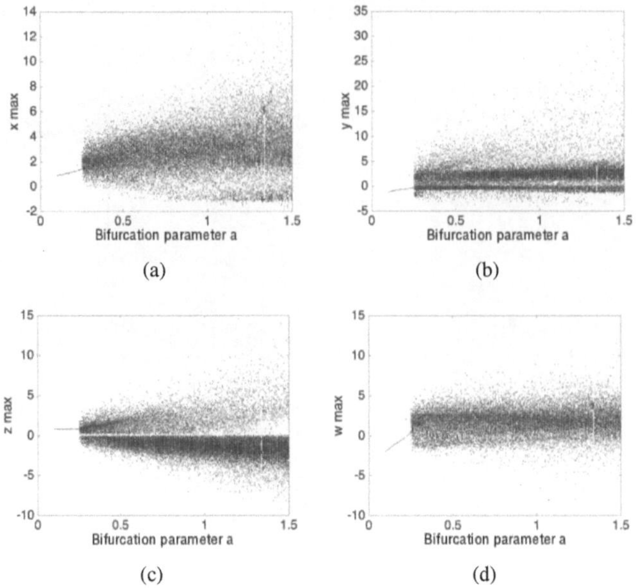

Figure 2.14: Bifurcation diagrams of the hyper-chaotic system introduced in [63] by varying parameter a vs. the state variables: (a) x, (b) y, (c) z, and (d) w.

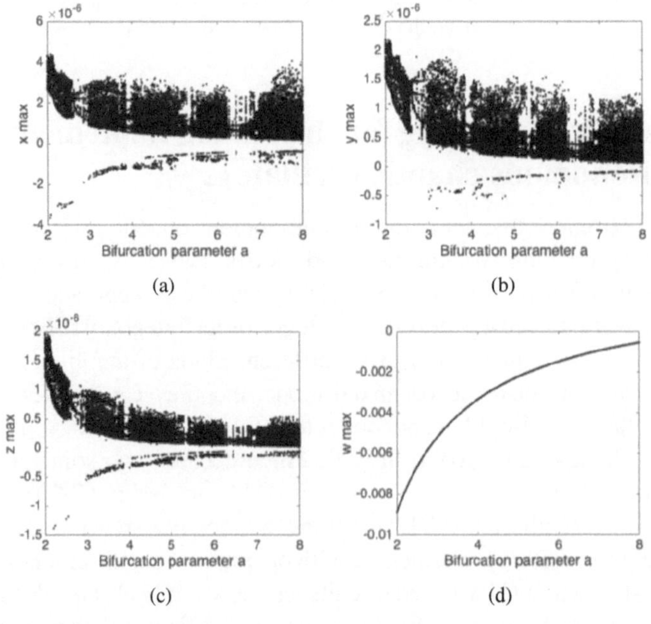

Figure 2.15: Bifurcation diagrams of the hyper-chaotic system introduced in [64] by varying parameter a vs. the state variables: (a) x, (b) y, (c) z, and (d) w.

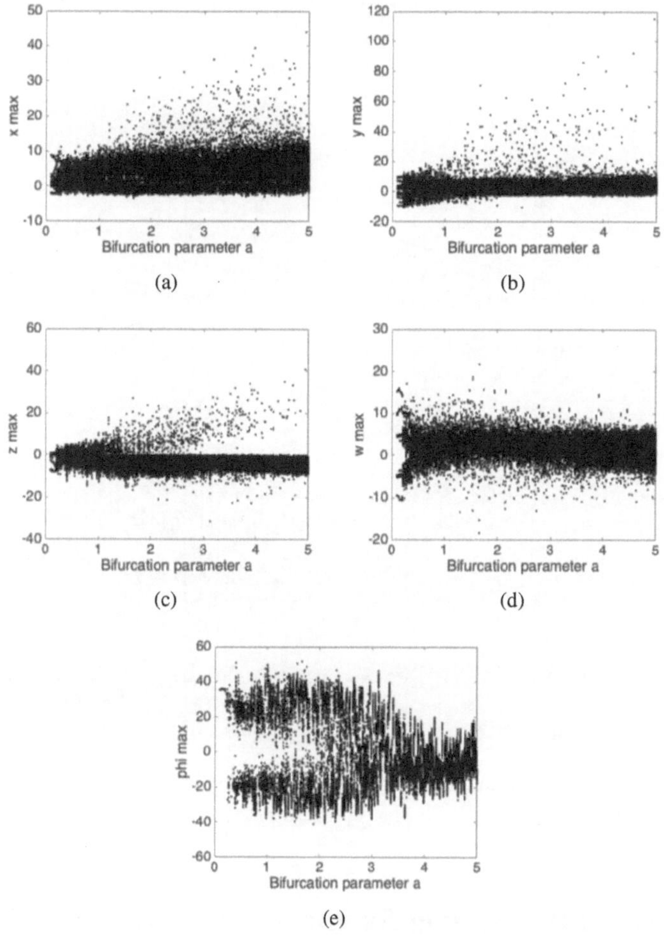

Figure 2.16: Bifurcation diagrams of the hyper-chaotic system introduced in [66] by varying parameter a vs. the state variables: (a) x, (b) y, (c) z, (d) w, and (e) φ.

amplitudes of the chaotic time series must be scaled to the ranges that the CMOS circuit or FPAA can drive. Lets us consider the Lorenz chaotic oscillator given in (2.1) with the parameters $\sigma = 10, \rho = 28$ and $\beta = 8/3$. The attractor and chaotic time series are shown in Fig. 2.17. As one sees, the amplitude ranges of the state variables x, y, z are between the ranges [-21.5, 21.5], [-29.44, 29.44], and [-54.26, 54.26], respectively. It is quite obvious that these ranges are higher than the voltage ranges supported by commercially available amplifiers, CMOS technology or FPAA devices, so that the mathematical equations should be down-scaled to allow their analog implementation.

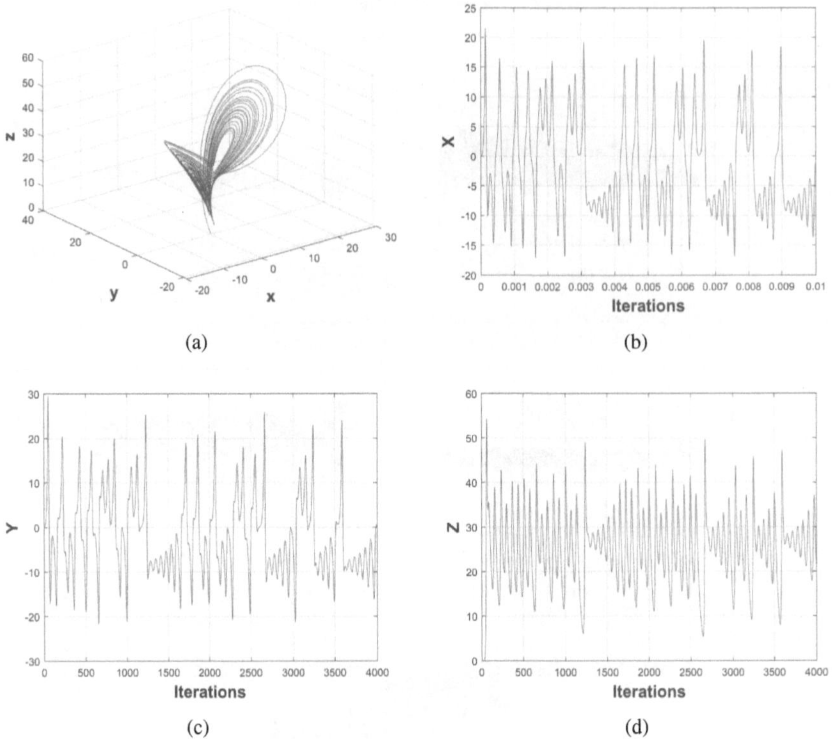

Figure 2.17: Lorenz chaotic attractor in (a), and its chaotic time series for each state variables: (b) x, (c) y and (d) z.

2.3.1 Amplitude scaling for the amplifier-based implementation of integer-order chaotic systems

By using operational amplifiers and multipliers, the circuit realization of the Lorenz chaotic oscillator is shown in Fig. 2.18, which consists of three commercially available amplifiers TL082 to perform the addition, subtraction and integration operations, and two multipliers AD633 to implement the nonlinear functions xz and xy. The values of the passive circuit elements are set after performing the downscaling as follows: The circuit elements associated to σ, ρ and β, are scaled to 1MΩ for the resistances in which $\sigma = 10$ is related to R_1 and R_2 for the state variables x and y, respectively. Defining $C_1 = C_2 = C_3 = 2$nf, the scaled R_1=1MΩ/100kΩ, and the time constant for the three integrators is established to be $1/R_1C_1 = 5 \times 10^3$.

The second equation of Lorenz oscillator $\dot{y} = x(\rho - z) - y$, has $\rho = 28$, and it is associated to R_3 and x, therefore the scaled value becomes R_3=1MΩ/35.7kΩ. The second input is associated with $-xz$ provided at the output of the mul-

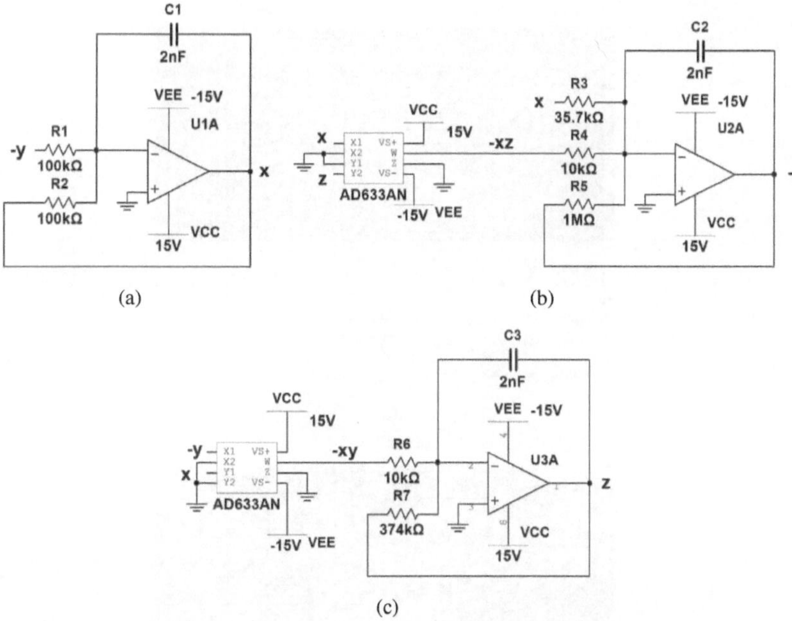

Figure 2.18: Analog implementation of Lorenz chaotic circuit showing the state variables: (a) x, (b) y, and (c) z.

tiplier AD633 and R_3 with unitary scaling, so that the equations are normalized to an amplitude of 0.1V, this is the scale factor 100 of the multiplier, i.e., R_4=1MΩ/10kΩ. The third input has y with unitary value R_5=1MΩ/1MΩ. For the equation $\dot{z} = xy - \beta z$, it is associated to the output $-xy$ of the multiplier AD633 and R_6 with unitary scaling, i.e., R_6=1MΩ/10kΩ. The second input is associated to z and β=8/3, leading to scale R_7=1MΩ/374kΩ. The scaling discussed above accomplishes that all the state variables have a similar dynamic range so that their voltage ranges are within the supplies of the amplifiers and multipliers. The equivalent equations replacing the coefficients by R and C elements is now given by (2.5), which are obtained analyzing Fig. 2.18. The SPICE simulation results are shown in Fig. 2.19, in which the passive circuit elements are set to: $R_1 = R_2 = 100k\Omega$, $R_3 = 35.7k\Omega$, $R_4 = R_6 = 10k\Omega$, $R_5 = 1M\Omega$, $R_7 = 374k\Omega$, and $C_1 = C_2 = C_3 = 2nF$.

$$
\begin{aligned}
\frac{dx}{dt} &= -\frac{1}{R_1C_1}y + \frac{1}{R_2C_1}x \\
\frac{dy}{dt} &= \frac{1}{R_3C_2}x - \frac{1}{R_4C_2}xz - \frac{1}{R_5C_2}y \\
\frac{dz}{dt} &= -\frac{1}{R_6C_3}xy + \frac{1}{R_7C_3}z
\end{aligned}
\tag{2.5}
$$

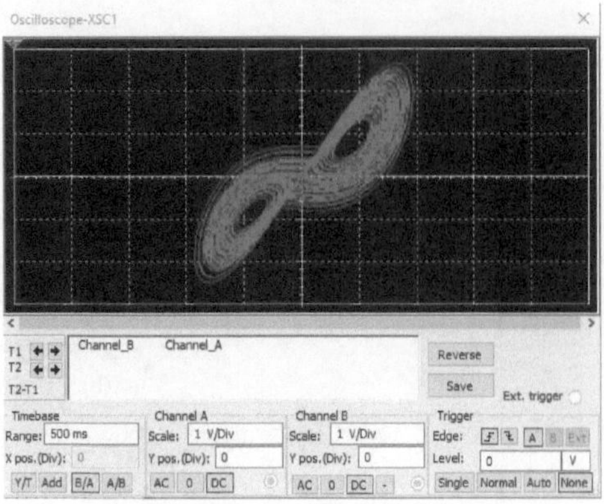

(a)

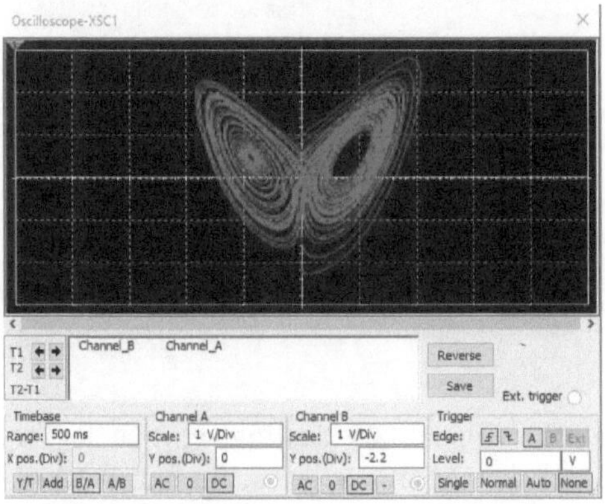

(b)

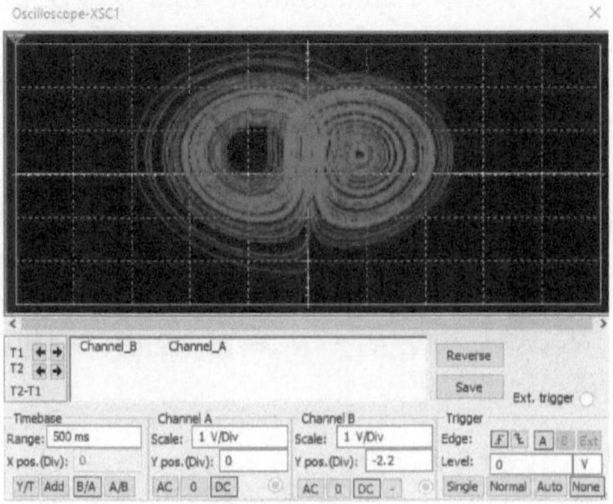

(c)

Figure 2.19: Attractors of Lorenz's chaotic circuit (1V/Div) in the: (a) $x-y$, (b) $x-z$, and (c) $y-z$ space portraits.

2.3.2 Amplitude scaling for the FPAA-based implementation of integer-order chaotic systems

An FPAA is an analog signal processor, equivalent to the digital one included within the field-programmable gate array (FPGA). The FPAA is an electronic device of specific purpose with the advantage of being reconfigurable electrically, and it can be used to implement a wide variety of analog functions, among them: integration, derivation, pondered sum/subtraction, filtering, rectification, comparator, multiplication, division, analog-to-digital conversion, voltage references, signal conditioning, amplification, nonlinear functions, generation of arbitrary signals, among others [72].

The Anadigm QuadApex development board is an easy-to-use platform designed to allow fast prototyping to implement and test analog designs on the FPAA that is biased with 3.3V and can be programmed using Anadigm Designer 2 EDA software [73], which generates C-code automatically. The AN231E04 FPAA consists of a 2×2 matrix of fully Configurable Analog Blocks (CABs), which are surrounded by programmable interconnect resources and analog input/output cells with active elements. The inclusion of an 8×256 bit look-up table (LUT) enables waveform synthesis and handling of several non-linear functions. The FPAA design process requires describing circuit functions that are represented as configurable analog modules (CAMs), which map onto portions of CABs. The EDA software and development board facilite instant prototyping of any circuit that is implemented in the EDA tool [74].

Lets us consider again the Lorenz system with the ranges shown in Fig. 2.17, which cannot be synthesized into an FPAA because the allowed range of the FPAA AN231E04 is between ± 3V. In this manner, the amplitude scaling can be performed by transforming the state variables to $x = 5X$, $y = 5Y$ and $z = 5Z$, thus updating the Lorenz equations to (2.6). In the FPAA a CAM Sum/Difference Integrator is used to perform addition, subtraction and integration operations. The CAM creates a summing integrator with up to three inputs that can be inverting or non-inverting topologies to create sums and differences in the transfer function. Each sampled input branch has a programmable integration constant. For example, the expression given in (2.7) has three variables K_1, K_2, K_3 associated to the input branches $V_{Input1}, V_{Input2}, V_{Input3}$. The third term in (2.7) will only be implemented if the corresponding CAM Option Input is turned on. The sign of each term is selected for each input branch in the CAM Options. Other terms are added for non-inverting inputs and subtracted for inverting inputs. For example, the transfer function for this CAM configured with only two non-inverting inputs becomes $\frac{\Delta V_{out}}{\Delta t} = \pm K_1 V_{Input1} \pm K_2 V_{Input2}$, where ΔV_{out} is the change in the output voltage during one clock period, and Δt is the length of one clock period. The equivalent transfer function in the Laplace domain is given in (2.8), where the constants can be replaced by capacitor values satisfying the following relations:

$K_1 = \frac{f_c C_{inA}}{C_{int}}$, $K_2 = \frac{f_c C_{inB}}{C_{int}}$ and $K_3 = \frac{f_c C_{inC}}{C_{int}}$.

$$\begin{aligned} \dot{X} &= \sigma(Y - X) \\ \dot{Y} &= \rho X - Y - 5XZ \\ \dot{Z} &= 5XY - \beta Z \end{aligned} \tag{2.6}$$

$$\frac{\Delta V_{out}}{\Delta t} = \pm K_1 V_{Input1} \pm K_2 V_{Input2} \pm K_3 V_{Input3} \tag{2.7}$$

$$V_{out}(s) = \frac{\pm K_1 V_{Input1}(s) \pm K_2 V_{Input2}(s) \pm K_3 V_{Input3}(s)}{s} \tag{2.8}$$

The integration constants are evaluated as $K = 1/RC$. In Anadigm Designer 2 the integration constant must be established considering the available values in $1/\mu S$, not in $1/S$. As an example, it can be established to $K = 1 \times 10^{-6}/RC$ to get the CAM-parameters shown in Table 2.4 for the blocks SumIntegrator. Figure 2.20 shows the seven configurable input/output structures IOCELL that implement Lorenz system, where n5, n6 and n7 are configured as outputs and bypass type, as shown in Table 2.4.

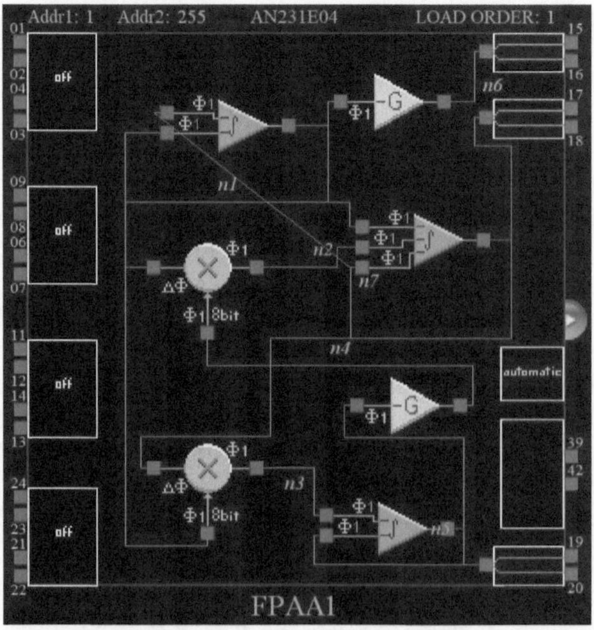

Figure 2.20: FPAA-based implementation of Lorenz chaotic oscillator.

Table 2.4: Configurable Analog Modules for the FPAA-based implementation of the Lorenz chaotic system.

Block Name	Options	Parameters	Clocks
SumIntegrator1 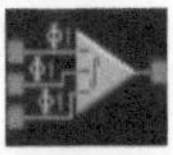	Output changes On: Phase 1 Input1: Non-Inverting Input2: Inverting Input3: Inverting	Int. Const. 1 (Upper)[1/us] 0.014 Int. Const. 2 (Middle)[1/us] 0.05 Int. Const. 3 (Lower)[1/us] 0.0005	ClockA 50 kHz (Chip Clock 1)
SumIntegrator2 	Output changes On: Phase 1 Input1: Inverting Input2: Inverting	Int. Const. 1 (Upper)[1/us] 0.005 Int. Const. 2 (Lower)[1/us] 0.005	ClockA 50 kHz (Chip Clock 1)
SumIntegrator3 	Output changes On: Phase 1 Input1: Inverting Input2: Inverting	Int. Const. 1 (Upper)[1/us] 0.005 Int. Const. 2 (Lower)[1/us] 0.00134	ClockA 50 kHz (Chip Clock 1)
Multiplier1 	Output changes On: Phase 1 Sample and Hold: Off	Multiplication Factor 1.00	ClockA 50 kHz (Chip Clock 1) ClockB 800 kHz (Chip Clock 0)
Multiplier2 	Output changes On: Phase 1 Sample and Hold: Off	Multiplication Factor 1.00	ClockA 50 kHz (Chip Clock 1) ClockB 800 kHz (Chip Clock 0)
GainHold1 	Input Sampling Phase 1 Sample and Hold: Off	Gain 1.00	ClockA 50 kHz (Chip Clock 1)
GainHold2 	Input Sampling Phase 1 Sample and Hold: Off	Gain 1.00	ClockA 50 kHz (Chip Clock 1)
IOCell5 	I/O Mode: Output Output of variable x	Output Type: Bypass	
IOCell6 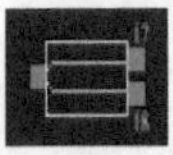	I/O Mode: Output Output of variable y	Output Type: Bypass	
IOCell7 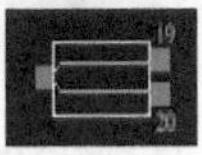	I/O Mode: Output Output of variable z	Output Type: Bypass	

The CAM is also used herein to perform the multiplication operations associated with the products of state variables in the Lorenz system, such as XZ and XY given in (2.6). The CAM Multiplier listed in Table 2.4 has the transfer function given in (2.9), where M is related to a multiplication factor and its value needed to be tuned to perform the multiplication of two variables in the correct way. This CAM of two inputs requires two frequencies to operate, and they are given by Clock A and Clock B, having a relation that requires that Clock B be 16 times Clock A [74]. Table 2.5 gives the FPAA configuration for the Clocks to be used in the implementation of (2.6).

$$V_{out} = M \cdot V_x \cdot V_y \tag{2.9}$$

Table 2.5: Frequencies of the Clocks in the FPAA.

Master Clock - ACLK (fc)	16 MHz		
System Clock 1 (sys1 = fc/20)	800 kHz	System Clock 2 (sys2=fc/1)	16 MHz
Clock 0 (sys1 / 1)	800 kHz	Clock 1 (sys1/16)	50 KHz

2.4 Simulating chaotic oscillators in Anadigm Designer tool

Anadigm Designer 2 is an EDA tool that can be used to quickly and easily construct complex analog circuits by selecting, placing and wiring building block like sub-circuits that are referred to as CAMs, and which can be synthesized into an FPAA. The virtual laboratory allows simulating the FPAA-based implementation and to view the results in the oscilloscope option [75], just before the physical implementation.

Figure 2.21 shows the four FPAA AN231E04 that are embedded into Anadigm QuadApex development board. To simulate the scaled Lorenz equa-

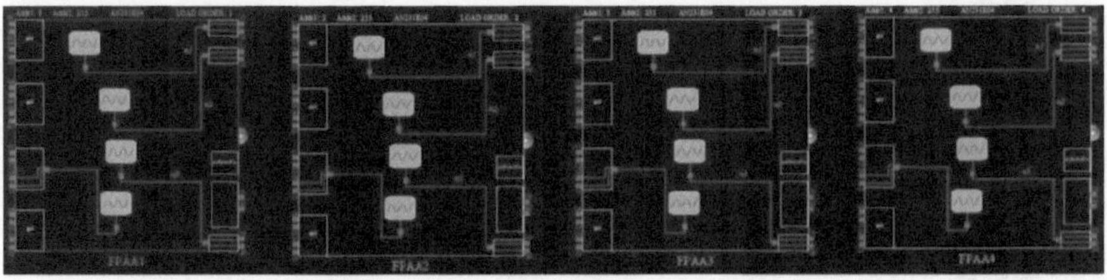

Figure 2.21: Sketch of the four FPAAs embedded into AN231E04 Anadigm QuadApex.

tions given in (2.6), the first step consists of estimating the required resources and to verify that the FPAA is suitable for the implementation. Afterwards, one can select, adjust and connect the CAMs and IO cells with wires, so that the Lorenz chaotic oscillator is implemented as shown in Fig. 2.20.

As the Lorenz chaotic oscillator is an autonomous circuit, one must insert a signal generator. It can be added as a dummy block such that the EDA tool can perform the simulation, whose results of the chaotic time series are shown in Fig. 2.22. In Fig. 2.22 it can also be observed that there is an external connection between the output O07 and input IO3, in which the cables are interchanged to invert the signal. This can be performed to emulate an inverter when the resources of the FPAA are limited. The simulation results provided by Anadigm Designer 2 are shown in Fig. 2.23.

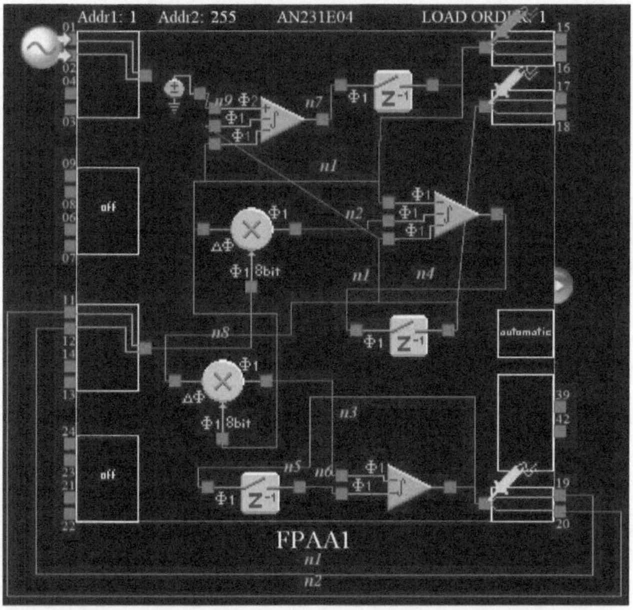

Figure 2.22: Description of Lorenz chaotic oscillator in Anadigm Designer 2 including delay blocks (z^{-1}) to observe the state variables x, y, z.

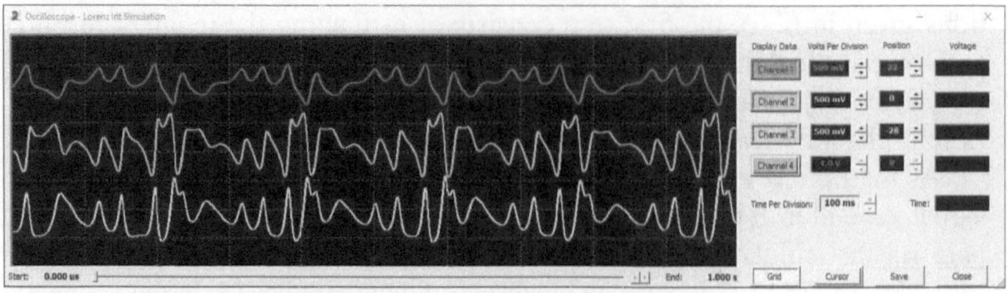

(a)

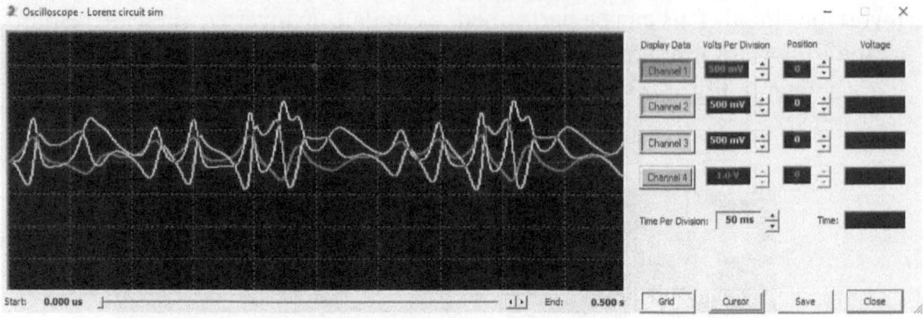

(b)

Figure 2.23: Simulation of Lorenz chaotic oscillator in Andigm Designer 2 to observe the state variables x, y, z.

2.5 Experimental results of the FPAA-based implementation of Lorenz system

The synthesis of (2.6) into the FPAA Anadigm QuadApex Development Board AN231E04, requires the configuration of the CAM shown in Table 2.4. Figure 2.20 shows the final design that does not include the delay blocks shown in Fig. 2.22. The FPAA-based implementation is measured using an oscilloscope to observe the experimental attractors and chaotic time series, which are shown in Fig. 2.24.

The same design process can be performed for chaotic oscillators of a higher-dimension, the unique drawback is the quantity of resources that are available in the FPAA.

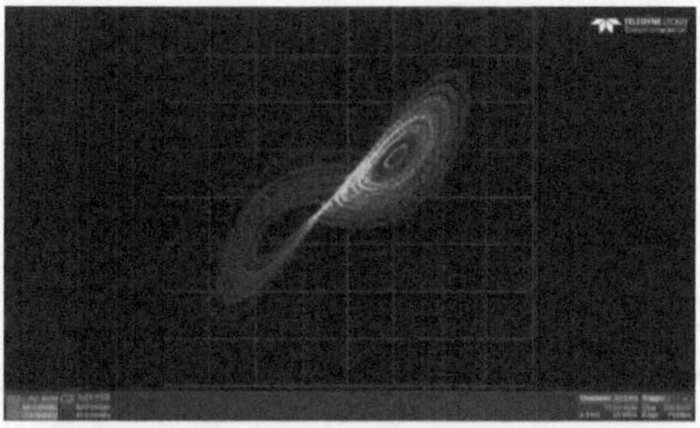

(a)

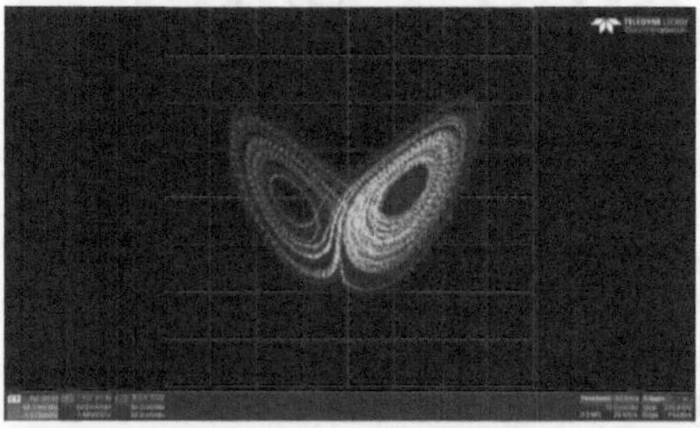

(b)

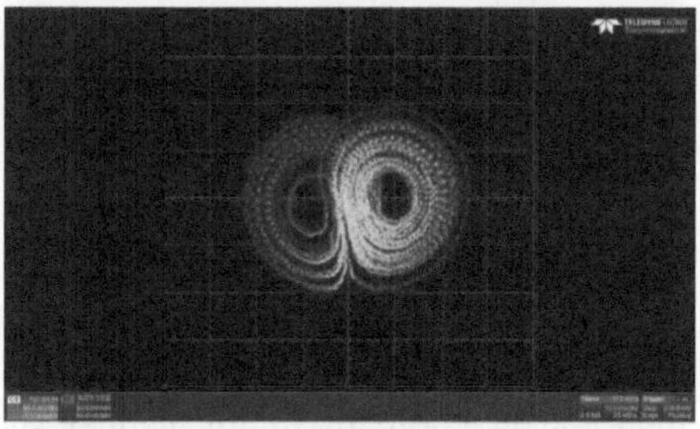

(c)

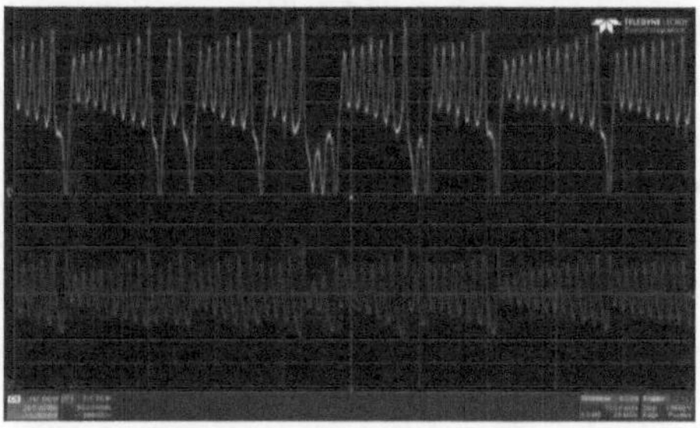

(d)

Figure 2.24: Experimental observation of the chaotic attractors of the FPAA-based implementation of Lorenz circuit (50 mV/Div): (a) portrait $x-y$, (b) portrait $x-z$, (c) portrait $y-z$, and (d) time series x (200 mV/Div) with amplitude of 1.518 V and y (50 mV/Div) with an amplitude of 1.666 V.

Chapter 3

Fractional-Order Chaotic/Hyper-chaotic Oscillators

Integer-order chaotic oscillators are modeled by ordinary differential equations (ODEs) whose derivatives have integer orders ($q = 1, 2, 3, \ldots$). Fractional-order chaotic oscillators are modeled by ODEs whose derivatives have fractional orders ($q = \frac{1}{2}, \frac{1}{4}, 0.8, 0.9, 0.9123, \ldots$). The special numerical methods described in Chapter 1 can be applied to solve fractional-order ODEs, and one can also apply approximations in the frequency domain. In this manner, such special definitions are described and herein divided in frequency and time-domain methods and applied to solve fractional-order chaotic and hyper-chaotic oscillators.

In 1695 Leibniz sent a letter answering L'Hopital's question about what could happen if the derivative has fractional-order $q = 1/2$. Thus they introduced fractional-order calculus. Leibniz said that if $q = 1/2$ then it will conduct a paradox from which one day it will conduct towards useful consequences. From the time fractional-calculus was developed as pure mathematics, and in the last decades and due to the availability of computing resources it has been found that fractional-derivatives are extremely useful to describe and/or model physical phenomena with more exactness than by applying integer-order calculus. Nowadays, a huge number of phenomena in the real engineering applications apply fractional calculus, for example in acoustic, thermal systems, mechanics, material science, signal processing, systems identification, reconfigurable hardware, etc. [35].

The nonlinear dynamical systems as fractional-order chaotic oscillators can be described by ODEs whose derivatives can have the same fractional-orders (called commensurate systems) or different fractional-orders (called incommensurate systems). In both cases, one applies special numerical methods to solve the fractional-order ODEs in which the challenge is the best adaptation of the method to solve the problem. Among the currently available methods to solve fractional-order chaotic oscillators, this book reviews the ones that can be implemented on embedded systems such as field-programmable analog arrays (FPAAs) and field-programmable gate arrays (FPGAs). The frequency domain methods are applied to implement fractional-order chaotic oscillators by using fractances or the fractional-order derivatives are approximated by a ratio of polynomials of integer order, and both cases can be implemented by using discrete electronic devices or FPAAs. The time domain methods are applied to implement fractional-order chaotic oscillators by using FPGAs. The special definitions that are applied herein are Grünwald Letnikov and Adams Bashfort-Moulton (predictor-corrector) methods, which require one to estimate an adequate step-size h and memory size [4]. In some cases one can reduce the memory size to diminish computing time [32]. Special attention is paid choosing an adequate h to reduce computing time and to guarantee the stability of the method and to avoid computational chaos and superestability [14].

3.1 Numerical simulation in the frequency domain

Fractional-order chaotic systems can be considered a generalization of integer-order ones, so that as already done for integer-order chaotic oscillators, the most common instance of a fractional-order system can be represented by a transfer function in the frequency domain. The authors in [34] showed the FPAA-based implementation of fractional-order chaotic oscillators using first-order active filter blocks. Basically, the fractional-order derivatives are approximated by a ratio of integer-order polynomials in the Laplace domain. The transfer function is decomposed into first-order power blocks that are multiplied and then those blocks are synthesized by active filter blocks. This process is described in the following sections.

Lets us consider the irrational transfer function also known as fractional-order integrator given in (3.1), where $s = j\omega$ is the complex frequency and the fractional-order is a positive real number such that $0 < \alpha < 1$.

$$H(s) = \frac{1}{s^{\alpha}} \qquad \alpha \in \mathbb{R}^{+} \tag{3.1}$$

The fractional-order integrator can be approached by different methods in the frequency domain and the resulting transfer function can be synthesized by integer-order blocks that can be implemented by commercially available amplifiers or into an FPAA. For instance, Charef approximated the derivatives and integrators of fractional-orders from 0.1 to 0.9 in steps of 0.1, as detailed in [76]. By applying Charef's approximations, several implementations of fractional-order chaotic oscillators have been introduced, for example: the authors in [77, 78] performed implementations of incommensurate fractional-order systems, and also fractional-order hyper-chaotic oscillators have been implemented in [79, 80].

The fractional-order Lorenz chaotic oscillator is given in [81], and modeled by (3.2), where it can be appreciated that just the derivatives have fractional-orders q_1, q_2, q_3. In its integer-order version it generates chaotic behavior by setting: $\sigma = 10, \rho = 28$ and $\beta = 8/3$, but in its fractional-order case, it can be of commensurate type when the derivatives have the same fractional orders, such as $q_1 = q_2 = q_3 = 0.9$. Otherwise, if the fractional-orders are different the dynamical system is incommensurate. A time simulation by applying Grünwald Letnikov method generates the phase-space portraits shown in Fig. 6.5, where the ranges of the state variables x, y, z are higher than that supported by commercially available amplifiers and FPAAs, therefore, the state variables will be down-scaled as shown in Chapter 2 to perform its electronic circuit implementation.

$$\begin{aligned} {}_0D_t^{q_1}x(t) &= \sigma(y(t) - x(t)), \\ {}_0D_t^{q_2}y(t) &= x(t)(\rho - z(t)) - y(t), \\ {}_0D_t^{q_3}z(t) &= x(t)y(t) - \beta z(t), \end{aligned} \tag{3.2}$$

The fractional-order derivatives can also be solved by in the frequency domain by using the fractional-order integrator $H(s) = \frac{1}{s^q}$, as detailed in [23]. It can be implemented by a fractance or fractor [44], which consists of arrays of passive circuit elements such as resistances and capacitances that are interconnected in series or parallel arrays to form ladder or tree topologies [82], as shown in Fig. 3.2, in which the symbol of the fractance has the form of an element similar to a resistance and a capacitance.

Basically, there are two types of circuit interconnections for the implementation of the fractor or fractance when using RC arrays, namely: ladder and tree topologies that can be cascaded to approximate complex fractional-orders. However, in electronics, yet one cannot find a pure fractance device that accomplishes the desired fractional-order being implemented. Henceforth, the fractance or fractor is emulated by combining RC ladder, cascade or tree topologies. Similar to the integer-order derivative and integrator blocks, by using an operational amplifier one can implement the fractional-order differentiation and integration operations, as shown in Fig. 3.3.

In the Laplace domain, the equivalent rational function of the RC circuit model between nodes A and B shown in Fig. 3.2(a) and 3.2(c), can be expanded to approximate with low error, and in a wide range of operation in the frequency domain, the integrator $\frac{1}{s^q}$. For instance, the transfer functions associated to the fractional-orders between $0.1 < q < 0.9$, and at steps of 0.1, are given in Table 3.1, which were approximated according to the frequency domain method detailed in [23, 83].

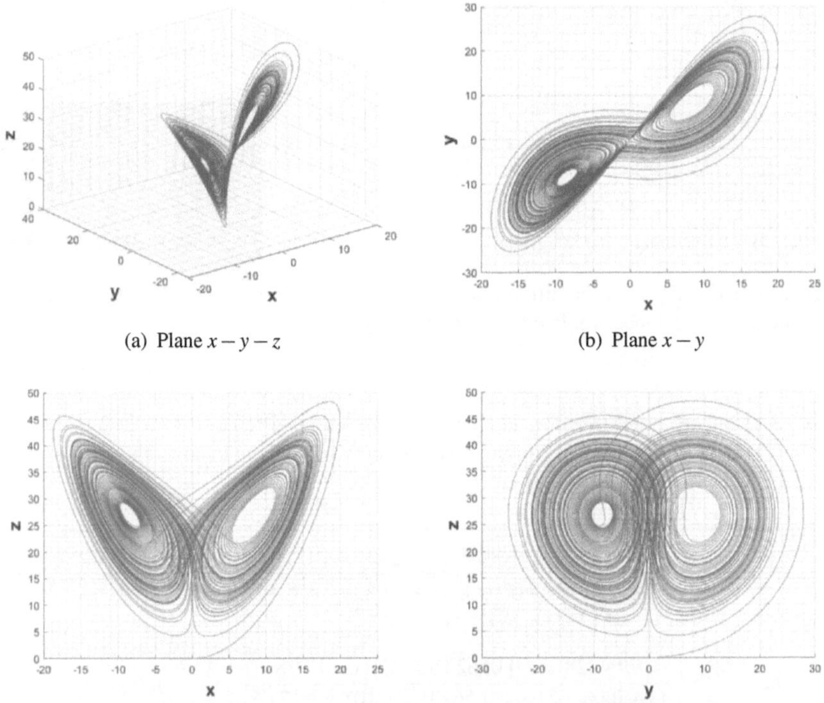

(a) Plane $x-y-z$ (b) Plane $x-y$

(c) Plane $x-z$ (d) Plane $y-z$

Figure 3.1: Time simulation of the fractional-order Lorenz chaotic oscillator by applying Grünwald-Letnikov method for $T_{sim} = 100$ seconds, $h = 5 \times 10^{-3}$, initial conditions of $(x_0, y_0, z_0) = (0.1, 0.1, 0.1)$, and the fractional-orders of the derivatives $q_1 = q_2 = q_3 = 0.9$.

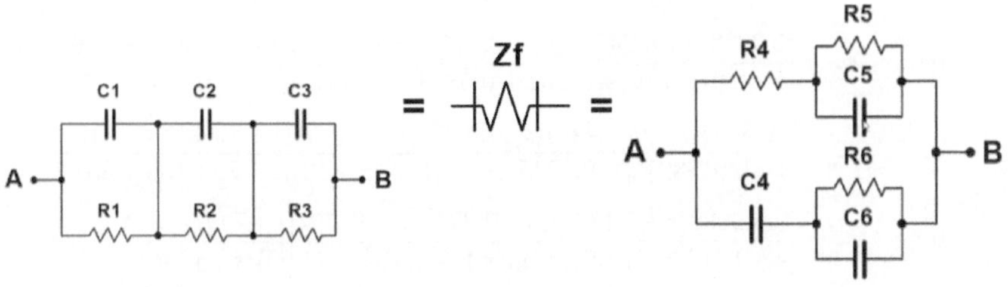

(a) Ladder topology (b) Fractance symbol (c) Tree topology

Figure 3.2: RC-circuit approximations of the fundamental fractance to implement the fractional-order of a dynamical system in the range $0 < q < 1$.

As one can infer, in the frequency domain the challenge is finding the best approximation of the fractional-order of the fractance [83]. Besides, the transfer

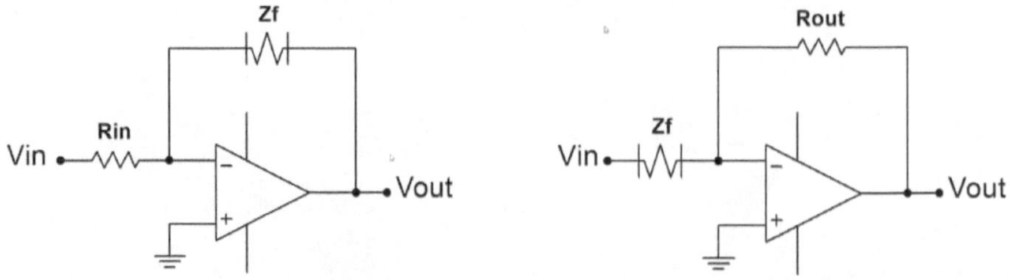

(a) Basic fractional-order integrator (b) Basic fractional-order differentiator

Figure 3.3: Fractional-order differentiator and integrator implemented with an operational amplifier and a fractance emulated with RC topologies.

Table 3.1: Integer-order transfer functions that approximate fractional-order integrators between $0.1 < q < 0.9$, with a maximum discrepancy $y = 2$dB and bandwidth of the system of $\omega_{max} = 10^3$ rad/s, taken from [23].

q-order		H(s) Transfer Function
0.1	$\approx$	$\frac{1584.8932(s+0.1668)(s+27.83)}{(s+0.1)(s+16.68)(s+2783)}$
0.2	$\approx$	$\frac{79.4328(s+0.05623)(s+1)(s+17.78)}{(s+0.03162)(s+0.5623)(s+10)(s+177.8)}$
0.3	$\approx$	$\frac{39.8107(s+0.0416)(s+0.3728)(s+3.34)(s+29.94)}{(s+0.02154)(s+01931)(s+1.73)(s+15.51)(s+138.9)}$
0.4	$\approx$	$\frac{35.4813(s+0.03831)(s+0.26.1)(s+1.778.)(s+12.12)(s+82.54)}{(s+0.01778)(s+0.1212)(s+0.8254)(s+5.623)(s+38.31)(s+261)}$
0.5	$\approx$	$\frac{15.8489(s+0.03.981)(s+0.2512)(s+1.585)(s+10)(s+63.1)}{(s+0.01585)(s+01)(s+0.631)(s+3981)(s+25.12)(s+158.5)}$
0.6	$\approx$	$\frac{10.7978(s+0.04642)(s+0.3162)(s+2.154)(s+14.68)(s+100.)}{(s+0.01468)(s+0.1)(s+0.6813)(s+4642)(s+3162)(s+2154)}$
0.7	$\approx$	$\frac{9.3633(s+0.06449)(s+0.578)(s+5.179)(s+46.42)(s+416)}{(s+0.01389)(s+0.1245)(s+1.116)(s+10)(s+89.62)(s+803.1)}$
0.8	$\approx$	$\frac{5.3088(s+0.1334)(s+2.371)(s+42.17)(s+749.9)}{(s+0.01334)(s+0.2371)(s+4.217)(s+74.99)(s+1334)}$
0.9	$\approx$	$\frac{2.2675(s+1.292)(s+215.4)}{(s+0.01292)(s+2.154)(s+359.4)}$

function in the Laplace domain that better approaches a given fractional-order should be implemented with standard values of the capacitors, while the resistors can be tuned to generate the lowest error to emulate the desired fractance. The values of the R and C elements can be determined according to [84, 85, 86], however, the selected values must generate low sensitivities. In this manner, and to avoid mismatches, it is much better to expand the transfer functions by first-order expressions, as shown in [34], to use FPAAs to complete the whole circuit implementation of a fractional-order chaotic oscillator.

3.1.1 Approximating a fractional-order integrator by a ladder fractance topology

The commensurate fractional-order Lorenz chaotic oscillator with $q = 0.9$, can be implemented by using the transfer function described by (3.3), which has a maximum discrepancy of $y = 2$dB [23, 82]. By using the ladder RC-topology shown in Fig. 3.2(a), and by performing an analysis considering $q = 0.9$ one can find the values of the circuit elements that are associated to the transfer function $H(s)$ between the nodes A and B from Fig. 3.2(a). The $H(s)$ is given in (3.4), where C_0 is a unitary parameter, and if its value is set to $C_0 = 1\mu$F, then $F(s) = H(s)C_0 = 1/s^{0.9}$, and the new expression of the ladder topology is given by (3.5). Finally, comparing (3.3) and (3.5), one can obtain the R and C values that are listed in Table 3.2.

$$H(s) = \frac{1}{s^{0.9}} = \frac{2.2675(s+1.292)(s+215.4)}{(s+0.01292)(s+2.154)(s+359.4)} \tag{3.3}$$

$$H(s) = \frac{\frac{C_0}{C_1}}{s+\frac{1}{R_1C_1}} + \frac{\frac{C_0}{C_1}}{s+\frac{1}{R_2C_2}} + \frac{\frac{C_0}{C_1}}{s+\frac{1}{R_3C_3}} \tag{3.4}$$

$$H(s) = \frac{1}{C_0}\frac{\left(\frac{C_0}{C_1}+\frac{C_0}{C_2}+\frac{C_0}{C_3}\right)\left[s^2+\frac{s\left(\frac{C_2+C_3}{R_1}+\frac{C_2+C_3}{R2}+\frac{C_1+C_2}{R_3}\right)+\frac{R_1+R_2+R_3}{R_1R_2R_3}}{C_1C_2+C_2C_3+C_1C3}\right]}{\left(s+\frac{1}{R_1C1}\right)\left(s+\frac{1}{R_2C_2}\right)\left(s+\frac{1}{R_3C_3}\right)} \tag{3.5}$$

Looking at the time simulation of applying Grünwald Letnikov method, one can see that the phase-space portraits shown in Fig. 6.5 exceeds the range of values supported by commercially available operational amplifiers, as well as of the required analog multipliers. In this manner, one must down-scale those state variable ranges to implement (3.6), and by applying circuit theory and the guidelines for down-scaling given in [61], then the equivalent circuit equations are given in (3.7). These equations can be implemented using the RC-approached fractance,

Table 3.2: Values of the R and C elements for the implementation of the fractional-order Lorenz chaotic oscillator.

Circuit Element	Element Value	Circuit Element	Element Value
$R_1 = R_{11} = R_{19}$	62.84 MΩ	$R_4 = R_{10} = R_{15}$	
$R_2 = R_{12} = R_{20}$	250 kΩ	$= R_{18} = R_{23}$	1 kΩ
$R_5 = R_8 = R_9 = R_{16}$		$R_3 = R_{13} = R_{21}$	2.5 kΩ
$= R_{24} = R_{25} = R_{26}$	10 kΩ	$R_6 = R_7$	2 kΩ
R_{17}	715 Ω	R_{14}	20 kΩ
R_{22}	7.5 kΩ	$C_1 = C_4 = C_7$	1.23 μF
$C_2 = C_5 = C_8$	1.84 μF	$C_3 = C_6 = C_9$	1.1 μF

commercial amplifiers like AD712 / LM324 / TL082, and the multiplier AD633, which needs an output coefficient value of $k = 0.1$ and a power supply of $\pm$ 12V.

$$\begin{aligned}
\frac{d^{0.9}x}{dt^{0.9}} &= 10(y-x), \\
\frac{d^{0.9}y}{dt^{0.9}} &= x(28-z)-y, \\
\frac{d^{0.9}z}{dt^{0.9}} &= xy-\frac{8}{3}z
\end{aligned} \tag{3.6}$$

$$\begin{aligned}
\frac{X(s)}{H(s)} &= \frac{R_5}{C_oR_4}\left[\frac{Y(s)}{R_7}-\frac{R_8X(s)}{R_9R_4}\right] \\
\frac{Y(s)}{H(s)} &= \frac{R_{13}}{C_oR_{14}}\left[\frac{X(s)}{R_{10}}-\frac{kR_{28}X(s)*Z(s)}{R_{27}R_{12}}-\frac{Y(s)}{R_7}\right] \\
\frac{Z(s)}{H(s)} &= \frac{R_{22}}{C_oR_{23}}\left[\frac{kX(s)*Y(s)}{R_{21}}-\frac{R_{25}Z(s)}{R_{27}R_{20}}\right]
\end{aligned} \tag{3.7}$$

Figure 3.4 sketches the implementation of state variable x by using a fractance, an inverter to get $-x$ and an adder. Figure 3.5 details the implementation of state variable y, in which the multiplier is required to solve the convolution between the variables $X(s) * Z(s)$, and Fig. 3.6 details the building blocks implementing state variable z, which also requires a multiplier to solve $X(s) * Y(s)$. Figure 3.7 shows the simulation results of the 2.7-order Lorenz chaotic oscillator that are in good agreement when applying the Grünwald-Letnikov method, as shown in Fig. 6.5. Thus demonstrating that the approximation of the fractional-order integrator in the frequency domain is also quite useful, and provides similar simulation results when applying time domain methods.

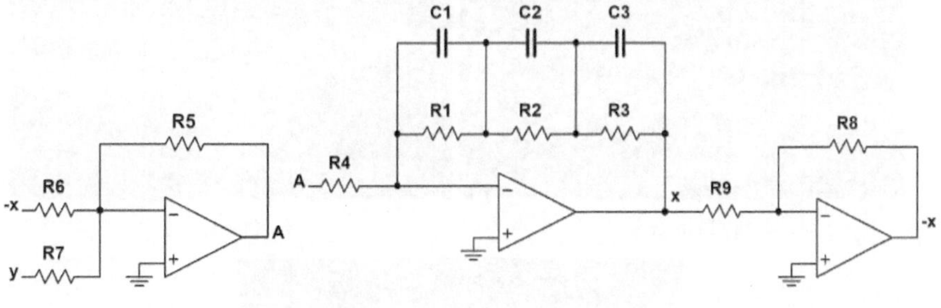

Figure 3.4: Ladder RC topology to implement $X(s)$ from (3.7).

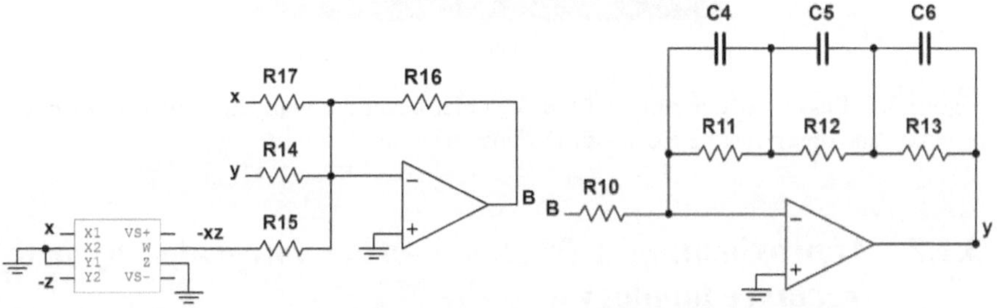

Figure 3.5: Ladder RC topology to implement $Y(s)$ from (3.7).

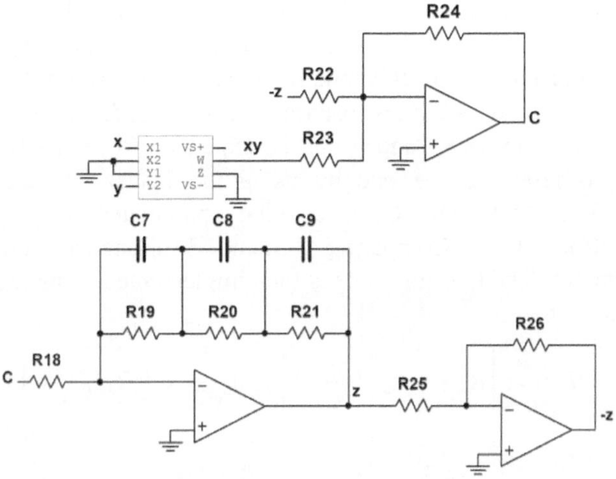

Figure 3.6: Ladder RC topology to implement $Z(s)$ from (3.7).

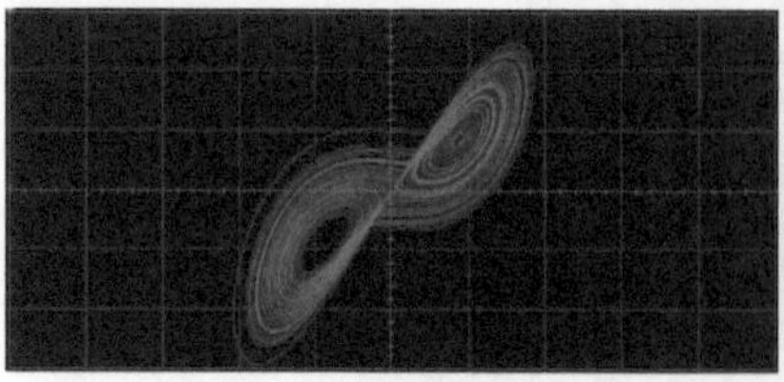

(a) Plane $x-y$

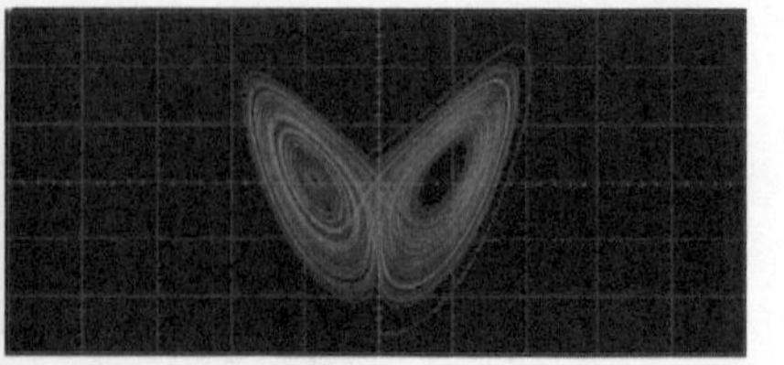

(b) Plane $x-z$

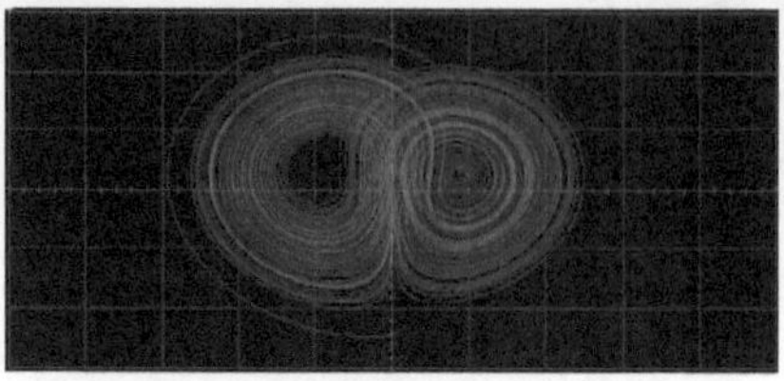

(c) Plane $y-z$

Figure 3.7: Phase-space portraits of the 2.7-order chaotic Lorenz attractor with $q_1 = q_2 = q_3 = 0.9$, and observing the plots with 1V/Div.

3.1.2 Approximating a fractional-order integrator by a tree fractance topology

Similar to the frequency domain approximation of the fractance by using RC-elements in ladder/cascade connection, this subsection shows the application of the tree topology to implement the fractances for the 2.7-order Lorenz chaotic oscillator. Again, the transfer function that approaches the commensurate fractional-orders of the integrators with $q = 0.9$ is taken from Table 3.1, which $H(s)$ is given in (3.3). It can be implemented by using the RC-based tree topology shown in Fig. 3.2(c), whose transfer function $H(s)$ between the nodes A and B is given by (3.8), and it can be expanded to get (3.9). This transfer function is different to the one by using the ladder fractance detailed in subsection 3.1.1, but again C_0 is a unitary parameter, thus $C_0 = 1\mu$F and $F(s) = H(s)C_0 = 1/s^{0.9}$. Comparing (3.3) and (3.9), one can find the values of the RC elements. The tree fractance is then implemented using the values listed in Table 3.3.

$$H(s) = \left[R_1 + \left(R_2//\frac{1}{sC_2}\right)\right] // \left[\frac{1}{sC_1} + \left(R_3//\frac{1}{sC_3}\right)\right] \tag{3.8}$$

$$H(s) = \frac{1}{C_0}\frac{\left(\frac{C_0}{C_1}+\frac{C_0}{C_3}\right)\left(s+\frac{R_1+R_2}{R_1C_2R_2}\right)\left(s+\frac{1}{C_1R_3+C_3R3}\right)}{s^3+s^2\left(\frac{R_1+R_2}{R_1C_2R_2}+\frac{1}{C_3R_3}+\frac{C_1+C_3}{C_1R_1C_3}\right)+s\left(\frac{R_1+R_2}{R_1C_2R_2C_3R_3}+\frac{1}{C_1R_1C_3R_3}+\frac{C_1+C_3}{C_1R_1C_2R_2C_3}\right)+\frac{1}{C_1R_1C_2R_2C_3R_3}} \tag{3.9}$$

Table 3.3: Values of the RC elements to implement the 2.7-order Lorenz chaotic oscillator.

Circuit Element	Element Value	Circuit Element	Element Value
$R_1 = R_{11} = R_{19}$	1.55 MΩ	$R_4 = R_{10} = R_{15}$	
$R_2 = R_{12} = R_{20}$	62.84 MΩ	$= R_{18} = R_{23}$	1 kΩ
$R_5 = R_8 = R_9 = R_{16}$		$R_3 = R_6 = R_7$	
$= R_{17} = R_{24} = R_{25} = R_{26}$	10 kΩ	$= R_{13} = R_{21}$	2.5 kΩ
R_{22}	33.5 kΩ	R_{14}	7 kΩ
$C_1 = C_4 = C_7$	0.73 μF	$C_3 = C_6 = C_9$	1.1 μF
$C_2 = C_5 = C_8$	0.52 μF		

The down-scaled equations are given in (3.7), where $H(s)$ is now synthesized by the tree fractance, and again k is the output coefficient of the multiplier AD633. Figure 3.8 shows the implementation of $X(s)$ by using the tree topology to implement the fractance, Fig. 3.9 shows the implementation of $Y(s)$, and Fig. 3.10 $Z(s)$. Their corresponding simulation results are shown in Fig. 3.11. Again, the simulation results are in good agreement when applying Grünwald-Letnikov method, as shown in Fig. 6.5.

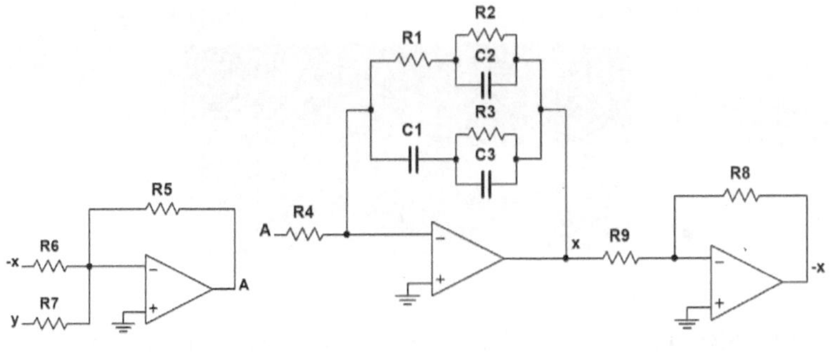

Figure 3.8: Implementation of variable $X(s)$ of the 2.7-order Lorenz chaotic circuit by using the tree fractance.

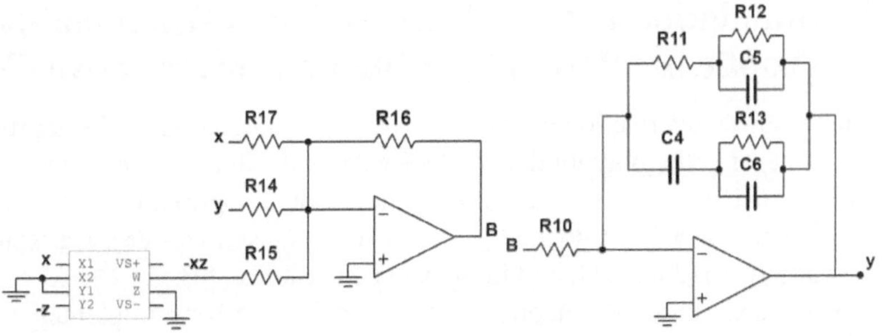

Figure 3.9: Implementation of variable $Y(s)$ of the 2.7-order Lorenz chaotic circuit by using the tree fractance.

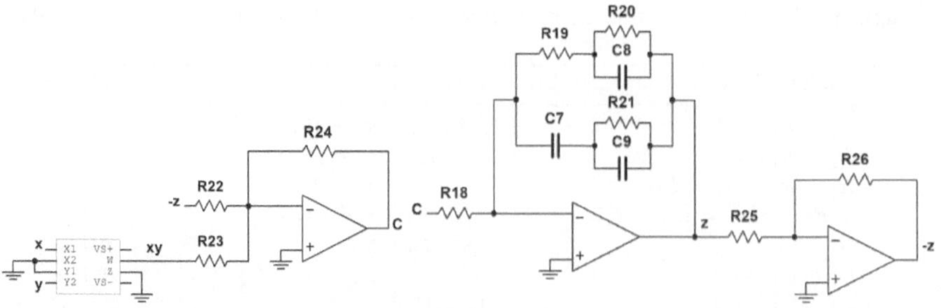

Figure 3.10: Implementation of variable $Z(s)$ of the 2.7-order Lorenz chaotic circuit by using the tree fractance.

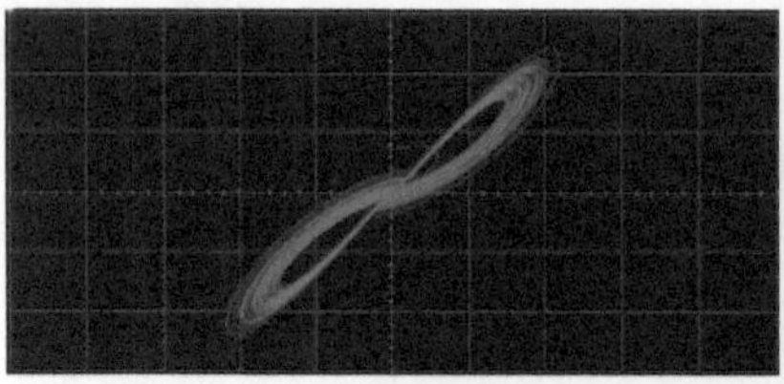

(a) Plane $x-y$

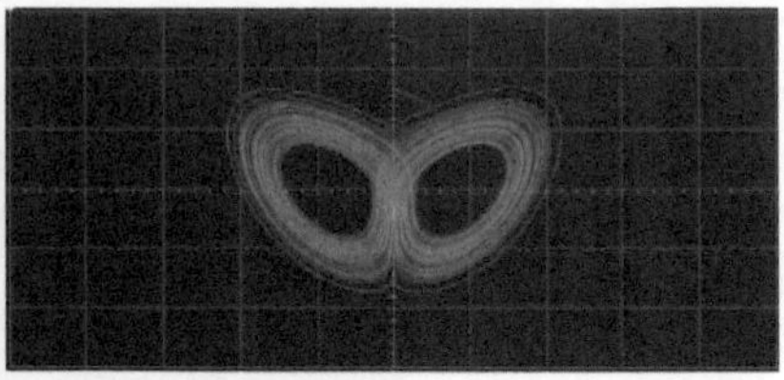

(b) Plane $x-z$

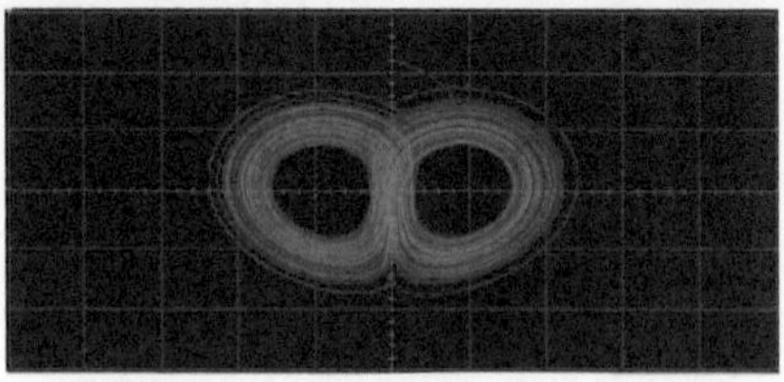

(c) Plane $y-z$

Figure 3.11: Simulation results of the 2.7-order Lorenz chaotic oscillator when $q_1 = q_2 = q_3 = 0.9$, and observed with 1V/Div.

3.2 Simulation and implementation of fractional-order chaotic oscillators by using first-order active filters

As shown above, the fractional-order derivatives or integrators can be approached by high integer-order polynomials in the Laplace domain and holding a constant phase within a chosen frequency of operation. Some approximations in the frequency domain were introduced by Chareff [87], Oustaloup [88], Carlson [89], Matsuda [90], Krishna [91], and Least Square methods [44].

Some fractional-order chaotic oscillators that have been implemented with analog electronics are given in [92, 78, 23, 93, 94, 77, 95, 96]. Other analog implementations and applications in synchronizing two fractional-order chaotic oscillators in a master-slave topology are given in [85, 97, 80, 98, 4]. However, they do not exploit the synthesis of the fractional-order derivatives and integrators by using integer integrators, as done in [34].

3.2.1 Approximation of $1/s^q$ by using first-order active filters

The fractional-order operators q_i in (3.2), can be described by $H(s) = \frac{1}{s^q}$ and approximated by Laplace polynomials as already done in [23] for $0.1 < q < 0.9$ in steps of 0.1. The transfer function of fractional-order 0.9 shown in (3.10), can be expanded in products of first-order polynomials to have $H_1(s)$, $H_2(s)$ and $H_3(s)$, given in 3.11, where $H_1(s)$ and $H_2(s)$ have one-pole and one-zero and can be implemented by the first-order filter topology shown in Fig. 3.12 [99], whose transfer function is given by (3.12).

$$H(s) = \frac{1}{s^{0.9}} \approx \frac{2.2675(s+1.292)(s+215.4)}{(s+0.01292)(s+2.154)(s+359.4)} \tag{3.10}$$

$$H_1(s) = \frac{s+215.4}{s+359.4} \qquad H_2(s) = \frac{s+1.292}{s+2.154} \qquad H_3(s) = \frac{2.2675}{s+0.01292} \tag{3.11}$$

$$H_{1,2}(s) = \frac{V_o}{V_i}(s) = -K\frac{s+z_1}{s+p_1} = -\frac{R_2}{R_1}\frac{s+\dfrac{1}{R_2C_2}}{s+\dfrac{1}{R_1C_1}} \tag{3.12}$$

The transfer function of $H_3(s)$ from (3.11) can be implemented by the first-order low-pass filter shown in Fig. 3.13, whose transfer function is given in (3.13).

$$H_3(s) = \frac{V_o}{V_i}(s) = \frac{-K}{s+p_1} = -\left(\frac{R_2}{R_1}\right)\left(\frac{\dfrac{1}{C_1R_2}}{s+\dfrac{1}{C_1R_2}}\right) \tag{3.13}$$

Figure 3.14 shows the whole implementation of the transfer function given in (3.10) by using the first-order active filters. The R and C values can be determined

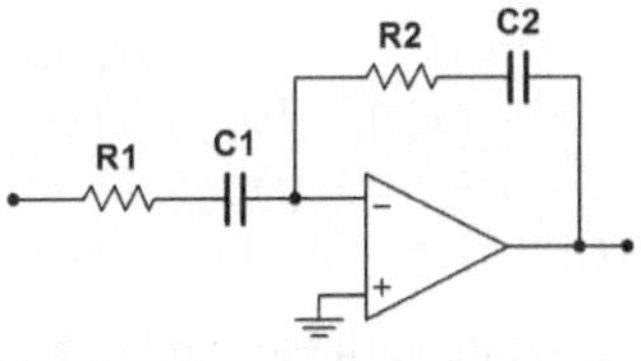

Figure 3.12: Implementation of one-pole and one-zero as $H_1(s)$ and $H_2(s)$ from (3.11).

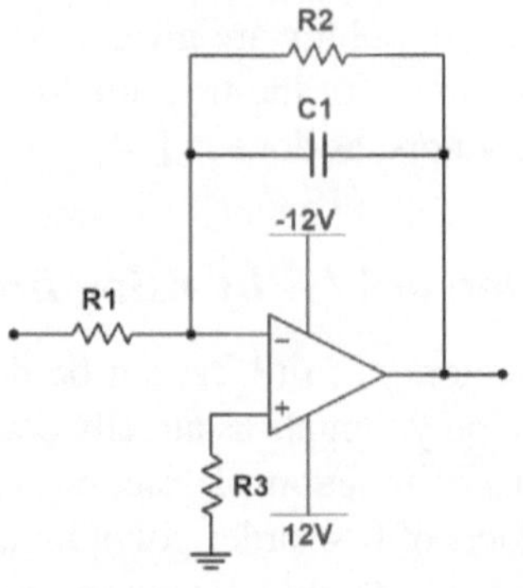

Figure 3.13: Implementation of $H_3(s)$ from (3.10).

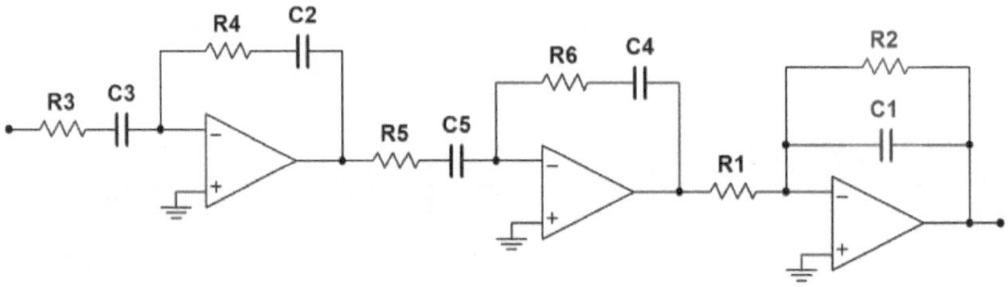

Figure 3.14: Analog implementation of the transfer function of the type given in (3.10) by using first-order active filter blocks.

according to [34], where one can find details on the implementation of fractional-order chaotic oscillators.

3.2.2 Implementation of a fractional-order hyper-chaotic oscillator

The implementation of the fractional-order hyper-chaotic oscillators can also be performed by using first-order active filter blocks. Lets us consider the fractional-order hyper-chaotic oscillator given in (6.14) [98], where the fractional-order of the derivatives $q = 0.9$ can be approximated by (3.10).

$$\begin{aligned}
{}_0D_t^{q_1}x(t) &= a(y(t)-x(t)) + by(t)z(t) + w(t) \\
{}_0D_t^{q_2}y(t) &= -cx(t) - dx(t)z^2(t) + gy(t) \\
{}_0D_t^{q_3}z(t) &= y^2(t) - kz(t) \\
{}_0D_t^{q_4}w(t) &= by(t) - w(t)
\end{aligned} \tag{3.14}$$

In (6.14) x, y, z, w are the state variables and a, b, c, d, g and k the coefficients. The chaotic behavior exists by using the initial conditions $[0.1, 0.1, 2.1, 0.1]$ and by setting the coefficient values to: $(a,b,c,d,g,k) = (35, 2.5, 7, 4, 28, 1/3)$. This fractional-order hyper-chaotic oscillator has seven equilibrium points given in Table 6.9. Its Jacobian matrix is given in (6.15) and the associated eigenvalues are given in Table 6.10.

$$J_{(x^*,y^*,z^*,w^*)} = \begin{bmatrix} -a & a+bx & by & 1 \\ -c-d & g & 2z & 0 \\ 0 & 2y & -k & 0 \\ 0 & b & 0 & -1 \end{bmatrix} \tag{3.15}$$

By using these eigenvalues in (6.16), one can determine the minimum commensurate fractional-order of the derivatives in order that the system generates chaotic behavior. In this case by using the real part as 26.323 and the imaginary part as $j0.1397$, the fractional-order is $q > 0.8915$. Therefore, the numerical

Table 3.4: Equilibrium points of the fractional-order hyper-chaotic oscillator given in (6.14).

$EP_0(0,0,0,0)$
$EP_1(-2.0430,-0.9925,0.81719,2.0034)$
$EP_2(2.0430,0.9925,0.81719,2.0034)$
$EP_3(-j2.2653,-j0.81142,-j0.90611,-2.4631)$
$EP_4(j2.2653,j0.81142,j0.90611,2.4631)$
$EP_5(-j5.5038,-j0.072293,-j2.2015,-14.540)$
$EP_6(j5.5038,j0.072293,j2.2015,-14.540)$

Table 3.5: Eigenvalues associated to the equilibrium points of the fractional-order hyper-chaotic oscillator given in (6.14).

$EP_0:$	$\lambda_1 = -30.847, \lambda_2 = 23.823, \lambda_3 = -0.97638, \lambda_4 = -0.33333$
$EP_1:$	$\lambda_1 = -29.546, \lambda_2 = 21.738, \lambda_3 = 0.4585, \lambda_4 = -0.98297$
$EP_2:$	$\lambda_1 = -28.786, \lambda_2 = 20.924, \lambda_3 = 0.51142, \lambda_4 = -0.98249$
$EP_3:$	$\lambda_1 = -32.849 - j0.1443, \lambda_2 = 26.323 + j0.1397,$
	$\lambda_3 = -0.8676 + j0.007746, \lambda_4 = -0.95049 - j0.003157$
$EP_4:$	$\lambda_1 = -32.849 + j0.1443, \lambda_2 = 26.323 - j0.1397,$
	$\lambda_3 = -0.8676 - j0.007746, \lambda_4 = -0.95049 + j0.003157$
$EP_5:$	$\lambda_1 = -41.265 + j0.9034, \lambda_2 = 34.349 - j0.8994,$
	$\lambda_3 = -1.0238 - j0.001251, \lambda_4 = -0.39305 - j0.002759$
$EP_6:$	$\lambda_1 = -41.265 - j0.9034, \lambda_2 = 34.349 + j0.8994,$
	$\lambda_3 = -1.0238 + j0.001251, \lambda_4 = -0.39305 + j0.002759$

simulation of the fractional-order hyper-chaotic oscillator can be performed by setting $q = 0.9$.

$$q \geq \frac{2}{\pi} \arctan \frac{|\text{Im}((\lambda)|}{|\text{Re}(\lambda)|} \tag{3.16}$$

Figure (6.6) shows the simulation results of (6.14) by applying FDE12 [47].

The fractional-order hyper-chaotic oscillator given in (6.14) has the block description given in 6.7, where $H(s)$ corresponds to the approximation of the fractional-order of $q = 0.9$ given in (3.10) and whose analog implementation is given in Fig. 3.14 by using first-order active filter blocks.

The synthesis of the blocks of the fractional-order hyper-chaotic oscillator shown in Fig. 6.7 can be performed by using amplifiers and multipliers as shown in Fig. 6.8. The amplifiers can drive up to $\pm$18V (AD712 / LM324 / TL082), the multipliers can be of type AD633 and $H(s)$ is implemented by the first-order active filters shown in Fig. 3.14. After performing the design process given in [34], the values of the circuit elements are set to: R_1=70 kΩ, R_2=R_5=2 kΩ, R_3=R_{10}=7 kΩ, R_6=28 kΩ, R_{13}=R_{19}=1 kΩ, R_9=17.5 kΩ, R_{11}=2.5 kΩ, R_{30}=10 kΩ, R_{21}=4 kΩ, R_4=R_7=R_{12}=R_{14}=R_{15}=R_{18}=R_{20}=R_2=R_{24}=10 kΩ. The SPICE simulation results of the fractional-order hyper-chaotic oscillator given in Fig. 6.8 are shown in Fig. 3.18 for the planes $x-y$, $x-z$, $x-w$ and $z-w$. As one sees, they are in good agreement with the numerical simulation results shown in Fig. 6.6.

3.2.3 Implementation of a fractional-order multi-scroll chaotic oscillator

The multi-scroll chaotic oscillator based on saturated nonlinear function (SNLF) series introduced in [100], can be transformed to its fractional-order version given in (3.17). The SNLF can be approximated by piecewise-linear (PWL) functions, and for two saturated levels with break points at ± 1, and saturation levels k, it can be described by (3.18). The fractional-orders of the derivatives q_1, q_2, q_3 can be commensurate and set to 0.9.

$$\begin{aligned}
{}_0D_t^{q_1} x(t) &= y(t), \\
{}_0D_t^{q_2} y(t) &= z(t), \\
{}_0D_t^{q_3} z(t) &= -ax(t) - by(t) - cz(t) + d_1 f_0(x(t), k).
\end{aligned} \tag{3.17}$$

$$f_0(x,k) = \begin{cases} k, & \text{if } \ x > 1, \\ kx, & \text{if } \ |x| \leq 1, \\ -k, & \text{if } \ x < 1. \end{cases} \tag{3.18}$$

The fractional-order chaotic oscillator based on SNLF series given in 3.17, can generate chaotic behavior by setting $a = b = c = d_1 = 0.7$, and using the initial conditions $(x(0), y(0), z(0)) = (0.2, 0, 0)$. But in this book with the generation of a 2-scroll attractor one can set $a = b = c = d_1 = 0.3$ and $k = 1$. The

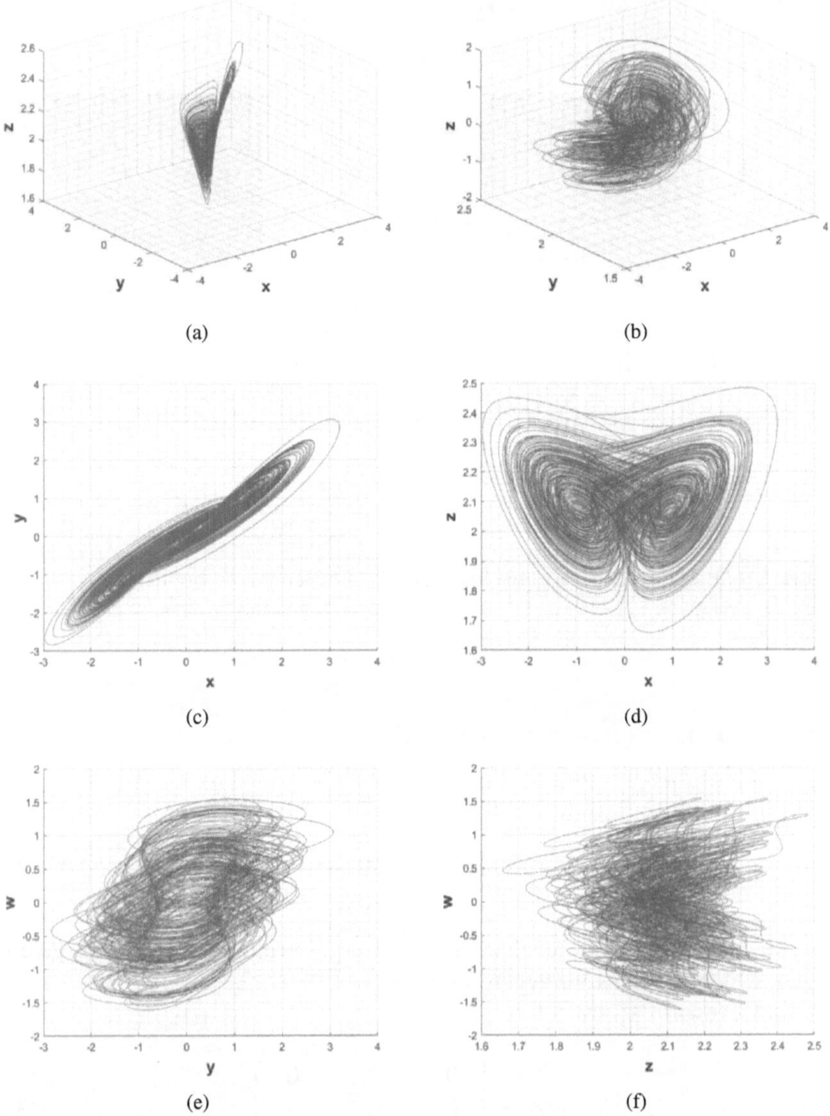

Figure 3.15: Simulation results of the fractional-order hyper-chaotic oscillator (6.14) by FDE12 [47]; by setting (a,b,c,d,g,k) = (35, 2.5, 7, 4, 28, 1/3), initial conditions $(x(0),y(0),z(0),w(0)) = (0.1,0.1,2.1,0.1)$, total time simulation $T_{sim} = 200s$, step-size $h = 0.005$, and $q_1 = q_2 = q_3 = q_4 = 0.9$.

equilibrium points are evaluated as: $EP_1 = (k_n d_1/a, 0, 0)$, $EP_2 = (0,0,0)$, and $EP_3 = (-k_n d_1/a, 0, 0)$. The Jacobian matrix evaluated at the equilibrium points

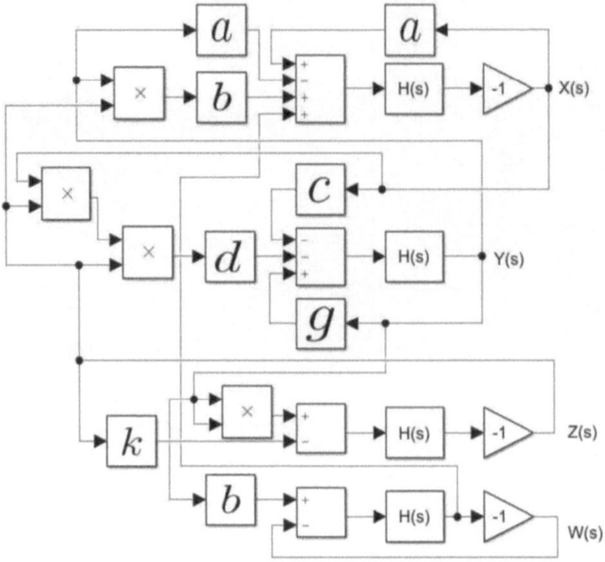

Figure 3.16: Block description of the fractional-order hyper-chaotic oscillator given in (6.14), where $H(s)$ is given in (3.10) and its analog implementation in Fig. 3.14 by using first-order active filters.

EP_1 and EP_3 is given in (3.19), which provides the eigenvalues: $\lambda_1 = -0.848$, $\lambda_2 = 0.074 + j0.9055$, and $\lambda_3 = 0.074 - j0.9055$.

$$J_{EP_{1,3}} = \begin{bmatrix} 0 & 1 & 0 \\ 0 & 0 & 1 \\ -0.3 & -0.3 & -0.3 \end{bmatrix} \tag{3.19}$$

The Jacobian matrix evaluated at the equilibrium point EP_2 is given in (3.20), which provides the eigenvalues: $\lambda_1 = 1.9309$, $\lambda_2 = -1.1154 + j1.6944$, $\lambda_3 = -1.1154 + j1.6944$.

$$J_{EP_2} = \begin{bmatrix} 0 & 1 & 0 \\ 0 & 0 & 1 \\ 3.15 & -0.3 & -0.3 \end{bmatrix} \tag{3.20}$$

The minimum fractional-order for which (3.17) generates chaotic behavior when setting $a = b = c = d_1 = 0.3$, and by using (6.16). If the fractional-order system is commensurate one gets $q_1 = q_2 = q_3 = q$, and $q \geq 0.873$. So that one can use $q = 0.9$, and the simulation results are shown in Fig. 3.19.

The block description of the fractional-order multi-scroll chaotic oscillator given in (3.17), is shown in Fig. 3.20, where again $H(s)$ is the approximation of the fractional-order integrators that are given in (3.10).

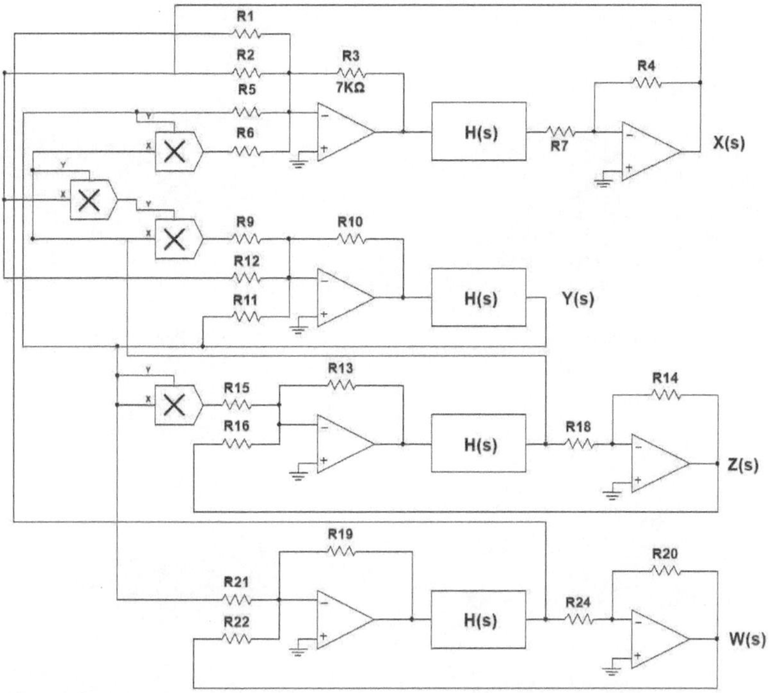

Figure 3.17: Whole circuit implementation of the fractional-order hyper-chaotic oscillator shown in Fig. 6.7 and described from (6.14) with commensurate fractional-order of 0.9.

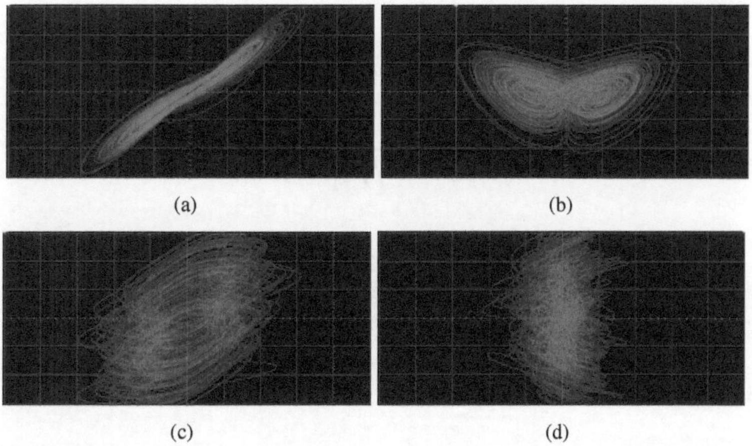

Figure 3.18: SPICE simulation results of the fractional-order hyper-chaotic oscillator shown in Fig. 6.8 with $q_1 = q_2 = q_3 = q_4 = 0.9$ and scale 500mV/Div for the planes: (a) $x - y$, (b) $x - z$, (c) $x - w$, and (d) $z - w$.

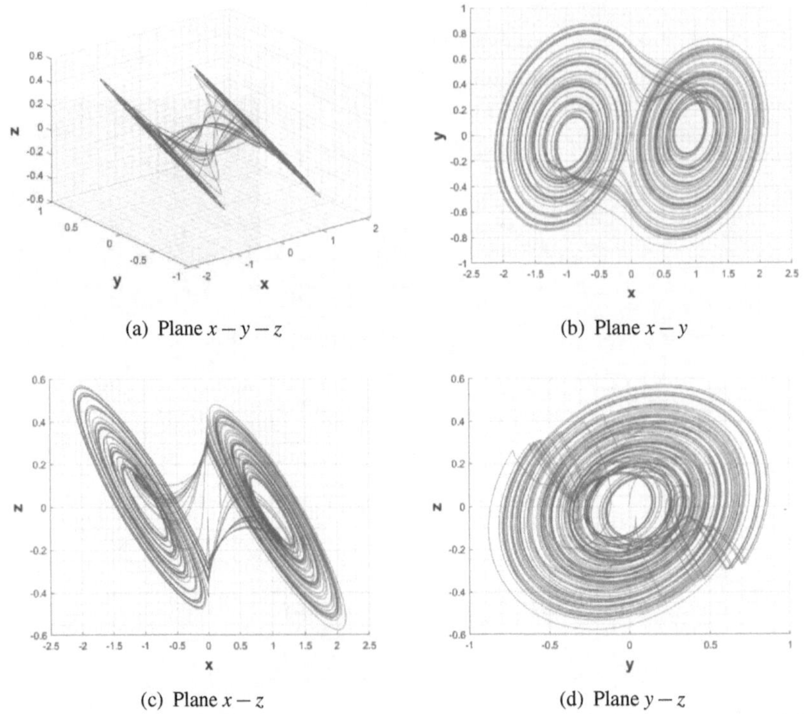

Figure 3.19: Numerical simulation of the fractional-order multi-scroll chaotic oscillator given in (3.17) with $a = b = c = d_1 = 0.3$, $k = 1$, time simulation $T_{sim} = 200s$, step-size $h = 0.005$, initial conditions $(x(0), y(0), z(0)) = (0.2, 0, 0)$, and $q_1 = q_2 = q_3 = 0.9$.

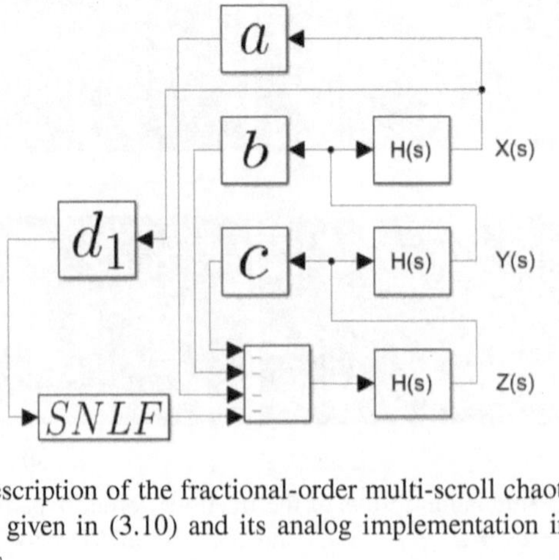

Figure 3.20: Block description of the fractional-order multi-scroll chaotic oscillator given in (3.17), where $H(s)$ is given in (3.10) and its analog implementation in Fig. 3.14 by using first-order active filters.

The whole circuit implementation of the block description of the fractional-order multi-scroll chaotic oscillator shown in Fig. 3.20, is given in Fig. 6.11, where $H(s)$ is given in (3.10) and its analog implementation in Fig. 3.14 by using first-order active filters. The amplifiers can be of type AD712 / LM324 / TL082, and the circuit elements are set to: R_1=R_2=R_{14}=R_{15}=R_{32}=R_{33}=100 kΩ, R_{27}=369.2 kΩ, R_{28}=R_{31}=400 kΩ, R_{29}=40 kΩ, R_{30}=950 kΩ. The SPICE simulation results are shown in Fig. 3.22.

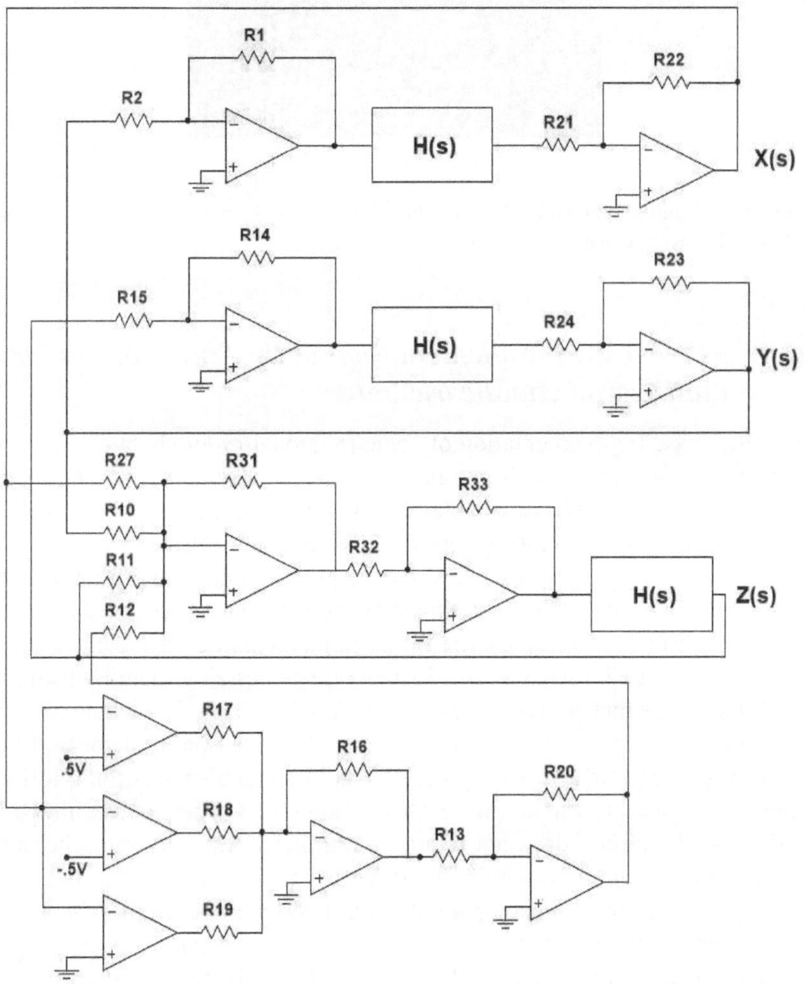

Figure 3.21: Circuit implementation of the fractional-order multi-scroll chaotic oscillator shown in Fig. 3.20.

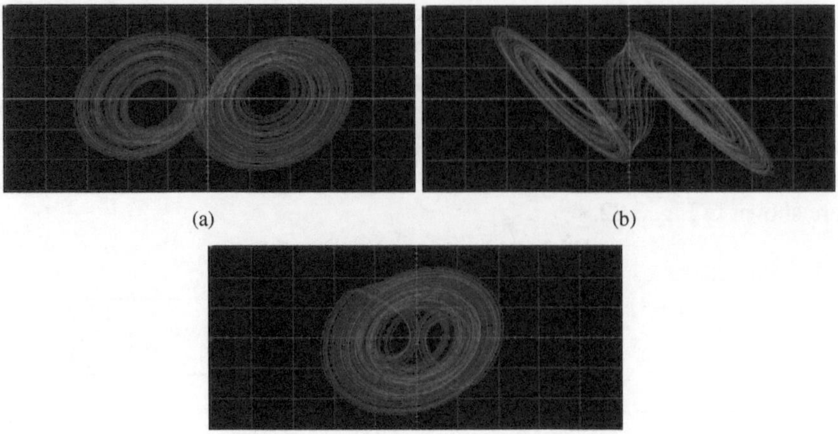

(a) (b) (c)

Figure 3.22: Simulation results of the fractional-order circuit shown in Fig. 6.11 with scales of 500mV/Div for the planes: (a) $x-y$, (b) $x-z$, and (c) $y-z$.

3.2.4 FPAA-based implementation of the fractional-order multi-scroll chaotic oscillator

The FPAA-based implementation of some fractional-order chaotic oscillators is given in [4], and more details can be found in [34]. By applying the amplitude scaling process described in Chapter 2, one can implement the fractional-order multi-scroll chaotic oscillator given in (3.17) by using FPAAs, which can drive voltages in the range between ± 3 V. The approximation of the fractional-order of 0.9 is again synthesized by using first-order active filters, as shown in the previous sections. For instance, the FPAA implementation of the block description shown in Fig. 3.20, can be performed by synthesizing them by using three AN231E04 chips that are connected as shown in Fig. 6.33. The whole fractional-order multi-scroll chaotic oscillator given in (3.17) can be implemented in one FPAA, where $H(s)$ that is given in (3.10), is implemented according to its amplifier topologies shown in Fig. 3.14 by using first-order active filters. This fractional-order circuit does not require multiplier CAMs. In total, the number of amplifiers that are used within the FPAA is equal to 20, and the SNLF is implemented by an array of comparators and inverters, and one CAM Sum/Diff is programmed to perform the sum $-ax(t)-by(t)-cz(t)+d_1 f_0(x(t),k)$ from (3.17) just as shown in Fig. 3.20. Figure 6.33 shows the whole FPAA-based implementation schematics using Andigm Desginer 2.

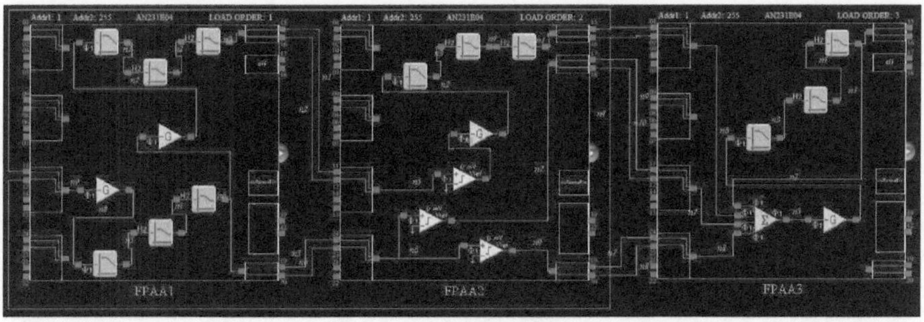

Figure 3.23: FPAA-based implementation of the fractional-order multi-scroll chaotic oscillator given in (3.17) by using the approximated transfer function given in (3.10) for the fractional-order integrator $1/s^{0.9}$.

3.3 Simulation of self-excited fractional-order chaotic oscillators by Gründwald-Letnikov method

Lets us consider the fractional-order Chen chaotic oscillator given in (3.21). It generates chaotic behavior by setting $a = 35, b = 3$, and $c = 28$ and initial condition $x_0 = y_0 = z_0 = 0.1$ [101]. Its numerical simulation is shown in Fig. 3.24 with $q_1 = q_2 = q_3 = 0.9$ and by applying the Grünwald-Letnikov definition with step-size $h = 0.001$. As shown in [4, 34], one can use the short-memory length concept to truncate the memory size, as discussed in Chapter 1. In this case, the memory length is set to 32, it means that the Grünwald-Letnikov definition will occupy 32 previous terms to calculate the new value in the execution of the numerical method.

$$\begin{aligned} D_t^{q_1} x &= a(y-x) \\ D_t^{q_2} y &= (c-a)x - xz + cy \\ D_t^{q_3} z &= xy - bz \end{aligned} \tag{3.21}$$

Not all the fractional-order chaotic oscillators can be simulated with short memory length, see for example the case of the oscillator given in (3.22) [102]. It generates chaotic behavior by setting $a = 2.05, b = 1.12$ and $c = 0.4$, initial conditions of $x_0 = y_0 = z_0 = 0.01$. Figure 3.25 shows the attractor generated when the fractional-order is $q_1 = q_2 = q_3 = 0.998$, and by applying the Grünwald-Letnikov definition with $h = 0.001$, but in this case the memory length L_m must be equal or higher than 256.

$$\begin{aligned} D_t^{q_1} x &= y \\ D_t^{q_2} y &= z \\ D_t^{q_3} z &= -ax - by - cz - x^2 \end{aligned} \tag{3.22}$$

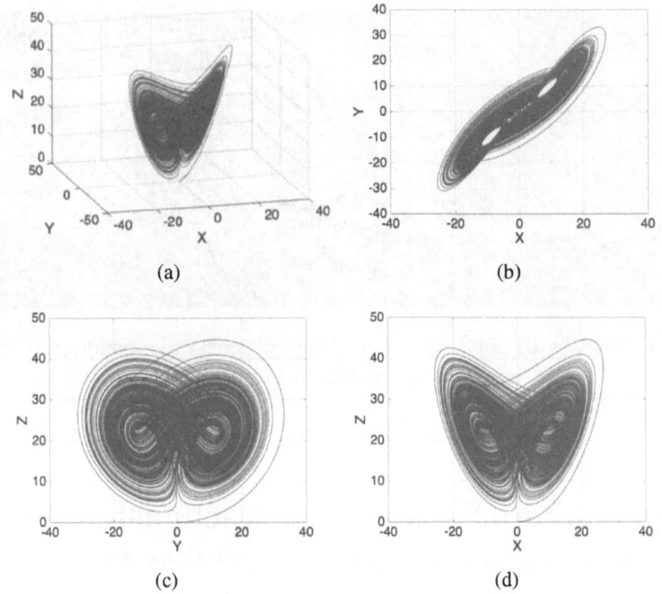

Figure 3.24: Simulation results of the fractional-order chaotic attractor based on (3.21) by applying Grünwald-Letnikov definition with $h = 0.001$ and memory length $L_m = 32$.

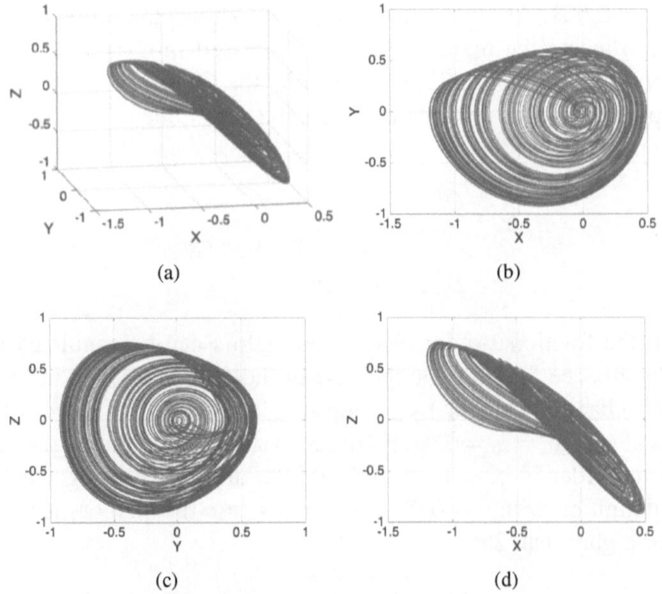

Figure 3.25: Simulation of the fractional-order chaotic oscillator given in (3.22) by applying Grünwald-Letnikov definition with $h = 0.001$ and memory length $L_m = 256$.

Other self-excited fractional-order chaotic oscillators are listed in Table 3.3, in which one can see the values of their coefficients, initial conditions, fractional-orders of the derivatives, step-size and memory length L_m by applying Grünwald-Letnikov. The numerical simulation results are shown in Fig. 3.26.

Table 3.6: Self-excited fractional-order chaotic oscillators.

System	Equations	Parameters
Rajagopal [103]	$D_t^{q_1} x = y$ $D_t^{q_2} y = -4x - 4yz$ $D_t^{q_3} z = -x^2 + y^2 + bz^2 - a$	$a = 1.493, b = 0.6$ $x_0 = y_0 = z_0 = 0.1$ $q_1 = q_2 = q_3 = 0.998$ $h = 0.001$, $L_m = 128$
Ruan [104]	$D_t^{q_1} x = y$ $D_t^{q_2} y = -\frac{1}{3}x + a(1 - z^2)y$ $D_t^{q_3} z = -y - 0.6z + yz$	$a = 0.5, x_0 = 0.1$ $y_0 = 0.2, z_0 = 0.3$ $q_1 = q_2 = q_3 = 0.97$ $h = 0.01$, $L_m = 512$
Niu [105, 106]	$D_t^{q_1} x = a(y - x) + yz$ $D_t^{q_2} y = cx - y - xz$ $D_t^{q_3} z = xy - bz$	$a = 35, b = 8/3, c = 55$ $x_0 = y_0 = z_0 = 0.01$ $q_1 = q_2 = q_3 = 0.96$ $h = 0.001$, $L_m = 64$

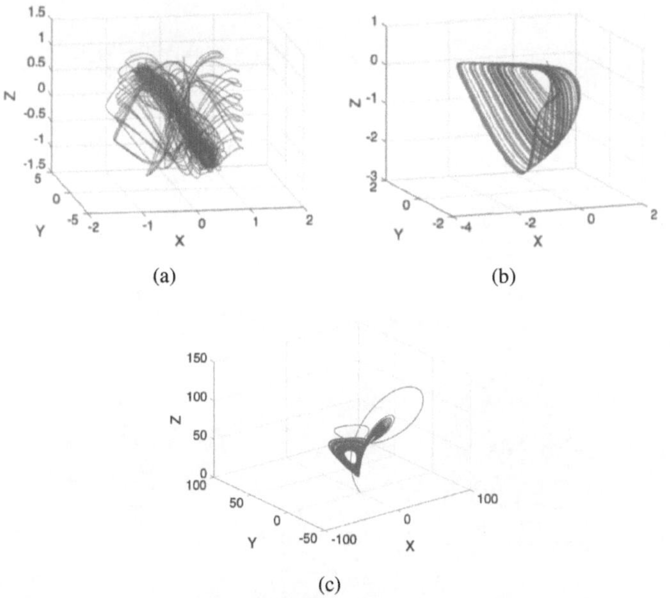

Figure 3.26: Simulation of self-excited fractional-order chaotic oscillators by applying Grünwald-Letnikov with the conditions given in Table 3.3. $x - y - z$ views of (a) Rajagopal [103], (b) Ruan [104] and (c) Niu [105, 106].

3.4 Simulation of self-excited multi-scroll and hyper-chaotic fractional-order chaotic oscillators by Grünwald-Letnikov method

The classification of self-excited fractional-order chaotic oscillators include multi-scroll and hyper-chaotic systems. In the case of simulating the multi-scroll fractional-order chaotic oscillator given in (3.17), where the PWL-function is described by (3.18), and by applying the Grünwald-Letnikov method, one must also estimate the memory length. As shown above, it generates chaotic behavior by setting $a = b = c = d_1 = 0.3$, the saturation $k = 1$, initial conditions $x_0 = y_0 = z_0 = 0.01$, and fractional-orders of the derivatives $q_1 = q_2 = q_3 = 0.95$. The time simulation is performed by setting $h = 0.005$, and the memory length is estimated as $L_m = 256$. The SNLF is increased by adding more PWL segments, so that one can generate more than 2-scroll attractors. For instance, Fig. 3.27 shows the simulation of a 4-scroll fractional-order chaotic attractor.

As mentioned above, the hyper-chaotic fractional-order chaotic oscillators are modeled by more than 3 ODEs, so that they can have more than one positive Lyapunov exponent. Equation 3.23 describes a self-excited fractional-order

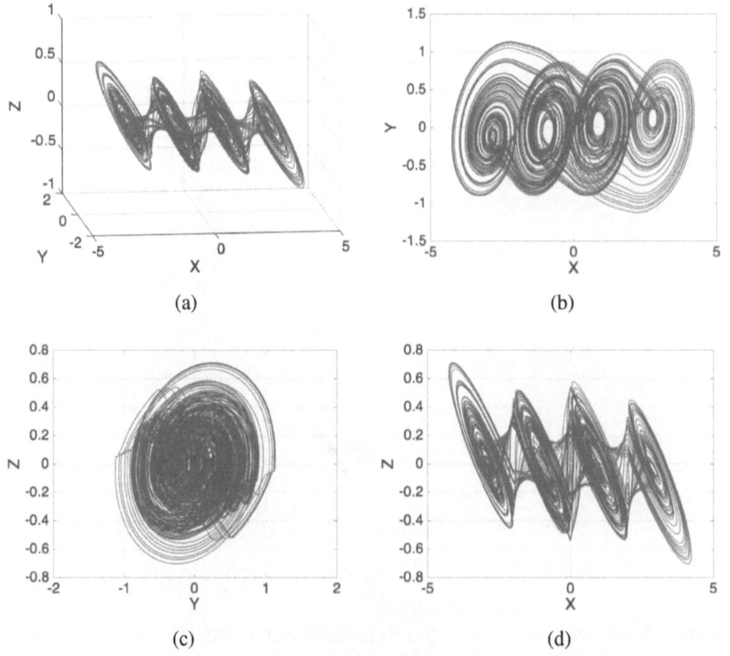

Figure 3.27: Simulation of a 4-scroll fractional-order chaotic attractor from (3.17) by applying the Grünwald-Letnikov method. Portrait views in the planes: (a) $x-y-z$, (b) $x-y$, (c) $y-z$ and (d) $x-z$.

chaotic oscillator that generates chaotic behavior if $a = 35, b = 2.5, c = 7d = 4, g = 28, k = 1/3$, and initial conditions $x_0 = y_0 = w_0 = 0.1$, and $z_0 = 2.1$ [98]. Figure 3.28 shows the views of the fractional-order attractor with $q_1 = q_2 = q_3 = q_4 = 0.9$ and solved by applying Grünwald-Letnikov with $h = 0.001$ and an estimated memory length of $L_m = 256$.

$$
\begin{aligned}
D_t^{q_1}x &= a(y-x)+byz+w \\
D_t^{q_2}y &= -cx-dxz^2+gy \\
D_t^{q_3}z &= y^2-kz \\
D_t^{q_4}w &= by-w
\end{aligned}
\tag{3.23}
$$

Other two 4D hyper-chaotic fractional-order chaotic oscillators are given in Table 3.4, where one can see the equations, their coefficient values, initial conditions, fractional-order derivatives, step-size and memory length by applying the Grünwald-Letnikov method. The simulation results are shown in Fig. 3.29.

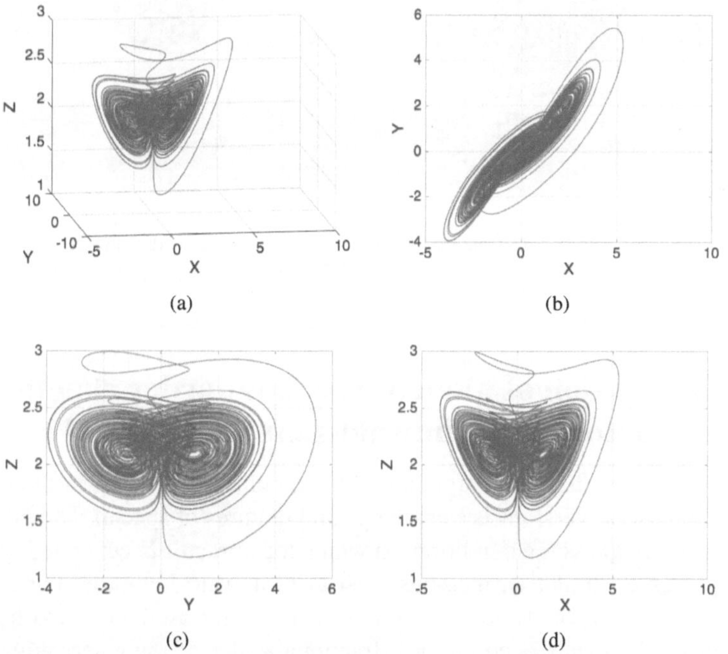

Figure 3.28: Simulation views of the fractional-order chaotic attractor based on (3.23) by applying Grünwald-Letnikov with $h = 0.001$ and memory length of $L_m = 256$.

Table 3.7: 4D hyper-chaotic fractional-order chaotic oscillators.

System	Equations	Parameters
Zhou [107]	$D_t^{q_1}x = ax - cyz$ $D_t^{q_2}y = xz - by + w$ $D_t^{q_3}z = xy - dz$ $D_t^{q_4}w = -kw - y$	$a = 10, b = 38$ $c = 1, d = 22.5, k = 3.08$ $x_0 = y_0 = z_0 = w_0 = 0.1$ $q_1 = q_2 = q_3 = q_4 = 0.98$ $h = 0.001, L_m = 256$
Dadras [108]	$D_t^{q_1}x = ax - yz + w$ $D_t^{q_2}y = xz - by$ $D_t^{q_3}z = xy - cz + xw$ $D_t^{q_4}w = -y$	$a = 8, b = 40, c = 14.9$ $q_1 = q_2 = q_3 = q_4 = 0.95$ $x_0 = y_0 = z_0 = w_0 = 0.01$ $h = 0.001, L_m = 256$

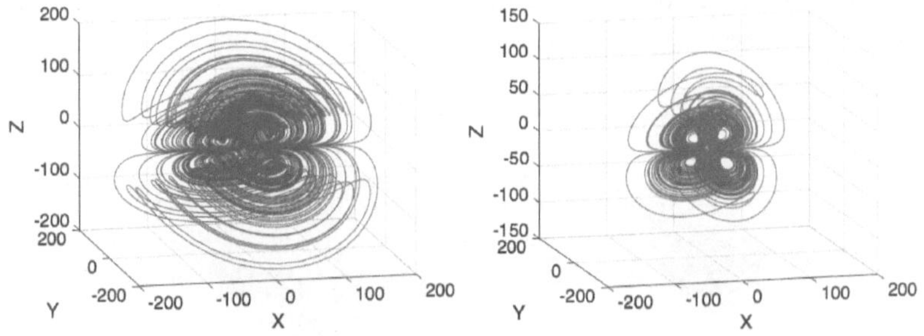

(a) (b)

Figure 3.29: Simulation of self-excited hyper-chaotic fractional-order attractors by applying Grünwald-Letnikov. (a) $x - y - z$ view of Zhou [107] and (b) $x - y - z$ view of Dadras [108].

3.5 Simulation of hidden fractional-order chaotic oscillators by Grünwald-Letnikov method

As mentioned in Chapter 1, simulating hidden chaotic systems in their integer or fractional-order versions is a challenge and requires of special methods. However, if one can guess the conditions in which the attractor is generated, then one can apply time simulation methods such as Grünwald-Letnikov. For instance, Table 3.5 shows three hidden fractional-order chaotic oscillators and lists their coefficient values, initial conditions, fractional-order of their derivatives, step-size and memory length to apply Grünwald-Letnikov. In this case, the time simulation results are shown in Fig. 3.30.

Table 3.8: Hidden fractional-order chaotic oscillators.

System	Equations	Parameters
Pham [109]	$D_t^{q_1}x = y,$ $D_t^{q_2}y = -x - yz,$ $D_t^{q_3}z = ax^2 + by^2 + cxy + d$	$a = 1.2, b = 1$ $c = 1, d = -1.3$ $x_0 = y_0 = z_0 = 0.01$ $q_1 = q_2 = q_3 = 0.992$ $h = 0.005, L_m = 256$
Li [110]	$D_t^{q_1}x = y - ax + bxz - w$ $D_t^{q_2}y = -cxz - dx + yz + 1$ $D_t^{q_3}z = 1 + xy$ $D_t^{q_4}w = kx$	$x_0 = -1.0, y_0 = z_0 = w_0 = 1.0$ $a = 3, b = 2, c = 10$ $d = 0.3, k = 1.3$ $q_1 = q_2 = q_3 = q_4 = 0.99$ $h = 0.001, L_m = 256$
Zhang [111]	$D_t^{q_1}x = a(y - x) + w$ $D_t^{q_2}y = -xz + w$ $D_t^{q_3}z = xy - b$ $D_t^{q_4}w = -cx$	$a = 12, b = 40, c = 3.6$ $x_0 = y_0 = z_0 = w_0 = 1.0$ $q_1 = q_2 = q_3 = q_4 = 0.98$ $h = 0.001, L_m = 256$

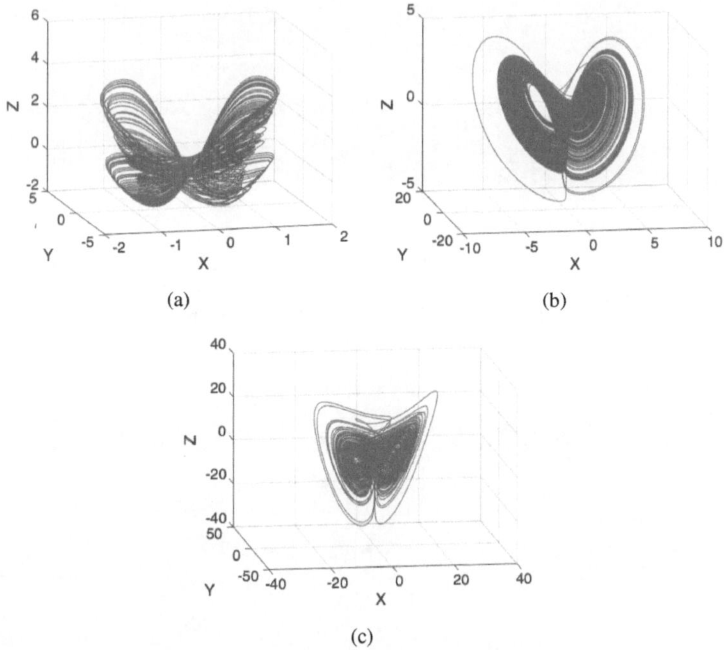

Figure 3.30: Time simulation of hidden fractional-order chaotic oscillators by applying Grünwald-Letnikov method and the parameters given in Table 3.5. (a) $x-y-z$ view of Pham [109], (b) $x-y-z$ view of Li [110] and (c) $x-y-z$ view of Zhang [111].

3.6 Simulation of fractional-order chaotic oscillators by applying Adams Bashforth-Moulton definition

The iterative equations of the predictor-corrector Adams Bashforth-Moulton method are given in Chapter 1. It will be detailed in Chapter 7 to perform the implementation of different fractional-order chaotic oscillators by using field-programmable gate arrays (FPGAs). The time simulation of the Chen fractional-order chaotic oscillator given in (3.21) with the fractional-orders of the derivatives $q_1 = q_2 = q_3 = 0.9$, and by applying Adams Bashforth-Moulton method is shown in Fig. 3.31. The time-step was evaluated to be $h = 0.001$. Details on the estimation of the memory length are given in Chapter 7. In the same manner, the time simulation of the fractional-order chaotic oscillator given in (3.22) is shown in Fig 3.32. The third fractional-order chaotic oscillator simulated by applying the Adams Bashforth-Moulton method is given in (3.23), and its simulation results are shown in Fig. 3.33. The simulation of these fractional-order chaotic oscillators confirms the suitability of the Adams Bashforth-Moulton method to solve fractional-order derivatives.

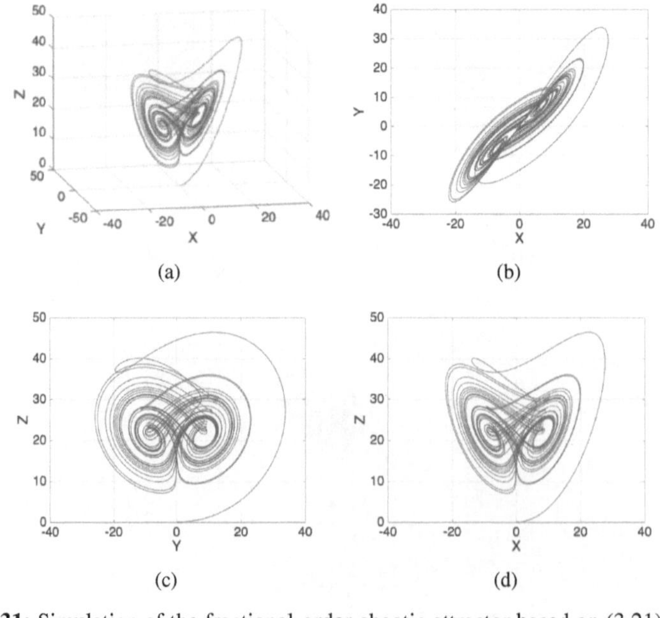

Figure 3.31: Simulation of the fractional-order chaotic attractor based on (3.21) by applying the Adams Bashforth-Moulton definition with $h = 0.001$. (a) $x-y-z$ view, (b) $x-y$ view, (c) $y-z$ view, and (d) $x-z$ view.

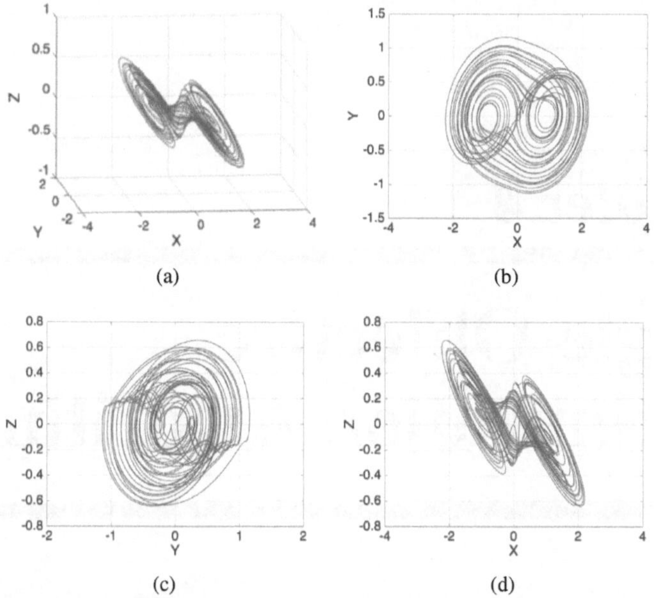

Figure 3.32: Simulation of the fractional-order chaotic attractor based on (3.22) by applying the Adams Bashforth-Moulton definition with $h = 0.001$. (a) $x - y - z$ view, (b) $x - y$ view, (c) $y - z$ view, and (d) $x - z$ view.

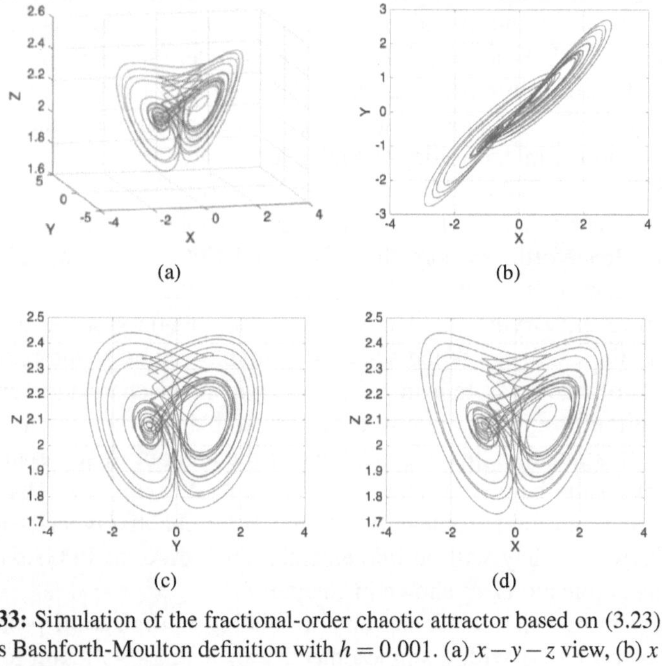

Figure 3.33: Simulation of the fractional-order chaotic attractor based on (3.23) by applying the Adams Bashforth-Moulton definition with $h = 0.001$. (a) $x - y - z$ view, (b) $x - y$ view, (c) $y - z$ view, and (d) $x - z$ view.

Chapter 4

Single-Objective Optimization Algorithms

Chapter 1 includes details of numerical methods to solve integer-order and fractional-order chaotic oscillators. Once solved and verified that the mathematical model generates chaotic behavior one can implement those chaotic oscillators by using analog and digital reconfigurable hardware. For instance, the implementation of self-excited and hidden integer-order chaotic/hyper-chaotic oscillators is described in Chapter 2 by using amplifiers and field-programmable analog arrays (FPAA). Later, Chapter 3 describes the implementation of self-excited and hidden integer-order chaotic/hyper-chaotic oscillators again by using amplifiers and FPAAs. Those integer and fractional-order chaotic oscillators are implemented by using field-programmable gate arrays (FPGAs) in Chapter 7. Henceforth, as both the FPAA and FPGA have the advantage of being reprogrammed/reconfigured, one can update the parameters of a chaotic system. However, varying the parameters of the chaotic systems is not a trivial task and the challenge is the search of suitable values of the parameters in order that the chaotic system improves its dynamical characteristics, such as maximizing the positive Lyapunov exponent, Kaplan-Yorke dimension D_{KY} or entropy. In this manner, this chapter focuses the attempt to solve an optimization problem using heuristics, and it will be applied to optimize characteristics of self-excited and hidden integer/fractional-order chaotic/hyper-chaotic oscillators. Afterwards, they will be implemented on FPAAs or FPGAs to develop engineering applications, as shown in Chapter 8.

In engineering, one can focus on formulating optimization problems into two classes: problems with a single-objective or a single function to optimize, and problems with two or more objectives. This chapter is dedicated to describing heuristics to solve single-objective problems, and Chapter 5 is devoted to describing some heuristics to solve multi-objective optimization problems. Both cases are applied in the optimization of chaotic systems, whose feasible solutions are implemented by FPAAs or FPGAs to develop applications in Chapter 8.

It is possible to apply a heuristic to a given problem without understanding why it is necessary to use that heuristic. But instead of guessing, this chapter provides details to understand what makes a problem a very hard problem to solve, and to recognize the conditions that justify the use of a heuristic to solve that problem. At the end of this chapter there are some examples of optimization problems that include constraints. Henceforth, since almost all practical problems in engineering and in optimizing chaotic systems have constraints, this chapter shows two forms two solve them.

Pseudo-codes of the algorithms, or heuristics, are provided along their description. In particular, some tests and results associated to Python programming language are given, while the entire codes are publicly available at `https://www.cs.cinvestav.mx/~fraga/OptCode.tar.gz`.

4.1 An optimization problem

The solution of an optimization problem is devoted to find its best solution, which is the maximum or the minimum of that given problem, within a certain bounds in the search space. Although in general there is not difference in adopting a maximization or minimization process, from now any problem herein will be treated as a minimization one. Therefore, if one has in hand a maximization problem, it can be transformed to a minimization problem basically by changing the sign of the function that defines the problem.

Formally, a single-objective optimization problem can be defined as follows:

$$\begin{aligned} \text{minimize: } & f(\mathbf{x}), \\ \text{subject to: } & \mathbf{g}(\mathbf{x}) \geq 0, \text{and} \\ & \mathbf{h}(\mathbf{x}) = 0, \\ & \mathbf{x} \in S \subset \mathbb{R}^n. \end{aligned} \tag{4.1}$$

where $f : \mathbb{R}^n \to \mathbb{R}$ is the function to optimize; $\mathbf{x} \in \mathbb{R}^n$, that is, the problem has n variables; and also one could have $\mathbf{g} : \mathbb{R}^n \to \mathbb{R}^p$, p inequality constraints; and $\mathbf{h} : \mathbb{R}^n \to \mathbb{R}^q$, q equality constraints. The solution to the problem $\mathbf{x}$ is in a subset S of the whole *search space* $\mathbb{R}^n$, where the constraints are satisfied, and this space S is called the *feasible space*.

The search space contains the feasible space and is defined by the *box constraints* given in (4.2), where each variable x_i is searched in the interval defined by the lower bound value l_i, and the upper bound value u_i, for $i = \{1,2,\ldots,n\}$.

$$x_i \in [l_i, u_i], \text{ for } i = \{1,2,\ldots,n\}. \tag{4.2}$$

One of the most difficult tasks in solving f in practical problems is to set the values for the box constraints in (4.2). The following sections discuss how to solve f in (4.1).

4.2 The Newton's method

If one has in hand the mathematical expression of function f in (4.1), the best that one can do is to apply the Newton's method to solve it. For instance, one can assume that f is a nonlinear function and multimodal, this is, it could have several local minima. As one will soon see, these kind of functions, nonlinear function and multimodal, are very hard to solve. Besides, if one has a nonlinear and multimodal problem but don't have its mathematical expression, it is well justified to use a heuristic.

Any function can be decomposed by its Taylor series expansion around a given point $\mathbf{a}$ as:

$$f(\mathbf{x}) = \sum_{i=0}^{\infty} \frac{(\mathbf{x}-\mathbf{a})^i f^{(i)}(\mathbf{a})}{i!} \tag{4.3}$$

By using the first two terms of (4.3), one obtains:

$$f(\mathbf{x}) \approx f(\mathbf{a}) + (\mathbf{x}-\mathbf{a})f'(\mathbf{a}) = f(\mathbf{a}) + \Delta\mathbf{a}f'(\mathbf{a}) \tag{4.4}$$

This is a linear approximation of function f around the point $\mathbf{a}$. In the regions where f is minimum, it is flat and then its derivative must be equal to zero. Thus, to solve the problem formulated in (4.1), it is necessary to calculate the derivative of function f, to equate the result with zero, and find the value of $\mathbf{x}$.

The lineal approximation of the derivative of f, using (4.4), becomes:

$$f'(\mathbf{x}) \approx f'(\mathbf{a}) + (\mathbf{x}-\mathbf{a})f''(\mathbf{a}) = f'(\mathbf{a}) + \Delta\mathbf{a}f''(\mathbf{a}) \tag{4.5}$$

Equaling (4.5) to zero, one obtains the expression for $\Delta\mathbf{a}$:

$$\Delta\mathbf{a} = \frac{-f'(\mathbf{a})}{f''(\mathbf{a})} \tag{4.6}$$

This is basically the Newton's method to solve a problem that is formulated as shown in (4.1). One needs to remember that it is necessary to give an initial value $\mathbf{a}_0$, in order to perform an iterative process as $\mathbf{a}_{i+1} = \mathbf{a}_i + \Delta\mathbf{a}$. To stop this iteration process, it is necessary to set a maximum number of iterations or it will stop until the value of $\Delta\mathbf{a}$ is lower than a given epsilon value. The initial value $\mathbf{a}_0$

must be very close to the solution, to be valid in the linear approximation around **a**. This is a contradiction: to solve a nonlinear problem, it is necessary to choose an initial point very close to the solution that one does not know!

An appropriate value of epsilon is associated to the *machine precision* value that is used practically in all personal computers. It is the smallest number such that one cannot distinguish $\varepsilon = \varepsilon + 1$. For a 32 bits, single precision real number (a `float` number in C/C++ programming language), it is approximately equal to 6×10^{-8}. For a double precision real number (a `double` or the real numbers used in python language) its value is approximately equal to 1×10^{-16}.

A pseudo-code in python of the Newton's method is given in the Algorithm 3.

Algorithm 3 Newton's method

Require: Initial value of **x**, functions f' and f''.
Ensure: Value of **x** that minimizes function f.

$i = 1$
while $i \leq 20$ **do** ▷ 20 maximum iterations
  $\Delta = -f'(x)/f''(x)$
  $x += \Delta$
  print(i, x)
  if $|\Delta| < 1e - 10$ **then**
    break
  end if
  $i += 1$
end while

4.2.1 One example

Lets us consider the function given in (4.7), which is plotted in Fig. 4.1. It is a nonlinear function, because it involves the sine function, and the figure also shows the derivative of f.

$$f(x) = (x-2)(x-5) + \sin(1.5\pi x) \tag{4.7}$$

By observing the plots in Fig. 4.1, one can see that in fact f is a multimodal function, and its derivative has 7 crosses with zero, 4 of them are local minima, and 3 of them are local maxima. The global minimum of f can be obtained by applying Newton's method, which pseudocode is shown in Algorithm 3. An execution of such implementation in python is shown in Fig. 4.2. The exact values of all the seven minima of maxima points can also be obtained by applying Newton's method. They are given in Table 4.1, where the iterations begin from the starting value for x_0 listed on the left column, the next column lists the number

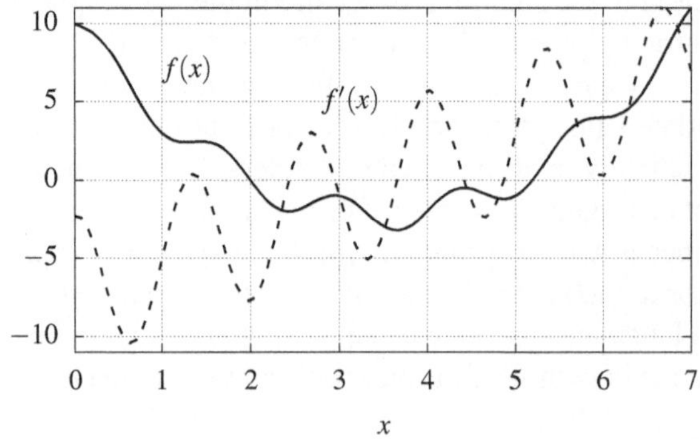

Figure 4.1: Graphs for the example function given in (4.7) and its derivative.

```
$ python f.py 3.5
1 3.688231755969256
2 3.6527624835611205
3 3.6528874443697874
4 3.6528874421621778
5 3.652887442162178
```

Figure 4.2: Execution of an implementation in python of the Newton's method. The starting value for x is set to 3.5.

Table 4.1: The 7 points where the derivative of function (4.7) is equal to zero.

x_0	Iterations	x	$f(x)$	Description
1.0	6	1.264980827648433	2.428748031158951	local minimum
1.4	6	1.441085745708321	2.475358840044047	local maximum
2.3	4	2.433055879794988	-2.003229585305734	local minimum
3.0	4	2.950004356645360	-0.975130079416381	local maximum
3.5	5	3.652887442162178	-3.224518019130687	**global minimum**
4.4	4	4.418288656759281	-0.485818390334725	local maximum
4.6	10	4.868491681705124	-1.191272590058309	local minimum

of iterations that are performed to find the minimum point, and the following columns show the values of x and $f(x)$. The starting values x_0 for the computation of the 7 minima points were obtained by observing the functions in Fig. 4.1. It is well-known that Newton's method is the fastest in finding a root, so that the number of iterations for all the roots-computations is less than 10, as appreciated in Table 4.1). However, without knowing the starting point value, one can search algorithms to find the global minimum, and it can be performed through the generation of a random value in the interval $[0,7]$ (as it is the range of values in the graph of f shown in Fig. 4.1) and it can be taken as the value for the starting point x_0 to apply Newton's method. 500 executions of this algorithm reached a 13.8% of success, which was counted if the result-value is different up to 0.001 of the optimum value.

4.2.2 Derivatives for optimization problems with several variables

In the pseudo-code of Newton's method listed in Algorithm 3, the problem to solve has a single variable. If the problem at hand increases in the number of variables, the optimization problem can be formulated as it was defined in (4.1), $f(\mathbf{x})$ with $\mathbf{x} \in \mathbb{R}^n$. Therefore, the derivatives of a problem with more than one variable can be expressed by the Jacobian by evaluating partial differential equations, as described in (4.8). $J(\mathbf{x})$ in this expression is known as the Jacobian of function f, and has the form of a column vector of size equal to the number of variables.

$$J(\mathbf{x}) = \begin{bmatrix} \frac{\partial f}{\partial x_1} \\ \frac{\partial f}{\partial x_2} \\ \vdots \\ \frac{\partial f}{\partial x_n} \end{bmatrix} \tag{4.8}$$

The second derivative of f, for this optimization problem with a single-objective function, has the form described by (4.9), in which H is well-known as the Hessian matrix of f, and has the size $n \times n$ of real numbers.

$$H(\mathbf{x}) = \begin{bmatrix} \frac{\partial^2 f}{\partial x_1 \partial x_1} & \frac{\partial^2 f}{\partial x_1 \partial x_2} & \cdots & \frac{\partial^2 f}{\partial x_1 \partial x_n} \\ \frac{\partial^2 f}{\partial x_2 \partial x_1} & \frac{\partial^2 f}{\partial x_2 \partial x_2} & \cdots & \frac{\partial^2 f}{\partial x_2 \partial x_n} \\ \vdots & \vdots & \ddots & \vdots \\ \frac{\partial^2 f}{\partial x_n \partial x_1} & \frac{\partial^2 f}{\partial x_n \partial x_2} & \cdots & \frac{\partial^2 f}{\partial x_n \partial x_n} \end{bmatrix} \tag{4.9}$$

By using both derivative matrices J and H, one can derive the iterative formulae to update the value of $\mathbf{x}$ in Algorithm 3, and it is given as:

$$\mathbf{x}_{i+1} = \mathbf{x}_i - H^{-1}(\mathbf{x}_i)J(\mathbf{x}_i) \tag{4.10}$$

As one can infer, the calculation of $\mathbf{x}$ for each iteration i has a very high cost, and in addition, the problem is more complex because it is necessary to have the expressions for J and H, and another computational issue is the high cost to invert the matrix H.

4.3 The two points gradient descent method

The application of Newton's method requires us to have the expression for the first derivatives to evaluate the Jacobian J, and the second derivative to evaluate the Hessian matrix. Besides, having only the expression for the first derivative, one could update iteratively the value of **x** (see Algorithm 3) by performing the following iterations:

$$\mathbf{x}_{i+1} = \mathbf{x}_i - \alpha J(\mathbf{x}_i), \tag{4.11}$$

In (4.11) the parameter α is associated to an scalar number, and J is again defined as the first derivative of f, and given in (4.8).

A good direction to search for the optimum value is in the negative of the derivative, as one could intuitively know from the characteristics of Newton's method. The optimization problem now is oriented to know how large that step should be made. The authors in [112] proposed to use two starting points instead of only one, as it is done in Newton's method, and then calculate the value of α in (4.11) by performing the following operations that are associated to the well-known two-points gradient descent method, and whose pseudo-code is given in Algorithm 4.

$$\begin{aligned} \Delta\mathbf{x} &= \mathbf{x}_2 - \mathbf{x}_1, \\ \Delta\mathbf{g} &= J(\mathbf{x}_2) - J(\mathbf{x}_1), \text{ and} \\ \alpha &= \frac{(\Delta\mathbf{x})^{\mathrm{T}}\Delta\mathbf{g}}{(\Delta\mathbf{g})^{\mathrm{T}}\Delta\mathbf{g}} \end{aligned} \tag{4.12}$$

By using the same example given in the previous Subsection 4.2.1, if it were executed for one hundred runs of an implementation of Algorithm 4, for a initial x_1 point generated randomly, and by setting the second initial point to $x_2 = x_1 + 0.05$. The result given a success rate of 14.8 % with an average of 11.26 iterations.

This two-points gradient descent method has the same inconveniences as Newton's method: it could lead to a local minimum or maximum, or could take more iterations, although it does not need the second derivative, and is a simpler algorithm compared to Newton's method.

4.4 The genetic algorithm

The Newton's method described in Section 4.2 and the two-point gradient method are based in local information. If the initial point (or points) are not near to the global optimum, their results could be a local minimum, a local maximum or even they could not converge in the maximum number of iterations. To cope with these problems, one can apply heuristics. For instance, a heuristic like the Genetic Algorithm (GA) uses a population, that is, a set of solutions that contain global information about the problem. The GA starts with a population of individuals (also called chromosomes) initialized randomly within the search bounds. The individuals are strings that represent possible solutions of

Algorithm 4 Two points gradient descent method

Require: Initial value of $\mathbf{x}_1$ and $\mathbf{x}_2$, expression for J
Ensure: Value of $\mathbf{x}$ that minimize function f.

```
f1 = f(x1); f2 = f(x2)
g1 = J(x1); g2 = J(x2)
if f1 < f2 then
    swap( x1, x2 )
    swap( g1, g2 )
end if
i = 1
while i ≤ 30 do                                        ▷ 30 maximum iterations
    Δx = x2 − x1
    Δg = g2 − g1
    α = transpose(Δx) ∗ Δg/(transpose(Δg) ∗ Δg)
    x1 = x2
    g1 = g2
    x2 = x2 − α ∗ g2
    g2 = J(x2)
    print(i, x2)
    if |g2| < 1e − 14 then
        break
    end if
    i += 1
end while
```

the optimization problem to be solved. One can compute the *fitness* of each individual, which is a measure of the "goodness" of the individual as a solution to the problem. These individuals are combined to produce a next generation of offspring, by performing operations of selection, cross-over, and mutation. This process is repeated to get subsequent generations. The general structure of a GA is described in Algorithm 5.

The coding of the representation is one of the most important issues in the GA design. One individual must code one solution of the problem, and for the general GA the code must be a binary string. In Table 4.2 it is shown the representation of the decimal numbers that can be coded with 4 binary digits. However, one problem can be visualized here: from one to other consecutive numbers the bits change in more that one place. For example, the number 7 in binary is 0111, and 8 is 1000, all four binary values change totally for the consecutive numbers 7 and 8. This problem can be solved by using Grey codes, which are represented by bit-strings and its main advantage is that any two consecutive numbers change only in one binary digit. In this manner, as in the GA each individual is initialized randomly, it will be the Grey code, and therefore it is only necessary to convert

Algorithm 5 Pseudo-code for a general Genetic Algorithm

Require: Population size μ, number of generations g, crossover probability p_c, mutation probability p_m. The fitness function f.
Ensure: A solution $\mathbf{x}$ that minimize $f(\mathbf{x})$

Randomly create an initial population $P = \{\mathbf{x}_1, \mathbf{x}_2, \dots, \mathbf{x}_\mu\}$.
Compute individuals' fitness.
for $i \in [1, g]$ **do**
    for $j \in [1, \mu/2]$ **do** ▷ μ must be an even number
        Select two individuals (called parents) to be reproduced.
        With probability p_c, apply crossover operator between parents.
        Apply the mutation operator to both children with a probability p_m.
        Compute children's fitness.
        Apply elitism mechanism.
    end for
    The new population of children replace to the parents.
end for
Report the individual with the best fitness.

the Grey code to a binary number, as shown in Fig. 4.3. The complete list of the 16 binary combinations is given in Table 4.2, showing their equivalents in decimal, binary and Grey code numbers.

The algorithm to convert any Grey code number to a binary number performs the operations given as follows: (1) the most significant bit (the most to the left) pass without change, (2) the next bits are calculated with the xor operation between bits g_{i+1} and b_i. Figure 4.3 shows an example of converting the Grey code number 1001 to the binary number 1110 (14 in decimal).

The natural representation of GAs as binary strings make them suitable to solve combinatorial problems. For solving continuous problems, the ones that use real numbers, one possible solution is to map linearly all binary numbers to the bounds of the search variable, using:

$$\frac{x-l}{u-l} = \frac{b}{2^q - 1} \tag{4.13}$$

where the variable x has the search range between the lower and upper limits $[l, u]$, and the binary variable b is represented with q bits (in our case, as each variable is represented as a string, with q binary numbers). The variable x will have a precision equal to $(u-l)/(2^q - 1)$ by using q bits in its representation.

After all individuals are evaluated, a selection process takes place to choose the individuals (presumably, the fittest) that will become the parents of the following generation. A variety of selection schemes exist and several of them are described in [113], including roulette wheel selection [114], stochastic remainder selection [115, 116], stochastic universal selection [117, 118], ranking selection

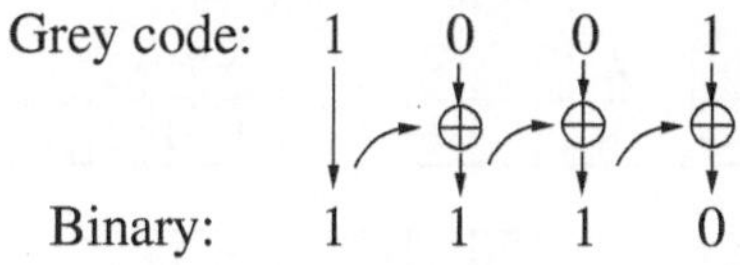

Figure 4.3: Example to convert Grey code 1001 to a binary number.

Table 4.2: The binary numbers from 1 to 15 and its Grey code representation.

Decimal Number	Binary Number	Grey Code
0	0000	0000
1	0001	0001
2	0010	0011
3	0011	0010
4	0100	0110
5	0101	0111
6	0110	0101
7	0111	0100
8	1000	1100
9	1001	1101
10	1010	1111
11	1011	1110
12	1100	1010
13	1101	1011
14	1110	1001
15	1111	1000

[119] and tournament selection. The tournament selection is easy to code: the population is shuffled randomly, then two consecutive parents are chosen, the one with better fitness win. In this selection step is where the problem is minimized (or maximized) in the comparison of both parents.

After being selected, crossover takes place. During this stage, the genetic material of a pair of individuals is exchanged in order to create the population of the next generation. This operator is applied with a certain probability p_c to pairs of individuals selected to be parents (p_c is normally set between 60% and 100%). When using binary encoding, there are three main ways of performing crossover:

1. Single-point crossover: A position of the chromosome is randomly selected as the crossover point as indicated in Fig. 4.4.

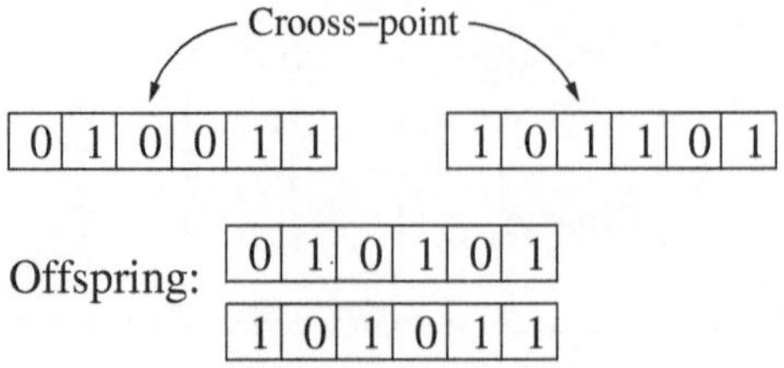

Figure 4.4: Use of a single-point crossover between two chromosomes. Notice that each pair of chromosomes produces two descendants for the next generation.

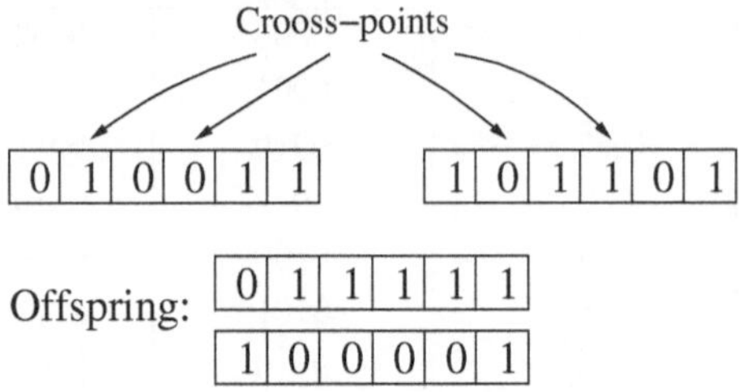

Figure 4.5: Use of two-point crossover between two chromosomes. In this case the genes at the extremes are kept, and those in the middle part are exchanged.

2. Two-point crossover: Two positions of the chromosome are randomly selected as to exchange chromosomic material, as indicated in Fig. 4.5.

3. Uniform crossover: This operator was proposed by Syswerda [120] and can be seen as a generalization of the two previous crossover techniques. In this case, for each bit in the first offspring it decides (with some probability p_c) which parent will contribute its value in that position. The second offspring would receive the bit from the other parent. This operator is illustrated in Fig. 4.6. Although for some problems uniform crossover presents several advantages over other crossover techniques [120], in general, one-point crossover seems to be a bad choice, but there is no clear winner between two-point and uniform crossover [121, 122].

The offsprings generated by the crossover operator are subject to mutation, which is a genetic operator that randomly changes a gene of a chromosome. As GA uses a binary representation, a mutation changes a 0 to 1 and viceversa. This operator is applied with a probability p_m to each binary variable in a chromosome (p_m normally adopts a low probability that goes from 1% up to 10% as a maximum). The use of this operator allows the introduction of new chromosomic material to the population and, from the theoretical perspective, it assures that–given any population–the entire search space is connected.

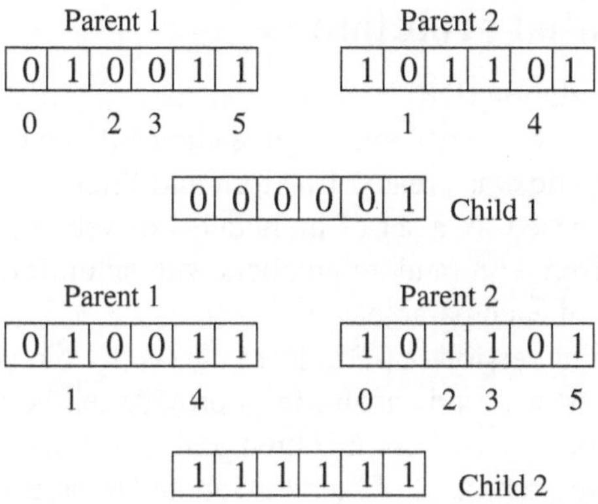

Figure 4.6: Use of 0.5-uniform crossover (i.e., adopting a 50% probability of crossover) between two chromosomes. The number below the parent's bits indicates to where the value at that position is moved into the child.

Finally, the individual with the highest fitness in the population is retained, and it passes intact to the following generation (i.e., it is not subject to either crossover or mutation). This operator is called elitism and its use is required to guarantee convergence of a simple GA, under certain assumptions.

4.4.1 One example

A code of the GA using tournament selection and two-point crossover was used to solve the problem described by the nonlinear function given in (4.7). The GA's parameters were set as: population size equal to 20 individuals, number of generations equal to 30, crossover and mutation probabilities equal to 0.7 and 0.1, respectively. 10 bits were used to code the single variable x. The search space was limited to be between the interval $[0,7]$. After 500 executions of this GA, it was get 97.2% of success. This is, in 487 of the 500 runs was get the global minimum value of $f(3.65395894428) = -3.22450414862$ (see Table 4.1) within an error of $7/(2^{10}-1) = 0.00684$ in the value for x. A success in this case was counted if $|x - 3.65288744| < 0.05$.

The high GA's success of 97.2% has a cost: each GA run takes $20 \times 30 = 600$ evaluations of the function f (one evaluation per individual in each iteration). Also, the main advantage compared to Newton's and the two-point gradient methods is that GA only requires the evaluation of f and not any information about the derivatives.

4.5 Differential evolution

The differential evolution (DE) heuristic uses naturally real numbers to code the problem's variables, thus it becomes a good choice if one has a continuous problem. The DE heuristic was created by Storn and Price in 1997 [123], where its population is composed by a set of individuals or vectors of real numbers. All vectors are initialized with random numbers with an uniform distribution within the search bounds of each parameter.

There are several versions of DE. Here the rand/1/bin version of DE is used because of its robustness and capability to provide the best results for different kinds of benchmarks and real optimization problems [124].

Nowadays there exists a lot of literature about DE, and a reader can use [125] for a good starting point about deeply DE details. The pseudo-code of DE is shown in Algorithm 6.

The core of DE can be identified from Algorithm 6, and it is in the loop on lines 8-13: a new individual is generated from three different individuals chosen randomly; each value of the new vector (represents a new individual) is calculated from the first father, plus the difference of the other two fathers multiplied by F, the difference constant; the new vector value is calculated if a random real number (between zero and one) is less than R, the DE's recombination constant. To prevent the case when the new individual could be equal to the current father i, at least one vector's component (a variable value) is forced to be calculated from their random fathers values: it is in line 9 of the pseudo-code, when $j = j_{\text{rand}}$, and j_{rand} is an integer random number between 1 and n. In lines 10-12 is verified if each combined variable values is within the search space. Then the new individual is evaluated, if it is better than the father (in lines 10-12), then the child replaces its father. The stop condition used here is: if the number of iterations is greater than a maximum number of iterations or when the difference in the objective function values of the worst and best individuals is less than s. This stop condition is called *diff* criteria in [126], and is recommended for a global optimization task.

According to the test in the CEC 2005 conference [127], DE is the second best heuristic to solve real parameter optimization problems, when the number of parameters is around 10.

A general form to set the parameters values for DE is: if n is the number of variables, the population size is set to $10n$, $F \in [0.5, 1.0]$, and $R \in [0.8, 1.0]$.

4.5.1 One example

An implementation of DE without the stop condition using s value was tested to solve the problem of the nonlinear function given in (4.7). The parameters used were: population size equal to 16 individuals, number of generations equal to 30, difference and recombination constants equal to 0.8 and 0.6, respectively. The

---

Algorithm 6 Differential evolution algorithm (rand/1/bin version)

Require: The search domain and the value s for the stop condition. The values for population size, μ; maximum number of generations, g; difference and recombination constants, F and R respectively.

Ensure: A solution of the minimization problem

initialize$(P = \{\mathbf{x}_1, \mathbf{x}_2, \ldots, \mathbf{x}_\mu\})$
evaluate(P)
$k = 0$
repeat
    for $i = 1$ to μ **do**
        Let r_1, r_2 and r_3 be three random integers in $[1, \mu]$, such that $r_1 \neq r_2 \neq r_3$
        Let j_{rand} be a random integer in $[1, n]$
        for $j = 1$ to n **do**
            $x'_j = \begin{cases} x_{r_3,j} + F(x_{r_1,j} - x_{r_2,j}) & \text{if } U(0,1) < R \text{ or } j = j_{\text{rand}} \\ x_{i,j} & \text{otherwise} \end{cases}$
            if $x'_j < l_i$ or $x'_j > u_i$ **then** ▷ Check bounds
                $x'_j = U(0,1)(u_i - l_i) + l_i$
            end if
        end for
        if $f(\mathbf{x}') < f(\mathbf{x}_i)$ **then**
            $\mathbf{x}_j = \mathbf{x}'$
        end if
    end for
    $\min = f(\mathbf{x}_1)$, $\max = f(\mathbf{x}_1)$
    for $i = 2$ to μ **do**
        if $f(\mathbf{x}_i) < \min$ **then**
            $\min = f(\mathbf{x}_i)$
        end if
        if $f(\mathbf{x}_i) > \max$ **then**
            $\max = f(\mathbf{x}_i)$
        end if
    end for
    $k \leftarrow k + 1$
until $(\max - \min) < s$ or $k > g$

---

search space was set to the interval $[0, 7]$. After 500 executions, DE got 100% of success. A success was counted if $|x - 3.65288744| < 10^{-4}$. A run of this DE takes $16 \times 30 = 480$ evaluations of the function f. The behavior of DE is better than the GA described in Section 4.4.1.

4.6 PSO and MOL algorithms

Particle swarm optimization (PSO) algorithm is another popular heuristic that can be used to solve single-objective continuous optimization problems. PSO is based on a mathematical model developed by Kennedy and Eberhart in 1995 [128, 129]. It is based in the metaphor of social behavior of birds or fishes. It also uses a population of candidate solutions (called *particles*), initialized randomly within the search bounds. These random values represent the initial positions for the particles. These particles have also a random initial velocity. Both, velocity and position, are updated as:

$$\begin{aligned} \mathbf{v}_i(t+1) &= \mathbf{v}_i(t) + c_1\text{rand}()(\text{pbest} - \mathbf{p}_i(t)) + c_2\text{rand}()(\text{gbest} - \mathbf{p}_i(t)) \\ \mathbf{p}_i(t+1) &= \mathbf{p}_i(t) + \mathbf{v}_i(t+1) \end{aligned} \tag{4.14}$$

where $\mathbf{v}_i(t+1)$ and $\mathbf{p}_i(t+1)$ represent the velocity and position of the particle in the $(t+1)$th iteration (or generation), respectively; rand() is a function that returns a uniform random real number within the interval $[0,1)$; pbest and gbest represent the best position of the particle and the best global position among all the particles; c_1 and c_2 are two parameters that represent the called confidence of the particle itself (cognition) and in the swarm (social behavior), respectively. These last two constants are the most relevant, the higher value for these constants, the faster the convergence will be. As mentioned in [129], the constants c_1 and c_2 can have values that may improve the convergence.

The expressions given in (4.14) suggest the movement of each particle $\mathbf{p}_i$ towards the best position (pbest) and the best global position (gbest), besides there are different variants using different update rules After each generation, the best particle position, and also the global best position, are calculated and stored.

The Many Optimizing Liaisons (MOL) algorithm [130], is a simplified version of PSO. The best known position of the particle, pbest in (4.14), is eliminated in MOL, and the velocity is update according to (4.15), which requires the values for parameters w, and c_2. This variant behaves similarly or better than the PSO algorithm according to [130].

$$\mathbf{v}_i(t+1) = w\mathbf{v}_i(t) + c_2\text{rand}()(\text{gbest} - \mathbf{p}_i(t)) \tag{4.15}$$

MOL is a purely social algorithm tending to follow the best swarm's particle (gbest), thus when the inertia coefficient $w = 1$, it restricts the particles exploring better solutions in the search space, so a challenge is finding the appropriate w value that allows the particles exploring different directions to find better solutions in the entire search space, in addition to maintain the velocity's limits in a previously defined range. The pseudo-code of MOL is given in Algorithm 7.

Algorithm 7 Pseudo-code for the Many Optimizing Liaisons (MOL) algorithm

Require: Swarm size μ, number of generations g, values for the parameters w and c_2. The fitness function f. Search range $[\mathbf{l}, \mathbf{u}]$.
Ensure: A solution $\mathbf{x}$ that minimizes $f(\mathbf{x})$

Randomly initialize the particles positions $P = \{\mathbf{p}_1, \mathbf{p}_2, \ldots, \mathbf{p}_\mu\}$.
Velocities range is set as $[-\mathbf{d}, \mathbf{d}]$, $d_i = u_i - l_i$, for $i \in \{1, 2, \ldots, n\}$
Randomly initialize the particles velocities $\{\mathbf{v}_1, \mathbf{v}_2, \ldots, \mathbf{v}_\mu\}$.
Compute individuals' fitness.
pbest$_i = \mathbf{p}_i$
Calculate gbest particle
for $i \in [1, g]$ **do**
    for $j \in [1, \mu]$ **do** ▷ For each dimension
        Update velocity for particle i ▷ Velocity bounds are check here
        Update position of particle i ▷ Position bounds are check here
        Update pbest and gbest particles
    end for
end for
Report the best particle position

4.6.1 One example

An implementation of MOL (in Algorithm 7) was used to solve the problem of the nonlinear function with a single variable x given in (4.7). The used parameters were: population size equal to 16 individuals, number of generations equal to 30, $w = -0.3328$, and $c_2 = 2.8446$. The search space was set to the interval $[0, 7]$. After 500 executions, MOL got 95% of success, if success is counted as $|x - 3.65288744| < 10^{-4}$, and it obtained 100% if success is calculated as $|x - 3.65288744| < 10^{-3}$.

4.7 Problems with constraints

Practical problems are subject to constraints. The algorithms showed in the previous sections do not handle constraints. In this section two methods are described to handle constraints in an optimization problem, namely: using a *penalty function*, and by modifying the selection step within a heuristic.

4.7.1 The penalty function

This is a very simple approach. It can be used without modifying the heuristic algorithm. It incorporates the constraint modifying the objective function as:

$$f_1(\mathbf{x}) = f(\mathbf{x}) + \alpha \sum_{i=1}^{p} \min[0, g_i(\mathbf{x})]^2 + \beta \sum_{i=1}^{q} h_i^2(\mathbf{x}) \tag{4.16}$$

And then now the f_1 is being optimized instead of f. α and β in (4.16) represent the penalty coefficients that weight the relative importance of each kind of constraints.

4.7.2 Handling constraints in the selection step

It is possible to handle the constraints in the operator of selection in any heuristic as the ones explained here (GA, DE, or MOL). Each time where two individuals are compared, the following mechanism is applied:

If the two compared individuals are feasible (constraints are satisfied), the individual that has the highest fitness value wins. If one of the individuals is infeasible and the other one is feasible, then the feasible individual wins. When two infeasible individuals are compared, the particle that has the lowest fitness value wins. The idea is to select leaders that, even when could be infeasible, lie close to the feasible region. The function to optimize is changed as:

$$f_2(\mathbf{x}) = \begin{cases} f(\mathbf{x}), & \text{if } \mathbf{x} \text{ is feasible} \\ \sum_{i=1}^{p} -g_i(\mathbf{x}) + \sum_{i=1}^{q} |h_i^2(\mathbf{x})|, & \text{otherwise} \end{cases} \tag{4.17}$$

This method can be incorporated into the GA in line 5 of Algorithm 5, in DE it can be incorporated in lines 14-16 of Algorithm 6, and in MOL in line 11 of Algorithm 7.

4.7.3 One example

To the problem of the nonlinear function with a single variable x given in (4.7), is added the constraint $g(\mathbf{x}) = y - 5 - 2.5(x-6) \geq 0$. This is a line and the feasible region is above the line showed in Fig. 4.1(a). In Fig. 4.1(b) is made a zoom within the range $x \in [2,4]$.

Both methods of penalty function and modified selection step were incorporated to GA, DE, and MOL algorithms. The results are shown in Table 4.3. The value for α in (4.16) was set equal to 20. The same values for all the parameters were used as it was described in the previous sections.

Observing the results in Table 4.3, the global optimum of the constrained problem is $f(2.43306) = -2.00323$ (see Fig. 4.7(b)). This global optimum is a local optimum for the unconstrained problem (see Table 4.1 in page 84). The

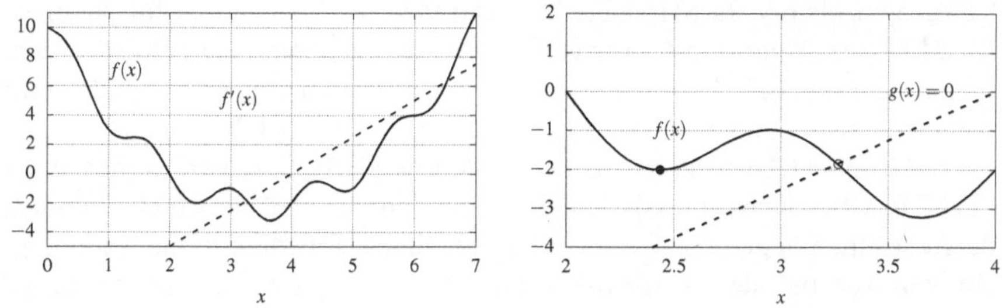

(a) (b)

Figure 4.7: Graph for the example function (4.7) and the constraint $y - 5 - 2.5(x-6) \geq 0$; the feasible region is above this line. At the right is shown the region within $x \in [2,4]$ of the same graph at the left. The point mark with '•' is the global optimum, the point '○' is a local optimum.

Table 4.3: Results of the penalty function and modifies selection step methods when are applied to the GA, DE, and MOL algorithms. Success was measured with 500 runs for each heuristic.

	Penalty Function Success	Modified Selection Step Success	x	$f(x)$
GA	95.4%	97.4%	2.43306	-2.00323
	4.4%	2.6%	3.25951	-1.85124
DE	100.0%	100.0%	2.43306	-2.00323
MOL	100.0%	99.8%	2.43306	-2.00323

GA also obtains a local minimum value in the intersection of the function f and the line that forms the constraint surface. This local optimum is marked as '○' in Fig. 4.7(b).

4.8 LSHADE

All the three heuristics studied in this chapter, GA in Section 4.4, DE in Section 4.5, and PSO and MOL in Section 4.6, require us to set values for all the parameters. To find the optimal values for a given problem is another optimization problem. One could do an exhaustive search for the optimal values within all the parameter's range, which is prohibitive if the execution time of the objective function is high.

The authors in [131] proposed a version of a self-adaptive DE, that was called SHADE (Success-History based Adaptive DE). Basically, it is an improved version of JADE [132], which uses a different parameter adaptation mechanism based on a history. In SHADE, the mean values of the successful parameters R and F (see Algorithm 6 in page 93), for each generation are stored in a historical memory. SHADE maintains a diverse set of parameters to guide control parameter adaptation as the search progresses for a more robust search.

LSHADE [133] is an improved version of SHADE, which further extends SHADE with a linear population size reduction mechanism, which continually decreases the population size according to a linear function. In the beginning of any heuristic the idea is to explore all the search space, then, once a promising area is found, this area must be exploited. In this second stage of the search not all the population is needed. LSHADE was used in the Congress on Evolutionary Computation (CEC) 2014 competition on real-parameter single-objective optimization. It obtained the first rank.

LSHADE code is publicly available [134] and is also used here for the solution of the nonlinear function with a single parameter x given in (4.7). LSHADE only requires the total number of functions evaluations, and it was set to 20% more than DE or MOL used, this is $16 \times 30 \times 1.2 = 576$. DE or MOL used 16 individuals and 30 generations. Recall that the problem given in (4.7) has a single variable x and it is used only for academic purposes. After 500 runs, LSHADE obtained a 100% of success with a precision of 10^{-8}, which uses internally LSHADE in the value of the objective function. The obtained precision was around 10^{-7} in the value of x. LSHADE also obtained a 100% of success with the constrained problem described in Section 4.7.3 by using the penalty function method (4.16) with $\alpha = 20$.

4.9 Optimization of D_{KY} of integer-order chaotic oscillators by DE and PSO

The application of the heuristics for the single-objective optimization of integer-order chaotic oscillators can be found in [135, 136, 68, 137, 138]. When they are optimized, their implementation can be performed by using modern complementary-metal-oxide-semiconductor (CMOS) technologies, as shown in [70, 69], or by using reconfigurable/reprogrammable devices such as FPAA and FPGA, as done herein in the following chapters.

In a dynamical system, Lyapunov exponents are good indicators of chaotic behavior, they are related to the exponentially fast divergence or convergence of nearby orbits in phase space, and they can be used to evaluate the Kaplan-Yorke dimension (D_{KY}). Since the existence of a positive Lyapunov exponent (LE+) is taken as an indication that chaotic behavior exists, it can be inferred that its maximization can generate more complex chaotic dynamics. For instance, the maximization of LE+ or the maximum Lyapunov exponent (MLE) by applying DE and PSO can be found in [135], where one can see the huge search spaces of the design variables of several chaotic oscillators of integer-order. As shown in

Chapter 1, integer-order chaotic oscillators are simulated by applying one-step, multi-step or special oscillatory numerical methods [17]. In those cases one challenge is the estimation of the step-size h to reduce CPU-time, i.e., the best numerical method is the one allowing the higher h and minimum error. For self-excited chaotic oscillators, the estimation of h can be associated to the eigenvalues of each equilibrium point, after guaranteeing the stability of the numerical method, as described in Chapter 1.

The optimization of D_{KY} by applying DE and PSO algorithms is performed in [136]. In that paper one can see the effect of the chaotic time series of the state variables associated to the highest D_{KY} when they are used to encrypt color images. One example to understand the very huge search spaces within the optimization of D_{KY} is by considering the Lorenz chaotic oscillator, who has three design parameters: σ, ρ, and β . These parameters can have different lengths in the integer and fractional values, for example: If one chooses two integer (10^2) and four fractional numbers (10^4), then the number of combi-nations becomes $10^6 \times 10^6 \times 10^6 = 10^{18}$. It is quite clear that the simulation of this huge number of cases can be unreachable in a couple of years, but the most important issue is that not all cases will generate chaotic behavior. This justifies the application of heuristics to search for the best coefficient values that provide high LE+ and high D_{KY}, for instance.

The maximization of D_{KY} by applying DE and PSO algorithms is performed by considering the chaotic system with infinite equilibria points [139], whose mathematical model is described by (4.18). Other two chaotic oscillators are considered, namely: Rössler system described by (4.19), and Lorenz chaotic system described in the previous chapters and given again by (4.20).

$$\begin{aligned} \dot{x} &= -z \\ \dot{y} &= xz^2 \\ \dot{z} &= x - ae^y + z(y^2 - z^2) \end{aligned} \tag{4.18}$$

$$\begin{aligned} \dot{x} &= -y - z \\ \dot{y} &= x + ay \\ \dot{z} &= b + z(x - c) \end{aligned} \tag{4.19}$$

$$\begin{aligned} \dot{x} &= \sigma(x - y) \\ \dot{y} &= x(\rho - z) - y \\ \dot{z} &= xy - \beta z \end{aligned} \tag{4.20}$$

The adaptation of DE algorithm to optimize D_{KY} is shown in Algorithm 16, and the adaptation of PSO is shown in Algorithm 17.

Algorithm 8 Differential Evolution

Require: D, G, N_p, CR, F, and $func(\bullet)$.

Initialize the population randomly (x)
Evaluate the population with the function associated to the chaotic oscillator $[func(x)]$
Save the evaluation results in *score*
for $(counter = 1; counter \leq G; counter{+}{+})$ **do**
  for $(i = 1; i \leq N_p; i{+}{+})$ **do**
    Select three different indexes randomly (a, b, and c)
    for $(j = 1; j \leq D; j{+}{+})$ **do**
      if $U(0,1) < CR \,||\, j = D$ **then**
        $trial_j \leftarrow x_{aj} + F(x_{bj} - x_{cj})$
      else
        $trial_j \leftarrow x_{ij}$
      end if
    end for
    $fx \leftarrow func(trial)$
    if fx is better than $score_i$ **then**
      $score_i \leftarrow fx$
      $x_i \leftarrow trial$
    end if
  end for
end forreturn x and *score*

The D_{KY} requires that the Lyapunov exponents be ordered from the most positive (MLE) to the most negative so that one can use (4.21), where k is an integer such that the sum of the Lyapunov exponents (λ_i) is non-negative. If chaotic behavior is guaranteed in Equations (4.18)–(4.20), then k = 2, so that λ_{k+1} is the third Lyapunov exponent (LE+), and therefore D_{KY} is higher than 2.

$$D_{KY} = k + \frac{\sum_{i=1}^{k} \lambda_i}{\lambda_{k+1}} \tag{4.21}$$

The Lyapunov exponents can be obtained by applying Wolf's method [140]. The initial conditions (x_0, y_0, z_0) matters to reduce computing time in computing Lyapunov exponents and D_{KY}. In this manner, the appropriate initial conditions are: $(ae^{0.5}, 0.5, 0.75)$ for the chaotic oscillator with infinite equilibria (4.18), $(0.5, 0.5, 0.5)$ for Rössler (4.19), and $(0.1, 0.1, 0.1)$ for Lorenz (4.20). The maximization of D_{KY} by applying DE and PSO algorithms was executed with the same conditions for the three chaotic oscillators, i.e., the same number of populations ($P = 30$) and maximum generations ($G = 20$). The numerical simulations were performed for 10,000 iterations, discarding the first 1000 iterations as they include the transient behavior. The optimization results were analyzed and the

Algorithm 9 Particle Swarm Optimization

Require: D, G, N_p, α, β, and $func(\bullet)$.
Initialize the position of the particles randomly (x)
Initialize the velocity of the particles (v)
Evaluate the position of the particles with your function $[func(x)]$
Save the evaluation results in *score* and $p \leftarrow x$
Find the best value from p and save it in g
for ($counter = 1; counter \leq G; counter$++) **do**
    for ($i = 1; i \leq N_p; i$++) **do**
        for ($j = 1; j \leq D; j$++) **do**
            $v_{ij} \leftarrow \alpha v_{ij} + U(0,\beta)(p_{ij} - x_{ij}) + U(0,\beta)(g_j - x_{ij})$
            $x_{ij} \leftarrow x_{ij} + v_{ij}$
        end for
        $f_x \leftarrow func(x_i)$
        if f_x is better than $score_i$ **then**
            $score_i \leftarrow f_x$
            $p_i \leftarrow x_i$
            if p_i is better than g **then**
                $g \leftarrow p_i$
            end if
        end if
    end for
end forreturn x, p, g, and *score*

highest values of D_{KY} are given in Tables 4.4 and 4.5, where one can see the values of the design parameters associated to the best five values of D_{KY} provided by DE and PSO, and it also shows their associated LE+.

The chaotic time series were tested in [136] for an image encryption system and the conclusion is that the best chaotic oscillator is Lorenz because its state variable x provided the lowest correlation between the original image and the chaotic channel, while the image was 100% successfully recovered. This process is shown in the following chapters but by using optimized fractional-order chaotic oscillators.

Table 4.4: Optimization results for the five highest values of D_{KY} provided by DE.

Oscillator	DE		
	Design Parameters	LE_+	D_{KY}
[139]	$a = 0.1006$	0.0753	2.0791
	$a = 0.0938$	0.0730	2.0789
	$a = 0.0939$	0.0726	2.0786
	$a = 0.0935$	0.0726	2.0785
	$a = 0.1007$	0.0747	2.0788
Rössler	$a = 0.3609$ $b = 0.1000$ $c = 11.3470$	0.2711	2.07890
	$a = 0.3947$ $b = 0.5490$ $c = 9.12060$	0.2600	2.07140
	$a = 0.3720$ $b = 0.2055$ $c = 12.0147$	0.2710	2.07870
	$a = 0.3930$ $b = 0.8505$ $c = 13.0501$	0.2645	2.07810
	$a = 0.3643$ $b = 0.1537$ $c = 12.7643$	0.2764	2.07880
Lorenz	$\sigma = 29.9226$ $\rho = 89.8095$ $\beta = 13.9727$	3.3129	2.08430
	$\sigma = 29.9388$ $\rho = 89.8923$ $\beta = 14.1895$	3.3122	2.08390
	$\sigma = 29.7786$ $\rho = 89.7268$ $\beta = 13.4876$	3.3168	2.07960
	$\sigma = 29.9222$ $\rho = 89.9781$ $\beta = 14.1956$	3.3149	2.07910
	$\sigma = 29.7066$ $\rho = 89.8899$ $\beta = 13.7180$	3.3199	2.07410

Table 4.5: Optimization results for the five highest values of D_{KY} provided by PSO.

Oscillator	PSO		
	Design Parameters	LE_+	D_{KY}
[139]	$a = 0.0937$	0.0753	2.079
	$a = 0.1007$	0.0788	2.0789
	$a = 0.1028$	0.0796	2.0796
	$a = 0.0935$	0.0726	2.0785
	$a = 0.1007$	0.0787	2.0791
Rössler	$a = 0.3609$ $b = 0.1000$ $c = 11.3470$	0.2711	2.02700
	$a = 0.3947$ $b = 0.5490$ $c = 9.12060$	0.2600	2.02690
	$a = 0.3947$ $b = 0.2055$ $c = 9.12060$	0.2600	2.02650
	$a = 0.3930$ $b = 0.8505$ $c = 13.0501$	0.2645	2.02350
	$a = 0.3643$ $b = 0.1537$ $c = 12.7643$	0.2764	2.02330
Lorenz	$\sigma = 30.0000$ $\rho = 90$ $\beta = 12.3872$	3.3129	2.07230
	$\sigma = 29.8297$ $\rho = 90$ $\beta = 13.7954$	3.3122	2.07290
	$\sigma = 29.9966$ $\rho = 90$ $\beta = 13.4876$	3.3168	2.07190
	$\sigma = 30.0000$ $\rho = 90$ $\beta = 14.1956$	3.3149	2.07197
	$\sigma = 29.8375$ $\rho = 90$ $\beta = 13.8036$	3.3199	2.07179

Chapter 5

Multi-Objective Optimization Algorithms

Single-objective optimization algorithms were described in Chapter 4, whose main characteristics are that they provide a single solution to a given problem. In a multi-objective optimization problem (MOOP), the heuristics can generate a set of solutions. This behavior is because the objetives are in conflict, that is: one could not improve one objective without deteriorating the other. In such a case, the set of solutions are the trade-offs among the objective functions. Three well-known heuristics that have been applied to solve MOOPs are described herein, namely: non-dominated sorting genetic algorithm (NSGA-II), multi-objective evolutionary algorithm with decomposition (MOEA-D), and MOMBI-III. This last one is an improvement of the original Many-Objective Meta-heuristic Based on the R2 Indicator (MOMBI [141]), which was developed by Raquel Hernández Gómez in 2013 as a result of her master thesis [142]. In the area of optimization with heuristics the term *multi-objective optimization* is associated to solve problems with two and three objectives, while for four and more objectives it is called *many-objective optimization.* In this book the integer/fractional-order chaotic/hyper-chaotic oscillators are optimized considered as objective functions the positive Lyapunov exponent (LE+), Kaplan-Yorke dimension D_{KY} and Kolmogorov-Sinai entropy (associated to the values of the LE+s). For this reason this chapter describes heuristics for MOOPs.

The NSGA-II algorithm was introduced in 2000 [143], and yet it is the reference algorithm to compare with. It can be applied to solve MOOPs without and with considering constraints. MOEA-D was introduced in 2004 [144], and MOMBI-II is an algorithm designed to solve many-objectives problems. The pseudo-codes of these algorithms, or heuristics, are described herein and they are already coded in C language. Some educational examples are publicly available at `https://www.cs.cinvestav.mx/~fraga/OptCode.tar.gz`.

5.1 Formulation of a multi-objective optimization problem

As mentioned in the previous chapter, a MOOP can also be formulated to maximize or minimize, but in this book the problems are oriented to formulate minimization problems. In this manner, and formally speaking, a multi-objective optimization problem is defined as:

$$\begin{aligned} \text{minimize: } & \mathbf{f}(\mathbf{x}), \\ & \mathbf{f}: \mathbb{R}^n \to \mathbb{R}^m \\ \text{subject to: } & \mathbf{g}(\mathbf{x}) \geq 0, \text{and} \\ & \mathbf{h}(\mathbf{x}) = 0, \\ & \mathbf{x} \in S \subset \mathbb{R}^n. \end{aligned} \tag{5.1}$$

where $\mathbf{f}(\mathbf{x}) = [f_1(\mathbf{x}), f_2(\mathbf{x}), \ldots, f_m(\mathbf{x})]^{\mathrm{T}}$ are the set of m functions to optimize, the problem has n decision variables, $\mathbf{x} \in \mathbb{R}^n$, and as in the case of single-objective problems, one could also have $\mathbf{g} : \mathbb{R}^n \to \mathbb{R}^p$, p inequality constraints; and $\mathbf{h} : \mathbb{R}^n \to \mathbb{R}^q$, q equality constraints.

The set of solutions to the problem $\{\mathbf{x}\}$ are in the *feasible region* S, which is a subset of the search space F, which at the same time is a subset of the whole space $\mathbb{R}^n$: $\{\mathbf{x}\} \in S \subset F \subset \mathbb{R}^n$.

The search space F is defined by the set of constraints called *box constraints*:

$$x_i \in [l_i, u_i], \text{ for } i = \{1, 2, \ldots, n\}, \tag{5.2}$$

where each variable x_i is searched in the interval defined by the lower bound l_i and the upper bound u_i values, for $i = \{1, 2, \ldots, n\}$.

5.2 Pareto dominance

In a single-objective optimization problem, if one has two solutions to the problem, for example x_1 and x_2, it is easy to see which solution is the best: if $f(x_1) < f(x_2)$, then x_1 is better than x_2. However, in a MOOP one cannot directly compare the set of solutions. Instead, the Pareto dominance is used, and it works as follows: one solution vector dominates (is better) another one, if all the values of the objective functions evaluated with this solution are less or equal to the corresponding values of the objective functions evaluated with the other solution vector, and in at least one value of the objective function is strictly lesser than the corresponding in the other vector.

Formally speaking, it is said that $\mathbf{x}_1$ dominates $\mathbf{x}_2$ if

$$\mathbf{x}_1 \prec \mathbf{x}_2 \text{ if } \begin{cases} f_j(\mathbf{x}_1) \leq f_j(\mathbf{x}_2), & \text{for } j \in \{1,2,\ldots,m\}, \text{ and} \\ f_j(\mathbf{x}_1) < f_j(\mathbf{x}_2), & \text{for at least one } j \in \{1,2,\ldots,m\}. \end{cases} \tag{5.3}$$

In (5.3) $\mathbf{x}_1, \mathbf{x}_2 \in S$.

All the non-dominated solutions form the set called *Pareto set*, and the corresponding values of the objective functions evaluated with that set form the *Pareto front*.

It is interesting to know how to compute in a computer program the Pareto dominance. For example, Algorithm 10 shows how to compute the dominance between two input vectors. The fist part in (5.3) is not used, instead only the second part is coded as shown in lines 5-12 in Algorithm 10. If a single element of vector "vfa" is less than the corresponding element in "vfb", "flag1" is activated (equaling it to 1). This "flag1" can be activated m times (the size of the vectors "vfa" and "vfb"). The same is done with vector "vfb" (and the corresponding elements in "vfa") but now the "flag2" is activated. If "vfa" dominates "vfb" this means that flag1 was activated and flag2 was not. If "vfb" dominates "vfa" this means that flag2 was activated and flag1 was not. If both flags were activated, then both vectors are non-dominated (and possibly both vectors are part of the Pareto front). In practically all heuristics, the whole population is evaluated at the same time, and the dominance is checked at a later step, therefore the inputs to Algorithm 10 are the values of the objective functions instead of the variable values.

5.2.1 Calculating the non-dominated solutions

Suppose we have a set with 5 solutions $\{\mathbf{x}_1, \mathbf{x}_2, \mathbf{x}_3, \mathbf{x}_4, \mathbf{x}_5, \}$, they can be compared between each pair of solutions to decide their dominance in a bi-objective problem. The 5 solutions have the values in the objective functions shown in Fig. 5.1, from which one can build the following matrix with the results of the comparison:

$$\begin{array}{cccccc} & \mathbf{f}(\mathbf{x}_1) & \mathbf{f}(\mathbf{x}_2) & \mathbf{f}(\mathbf{x}_3) & \mathbf{f}(\mathbf{x}_4) & \mathbf{f}(\mathbf{x}_5) \\ \mathbf{f}(\mathbf{x}_1) & - & 0 & 0 & 0 & 0 \\ \mathbf{f}(\mathbf{x}_2) & & - & -1 & 0 & 0 \\ \mathbf{f}(\mathbf{x}_3) & & & - & 1 & 0 \\ \mathbf{f}(\mathbf{x}_4) & & & & - & 0 \\ \mathbf{f}(\mathbf{x}_5) & & & & & - \end{array} \tag{5.4}$$

In (5.4) each column and row are labeled with $\mathbf{f}(\mathbf{x}_i)$, for $i \in \{1,2,\ldots,5\}$. This matrix includes only the dominance results between $\mathbf{f}(\mathbf{x}_i)$ and $\mathbf{f}(\mathbf{x}_j)$, for $j > i$,

Algorithm 10 A function to calculate the dominance between two solution vectors

Require: Vector va, vb, number of objectives m.
Ensure: Value of dominance between va and vb

```
function DOMINANCE(m, va, vb)
    vfa = evaluate( va )
    vfb = evaluate( vb )
    flag1 = 0
    flag2 = 0
    i = 0
    while i < m do
        if vfa[i] < vfb[i] then
            flag1 = 1
        else if vfa[i] > vfb[i] then
            flag2 = 1
        end if
        i += 1
    end while
    if flag1==1 and flag2==0 then
        return 1                                  ▷ va dominates vb
    else if flag1==0 and flag2==1 then
        return -1                                 ▷ vb dominates va
    else                                          ▷ Both are non-dominated
        return 0
    end if
end function
```

and $i, j \in \{1,2,\ldots,5\}$. It is not necessary to calculate all the values for $j < i$, and of course the dominance is not defined for a self-point. One doesn't know the function $\mathbf{f}$, and the $\mathbf{x}_i$ values, this means that dominance is computed by using only the values of the objective function, as in the example given in Fig. 5.1.

The problem now is how to calculate points in Fig. 5.1 that belong to the set of non-dominated solutions (if this set is optimal then it is the Pareto front), from the values given in matrix (5.4). By defining an auxiliar vector, initialized with zeros, one can mark in this auxiliar vector when a point dominates another (with any other value different than zero). At the end, all the elements in the auxiliar vector different than zero will be the indexes of the points in the Pareto front. One can use the pseudocode given in Algorithm 11 to solve this kind of problem. Algorithm 11 uses the function dominance() already described in Algorithm 10.

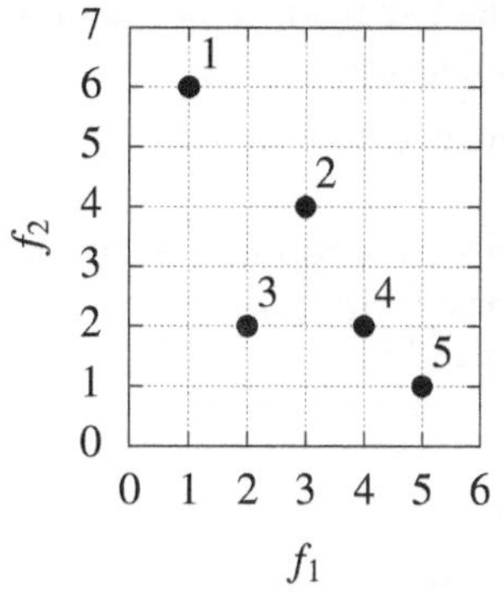

Figure 5.1: An example with 5 points in the objective space.

For this example, lets us suppose that one has the initialized array "solutions" given as:

$$\text{solutions} \rightarrow [0,0,0,0,0]$$

in line 1 of the pseudocode in Algorithm 11. The two comparisons that result different than zero (see matrix (5.4)) are: dominance($\mathbf{f}(\mathbf{x}_2), \mathbf{f}(\mathbf{x}_3)$), which is equal to -1; and dominance($\mathbf{f}(\mathbf{x}_3), \mathbf{f}(\mathbf{x}_4)$), which is equal to 1. Thus $\mathbf{f}(\mathbf{x}_3)$ dominates $\mathbf{f}(\mathbf{x}_2)$ in this call to "dominance" function, and afterwards the array called "solutions" is modified as:

$$\text{solutions} \rightarrow [0,1,0,0,0].$$

In the second call, $\mathbf{f}(\mathbf{x}_3)$ dominates $\mathbf{f}(\mathbf{x}_4)$, so that the array called "solutions" is now updated to:

$$\text{solutions} \rightarrow [0,1,0,1,0].$$

The indexes to the elements of this array equal to zero: 1, 3, 5, are the indexes to the elements that form the set of non-dominated solutions in the objective space (here it is supposed that the elements in the array are indexed starting with 1, while in Algorithm 11 the array indexing starts with 0).

5.3 NSGA-II

The Non-dominated Sorting Genetic Algorithm (NSGA-II) was proposed by Deb et al. in 2002 [143], and is still considered to be the state-of-the-art algorithms to solve MOOPs. The main goal for which NSGA-II was designed is to give and to spread all solutions on the Pareto front. As any evolutionary algorithm, it uses a set of solutions (a population) that is evolved by applying genetic operators (selection, mutation and crossover) to generate new solutions (or individuals); and only the best solutions survive to the next iteration of the algorithm. NSGA-

Algorithm 11 A pseudocode to calculate the set of non-dominated solutions for a set of already evaluated solutions

Require: Input set of evaluated solutions in a matrix V, of size $n \times m$, n row vectors of size m.

Ensure: Indexes to the points that form the set of non-dominated solutions

```
Initialize vector solutions[] with zeros
i = 0
while i < n do
    j = i + 1
    while j < n do
        d = dominance( m, V[i], V[j] )
        if d == 1 then                          ▷ V[i] dominates V[j]
            solutions[j] = 1
        else if d == −1 then                    ▷ V[j] dominates V[i]
            solutions[j] = 1
        end if
        j += 1
    end while
    i += 1
end while
i = 0                                           ▷ Print the indexes to nondominated solutions
while i < n do
    if solutions[i] == 0 then
        i += 1
        print( i )
    end if
end while
```

II presents two ideas that try to improve the common performance of traditional multi-objective algorithms: the *non-dominated sorting* and the *crowding distance*

The non-dominated sorting separates all the population in sub-fronts according to Pareto non-domination. A rank value is created according to the number of individuals that dominated each solution. The non-dominated solutions receive a rank equal to one, and the others receive a rank value according to how many subsets they dominate.

Also, to preserve a good spread of the final solutions, NSGA-II uses the second mechanism called crowding distance. This distance measures the average size of the cuboid formed with the points that enclose a solution in the population.

The main idea behind NSGA-II is to use a selection that preserves the individuals with the lowest rank value. When almost all the solutions exist in the first rank, the algorithm selects the solutions with the highest value of the crowding distance. Algorithm 12 describes how NSGA-II evolves the solutions. The genetic operators of NSGA-II, its two used mechanisms, and also how it handles the constraints, will be explained in detail in the following subsections.

Algorithm 12 Pseudocode of NSGA-II

Require: Number of generations g, number of individuals n

Randomly generate the population P_1
Evaluate the population
for $i = 1 : g$ **do**
    Apply genetic operators in P_i to generate Q_i
    Set $R_i = P_i \cup Q_i$
    Calculate the rank values for R_i
    Calculate the crowding distance of R_i
    $P_i \leftarrow$ select the n individuals with the lowest rank and highest crowding distance
end for

5.3.1 Genetic operators

NSGA-II uses the usual binary tournament for selection (see Section 4.4, on page 86). NSGA-II can use binary real variables. For binary variables, it uses single-point crossover and bitwise mutation (see Section 4.4). For real variables, it uses the simulated binary crossover (SBX) operator and polynomial mutation. SBX was invented also by Deb [145], he studied the working principle of single-point crossover on binary strings and proposed the probabilities distributions given in (5.5) that simulates the same crossover on real variables.

$$d(\beta, \eta_c) = \begin{cases} \frac{\eta_c+1}{2} \beta^{\eta_c} & \text{for } \beta \in [0:1] \\ \frac{\eta_c+1}{2} \frac{1}{\beta^{\eta_c+2}} & \text{if } \beta > 1 \end{cases} \tag{5.5}$$

A graph of $d(\beta)$ for several values of η_c is shown in Fig. 5.2. This operator is convenient, because the spread of children solutions around parent solutions can be controlled by using the η_c distribution index. A large value of this index allows only near parent solutions to be created, whereas a small value of the index allows distant solutions to be created. Another aspect of this crossover operator is that it allows us to have a more focused search when the population is converging. Similarly, for polynomial-based mutation, the amount of perturbation in a variable can also be controlled by the η_m distribution index.

NSGA-II crossover code does much more than only the direct implementation of (5.5). First, it uses a crossover probability, by default equal to 0.9 (it can be changed). Second, it applies SBX only 50% of the rest of the time for each variable, (exactly $0.9 \times 0.5 = 0.45$, or 45% of the time SBX is applied). This means

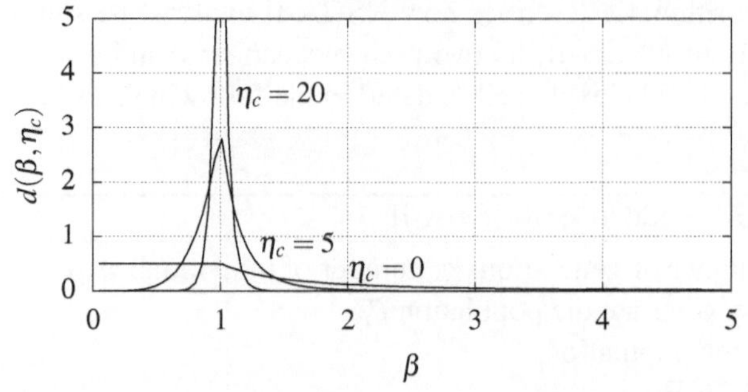

Figure 5.2: Graphs for the SBX distribution probabilities in (5.5) for $\eta_c = 0$, 5, and 20.

that $0.1 + 0.9 \times 0.5 = 0.55$, or 55% of the times the children are equal to the parents. And finally, once SBX is applied, the 50% of the rest of the times NSGA-II interchange the values of the modified variables (exactly it is $0.45 \times 0.5 = 0.225$, or 22.5% of the times). This crossover process can be visualized with three examples as shown in Fig. 5.3 with a single variable, with two fathers with values 2 and 8, and a search space in the interval $[0, 10]$.

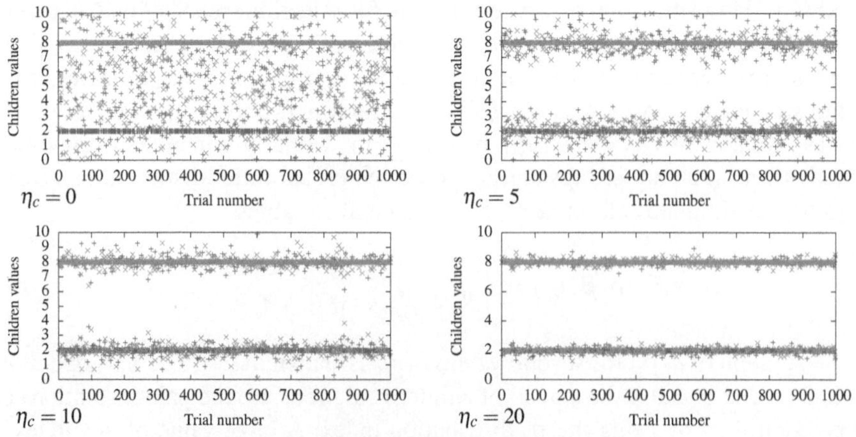

Figure 5.3: Four examples of NSGA-II crossover operator. In each graph it is shown the children values of 1000 trials of crossover operator on a single variable with father's values 2 and 8, and search space in the interval $[0, 10]$.

For the mutation operation, NSGA-II uses a mutation probability of $p_m = 1/n$ or $1/l$, where n is the number of real variables and l is the string length used for binary variables. The mutation operator uses a η_m (instead of η_c) values using

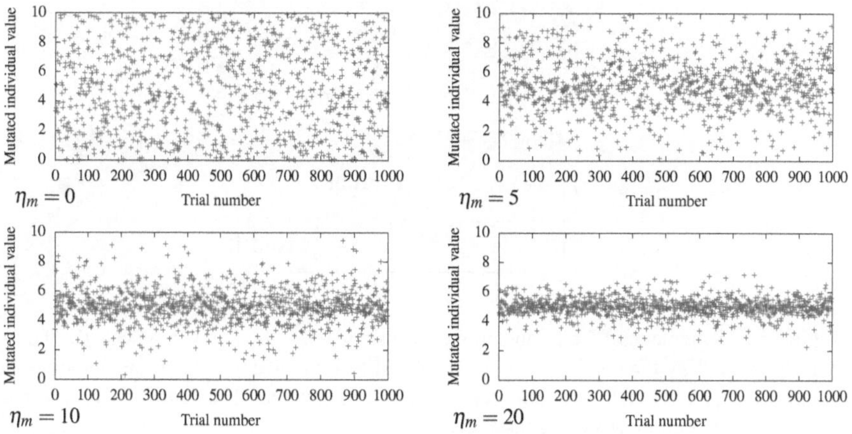

Figure 5.4: Four examples of NSGA-II polynomial mutation operator at different η_m values. It is shown in each graph the values of 1000 trials of the mutation operator applied to a varible equal to 5 and search space in the interval $[0, 10]$.

the second part of (5.5). For real variables, NSGA-II uses distribution indexes for crossover and mutation operators as $\eta_c = 20$ and $\eta_m = 20$, respectively. Figure 5.4 shows several examples of the mutation operator over one variable with value of 5 within a search space of $[0, 10]$.

To generate the simulations in Figs. 5.3 and 5.4 original NSGA-II code already available in [146] and also available in the code package here [147] was used.

5.3.2 Ranking and crowding distance

Elitism is introduced in the NSGA-II when both populations of parents and offspring are sorted with respect to its rank (line 6 in Algorithm 12). Ranking could be performed by the procedure described in Section 5.2.1: the first non-dominated solutions are obtained and removed from the population, to the rest the same procedure it is applied successively. It is not necessary to sort the whole joint new population, the sorting is stopped when the number of individuals in the obtained fronts is equal or greater than the population size. The individuals in the first front have a rank value of one. Each consecutive front increases its rank value in one. Figure 5.5 illustrated the front building process, as well as the crowding distance for a hypothetical population.

Crowding distance is illustrated in Fig. 5.5: it estimates the density of solutions surrounding a particular solution in the population, its value corresponds to the average distance of two points on either side of this point along each of the objectives. To calculate this distance, the population within a given front is

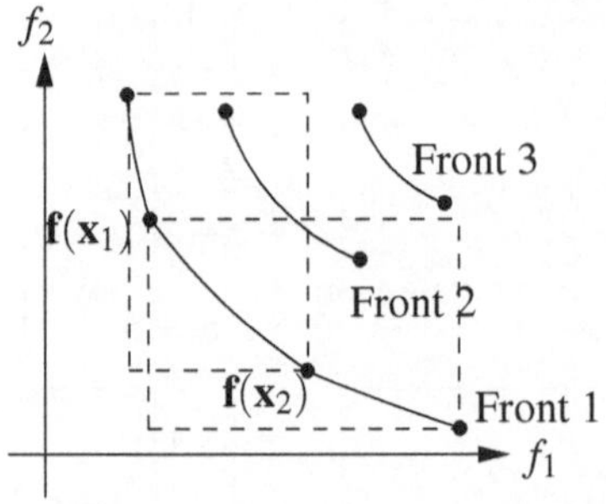

Figure 5.5: Schematic of fronts and crowding distance of individuals in NSGA-II. The fronts are represented by the continuous lines. The dashed regions depict the crowding distance of solutions $\mathbf{f}(\mathbf{x}_1)$ and $\mathbf{f}(\mathbf{x}_2)$. Here, solution $\mathbf{x}_2$ is preferred over $\mathbf{x}_1$. The boundary solutions of the first front have infinite crowding distance.

sorted along each objetive value and after each sort the hyperboxes lengths are computed and summed up.

In line 8 in Algorithm 12, when the new population is selected, if the last front is larger than the population size, then its solutions are sorted according to their crowding distance. The best solutions fill the remaining population.

The overall complexity of the NSGA-II algorithm is $O(mn^2)$, which is governed by the non-dominated sorting part of the algorithm.

It is worth noting that when two solutions in the tournament selection are compared or sorting of the population, it is preferred the solution with the lower (better) rank. Otherwise, if both solutions belong to the same front, then the preferred solution is the one that is located in a less crowded region.

5.3.3 *Constraints handling*

NSGA-II can also handle constraints, the process that it uses is similar to the one explained for the single-objective problems in Section 4.7.2, on page 96: when two solutions are compared, if both are unfeasible then the preferred one is with the lesser number of constraint violations; if one solution is feasible and the other is unfeasible, the feasible one is preferred; if both solutions are feasible the Pareto dominance is used to decide which is better.

5.4 MOEA-D

The Multi-Objective Evolutionary Algorithm based on Decomposition (MOEA-D) proposed by Zhang in 2007 [144], transforms an optimization problem into a number of single-objective optimization subproblems that are simultaneously optimized by evolving a population. Each individual solution in the population is associated with a subproblem. Three characteristics are crucial for MOEAD: a set of uniformly distributed weight vectors that cover the search space, an utility function, and a reference point. These three characteristics are used to solve each subproblem.

The set of weight vectors can be generated by using the simplex-lattice design approach, that will be described later in Subsection 5.4.1.

Let $\boldsymbol{\lambda} = [\lambda_1, \lambda_2, \ldots, \lambda_m]^{\mathrm{T}}]$ be a weight vector, i.e., $\lambda_i \geq 0$, for $i \in \{1, 2, \ldots, m\}$ and $\sum_{i=1}^{m} = 1$. The simplest form to transform the multi-objective problem in the form given in (5.1) to a single-objective problem is:

$$\begin{aligned} \text{minimize: } & f_{\text{ws}}(\mathbf{x}, \boldsymbol{\lambda}) = \sum_{i=1}^{m} \lambda_i f_i(\mathbf{x}), \\ \text{subject to: } & \mathbf{g}(\mathbf{x}) \geq 0, \text{and} \\ & \mathbf{h}(\mathbf{x}) = 0, \\ & \mathbf{x} \in S \subset \mathbb{R}^n. \end{aligned} \tag{5.6}$$

The problem formulation form given in (5.6) is called the *weighted sum* approach. Using different weight vectors one can generate different solutions to cover the Pareto front (PF). If the PF is convex (concave in the case of maximization), this approach could work well. However, not every Pareto optimal vector can be obtained by this approach in the case of non-convex PFs.

Here is where the utility functions are used. By using the Tchebycheff utility function, each single-objective problem solved by MOEA-D will be described by:

$$\begin{aligned} \text{minimize: } & f_{\text{tch}}(\mathbf{x}, \boldsymbol{\lambda}, \mathbf{z}^*) = \max_{i \in [1,m]} \{\lambda_i |f_i(\mathbf{x}) - z_i^*|\}, \\ \text{subject to: } & \mathbf{g}(\mathbf{x}) \geq 0, \text{and} \\ & \mathbf{h}(\mathbf{x}) = 0, \\ & \mathbf{x} \in S \subset \mathbb{R}^n, \end{aligned} \tag{5.7}$$

where $\mathbf{z}^* = [z_1^*, z_2^*, \ldots, z_m^*]^{\mathrm{T}}$ is the reference point, i.e., $z_i^* = \min\{f_i(\mathbf{x}) | \mathbf{x} \in S\}$ for each $i = \{1, 2, \ldots, m\}$. For each Pareto optimal point $\mathbf{x}^*$ there exists a weight vector $\boldsymbol{\lambda}$ such that it is the optimal solution of (5.7) and each of those optimal solutions is a Pareto optimal solution of (5.1). Therefore, one is able to obtain different Pareto optimal solutions by altering the weight vector [144].

As the weight vectors are generated linearly, the unities of the objectives must be also linear. This is a big difference between MOEA-D and NSGA-II algorithms that one must take it into account.

In Algorithm 13, the pseudocode of M is shown. It is supposed implicitly that the population size is equal to the number of weight vectors. The use of the external population (in lines 1 and 18) is optional [144], if it is not used, then the final population could have dominated solutions. In line 10, Differential Evolution (DE) could be used as the genetics operators as it was used in the CEC

Algorithm 13 Pseudocode of MOEA-D

Require: MOOP, stop condition, set of n weight vectors $L = \{\boldsymbol{\lambda}_1, \boldsymbol{\lambda}_2, \ldots, \boldsymbol{\lambda}_n\}$, neighborhood size k, utility function u.
Ensure: An approximation to the Pareto set in the external population (EP),
1: EP $\leftarrow \emptyset$
2: $(\forall \boldsymbol{\lambda}_i \in L)\ B(i) \leftarrow$ k-nearest-neighbors$(\boldsymbol{\lambda}_i, L, k)$
3: ▷ $B(i)$ stores the indexes of k nearest-neighbors weight vectors to each $\boldsymbol{\lambda}_i$
4: Initialize the population $P = \{\boldsymbol{p}_1, \boldsymbol{p}_2, \cdots, \boldsymbol{p}_n\}$
5: Evaluate the population P
6: Calculate reference point $\mathbf{z}^*$ as the minimum value for each objective
7: **while** stop condition is not fulfilled **do**
8:     **for** $i \in [1, n]$ **do** ▷ There are n subproblems to solve
9:         Let j_1 and j_2 two indexes randonly selected from $B(i)$
10:         Generate the child $\boldsymbol{c}$ from $\boldsymbol{p}_{j_1}$ and $\boldsymbol{p}_{j_2}$ using genetic operators
11:         Evaluate the individual $\boldsymbol{c}$
12:         Update reference point $\mathbf{z}^*$
13:         **for all** $j \in B(i)$ **do**
14:             **if** $u(\boldsymbol{c}, \mathbf{f}, \mathbf{z}^*, \boldsymbol{\lambda}_j) \leq u(\boldsymbol{p}_j, \mathbf{f}, \mathbf{z}^*, \boldsymbol{\lambda}_j)$ **then**
15:                 Replace individual $\boldsymbol{p}_j$ by $\boldsymbol{c}$
16:             **end if**
17:         **end for**
18:         Update EP: remove from EP all individuals dominated by $\mathbf{f}(\mathbf{c})$, otherwise add $\boldsymbol{c}$ to EP.
19:     **end for**
20: **end while**
21: **return** P

2009 competition [148]. In line 14, other utility functions could be used instead of the Tchebycheff one given in (5.7). Several studies have demonstrated that MOEA-D improves the NSGA-II results [149, 150].

5.4.1 Simplex-lattice design

A simplex in 2D is a line, and in 3D is a triangle. Simplex-lattice design is an algorithm to distribute weight vectors equally spaced over the simplex, forming an ordered arrangement called $\{m, h\}$ simplex-lattice, where m is the number of objectives and h is a parameter of proportion. This structure consists of all possible combinations of m proportions such that they sum 1. There exits $h+1$ proportions values equally spaced from 0 to 1, that is:

$$\left[\frac{0}{h}, \frac{1}{h}, \frac{2}{h}, \ldots, \frac{h}{h}\right]^{\mathrm{T}}$$

The total number of weight vectors in the $\{m,h\}$ simplex-lattice is given by the combinatorial number:

$$C_{m-1}^{h+m-1} = \frac{(h+m-1)!}{(m-1)!h!}. \tag{5.8}$$

There are several proposed algorithms that implement the simplex-lattice design. The code in python given in Listing 5.1 is an example of the implementation of the efficient algorithm described by Chasalow and Brand in 1995 [151]. Figure 5.6 shows the weight vectors obtained for $\{2,10\}$, for 2 objectives (11 vectors); and $\{3,13\}$, for 3 objectives (105 vectors). Each vector is represented in Fig. 5.6 as a dot.

Listing 5.1: Simplex-lattice design program in python.

```
import sys
import math
n = len( sys.argv )
if n != 3 :
        print( "Args: m h" )
        sys.exit(1)
m = int(sys.argv[1])
h = int(sys.argv[2])
# The number of vectors is calculated
c = math.factorial( m+h-1)/(math.factorial( m-1 )*math.factorial( h ))
print( "# ", c, "vectors with m =", m, ", h =", h )
# A vector is calculated each time
x = [0]*m    # It uses integer numbers
x[0] = h
j = 0
while x[m-1] < h :
        i = 0   # print the vector
        while i < m :
                v = float(x[i])/h
                print( v, end=' ' )
                i += 1
        print( )
        x[j] -= 1
        if j < m-2 :
                v = 0
                i = 0
                while i <= j :
                        v += x[i]
                        i += 1
                x[j+1] =  h-v
                i = j+2
                while i < m :
                        x[i] = 0
                        i += 1
                j += 1
```

```
    else :
        x[m-1] += 1
    k = -1
    i = 0
    while i < m - 1 :
        if x[i] > 0 :
            k = i
        i += 1
    if k >= 0 :
        j = k
i = 0   # Print the last vector
while i < m :
    v = float(x[i])/h
    print( v, end='␣' )
    i += 1
print( )
```

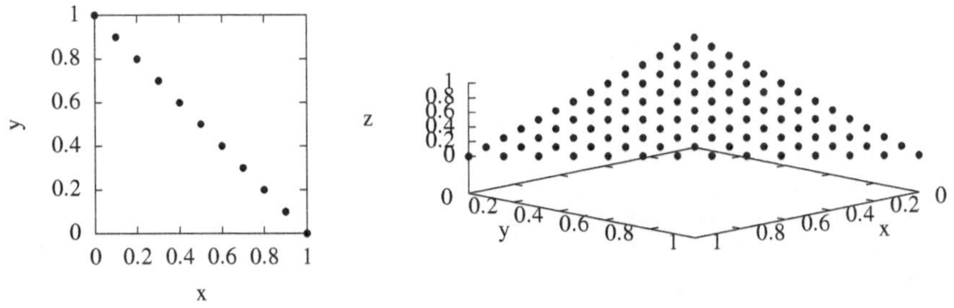

Figure 5.6: Example of two simplex-lattice design. In the left built with $\{2,10\}$ (2 objectives, 11 vectors are generated), and in the right built with $\{3,13\}$ (3 objectives, 105 vectors are generated).

5.5 MOMBI-II

The original Many-Objective Meta-heuristic Based on the R2 Indicator (MOMBI) was developed by Raquel Hernández Gómez in 2013 as a result of her master thesis [142, 141].

The Pareto dominance, explained in Section 5.2, has a problem when the number of objectives is increased: the space grows and any solution created randomly becomes non-dominated, thus Pareto dominance lost its effectivity. MOMBI is an algorithm developed to solve this problem. Indeed, MOMBI is an algorithm that can be used to solve a many-objective problems (with more than three objectives). Instead of using the Pareto dominance, MOMBI uses the R2 indicator to rank the population. The base of MOMBI is like the NSGA-II algorithm but now by using the R2 indicator for the selection mechanism.

The R2 indicator requires a set of utility functions and a set of weighting vectors (just as they were used in the MOEA-D algorithm). The R2 indicator maps an objective vector into a scalar value using a weight vector. The original MOMBI algorithm used the Tchebycheff utility function, and for generating the weight vectors the Simplex-Lattice Design method was adopted (see Subsection 5.4.1, in page 114). However, in some tests instances for many-objective optimization, MOMBI experimented loss of diversity.

MOMBI-II algorithm updated MOMBI and basically it was changed by using the utility function called achievement scalarezing function (ASF), which is defined as

$$\min f_{\text{asf}}(\mathbf{x}, \boldsymbol{\lambda}, \mathbf{z}^*) = \max_{i \in [1,m]} \left\{ \frac{|f_i(\mathbf{x}) - z_i^*|}{\lambda_i} \right\}, \tag{5.9}$$

where $\mathbf{x}$ is a solution, $\boldsymbol{\lambda}$ is a weight vector, and $\mathbf{z}^*$ is the reference point.

ASF is very similar to the Tchebycheff utility function used in (5.7) with respect to their mathematical expression; in ASF the weight component is divided instead of multiplied as in Tchebycheff function.

The R2 unary indicator for a constant reference set [152], is defined as:

$$R2(X,U) = -\frac{1}{|U|} \sum_{u \in U} \max_{\mathbf{x} \in X} \{u(\, \mathbf{f}(\mathbf{x})\,)\}. \tag{5.10}$$

Using the Tchebycheff utility function in (5.7), (5.10) could be rewritten as

$$\begin{aligned} R2(X,U) &= \frac{1}{|U|} \sum_{u \in U} \max_{\mathbf{x} \in X} \left\{ - \max_{i \in [1,m]} [\lambda_i |f_i(\mathbf{x}) - z_i^*|] \right\} \\ &= \frac{1}{|U|} \sum_{u \in U} \min_{\mathbf{x} \in X} \max_{i \in [1,m]} [\lambda_i |f_i(\mathbf{x}) - z_i^*|] \end{aligned} \tag{5.11}$$

Note that the problem in (5.7) is formulated as a minimization one, and therefore it can be transformed to maximization in (5.11) by using the dual property $\min f_{\text{tch}} = -\max(-f_{\text{tch}})$.

The R2 indicator by using ASF utility function is equal to:

$$R2(X,U) = \frac{1}{|U|} \sum_{u \in U} \min_{\mathbf{x} \in X} \max_{i \in [1,m]} \frac{|f_i(\mathbf{x}) - z_i^*|}{\lambda_i (z_i - z_i^*)}. \tag{5.12}$$

Here $\mathbf{z}^* = \min f_i(\mathbf{x})$ is the ideal vector, and $\mathbf{z} = \max f_i(\mathbf{x})$ is the nadir vector. Both vectors are used in (5.12) to normalize the objective function values. This normalization is a crucial task in MOMBI-II algorithm.

The pseudocode of MOMBI-II is shown is Algorithm 14. The reference points calculated in lines 3 and 10 are the ideal and the nadir vectors. MOMBI-II uses the same NSGA-II genetic operators: SBX for crossover, and polynomial mutation to generate the new population Q (in line 8). The initial population (line

5), and in each iteration the whole population (line 12), is ranked according to each weight vector by using the pseudocode given in Algorithm 15. The population is reduced by using their rank values: the individuals with lesser rank values, until the given value for the population size pass to the next generation.

Algorithm 14 MOMBI pseudocode

Require: MOOP, termination condition g, utility function U, set of weight vectors
Ensure: Approximation set to the Pareto optimal front
1: Initialize population P_0
2: Evaluate population P_0
3: Calculate reference points
4: Set $p.rank \leftarrow p.u^* \leftarrow \infty$ for each individual in P_0
5: Execute R2 ranking algorithm
6: **for** $i = 1 : g$ **do** ▷ g is the maximum number of generations
7:     Perform tournament selection
8:     Generate offspring Q_i using variation operators
9:     Evaluate population Q_i
10:     Update reference points
11:     Set $p.rank \leftarrow p.u^* \leftarrow \infty$ for each individual in P_i and Q_i
12:     Execute R2 ranking algorithm
13:     Reduce population $P_{i+1} \leftarrow P_i \cup Q_i$ according to their rank
14: **end for**
15: **return** P_g

Note in pseudocode 15 that the population P is ranked each time for each weighted vector $\boldsymbol{\lambda} \in L$. Also, when the R2 values are equal, the norm of the individuals are used to sort the whole population.

A constraint handling mechanism to MOMBI-II has been aggregated in [153].

MOMBI-III algorithm developed in [154] uses a set of utility functions, instead of a single one, and also incorporates an improved constraint handling mechanism. MOMBI-III first generates a sufficient number of feasible individuals until the R2 indicator can be applied to the population. This idea requires another parameter to be set which is the adequate number of feasible individuals.

Raquel created a project with all the source code in C/C++ of all her results. This framework includes the source code of algorithms NSGA-II, NSGA-III, MOEA-D, MOMBI, MOMBI-II, MOMBI-III, and some others more [155].

Algorithm 15 Ranking procedure

for all $\boldsymbol{\lambda} \in L$ **do**
  for all $p \in P$ **do**
    $p.\mu = f_{\text{asf}}(, \boldsymbol{\lambda}, \mathbf{z}^*)$
  end for
  Sort P w.r.t. μ and $||p||$ in increasing order
  rank $\leftarrow 1$
  for all $p \in P$ **do**
    if rank $<$ p.rank **then**
      p.rank $\leftarrow$ rank
    end if
    rank $\leftarrow$ rank $+1$
  end for
end for

5.6 Indicators

The indicators are used to compare the results of two or more multi-objective algorithms. Remember, the results of a MOOP is a set of solutions, thus it is necessary that the indicators can give a single number for the comparison of two sets of solutions. Also, the indicators help to evaluate the quality of an approximation to the Pareto optimal front, regarding to three main characteristics: *convergence*, *distribution*, and *diversity*. These characteristics are illustrated in Fig. 5.7, and their definitions are given as follows:

- Convergence: The distance of the resulting non-dominated front to the Pareto optimal front should be minimized.
- Distribution: A good (in most cases uniform) distribution of the solutions found is desirable.
- Spread: The extent of the obtained non-dominated front should be maximized, i.e., for each objective, a wide range of values should be covered by the non-dominated solutions.

In the literature, the term proximity is used instead of convergence, and the term diversity encompasses both distribution and spread. Good diversity is commonly of interest in objective space, but may also be required in decision-space. The case I shown in Fig. 5.7 shows solutions with good convergence but bad diversity. In the second case shown in Fig. 5.7, a diverse set of well-spread solutions is obtained, although these do not have convergence. The solutions in the third case in the same figure have convergence and diversity; however, the edges of the Pareto optimal front are not explored. Finally, the fourth case shows the solution of an ideal optimizer: good convergence and good diversity.

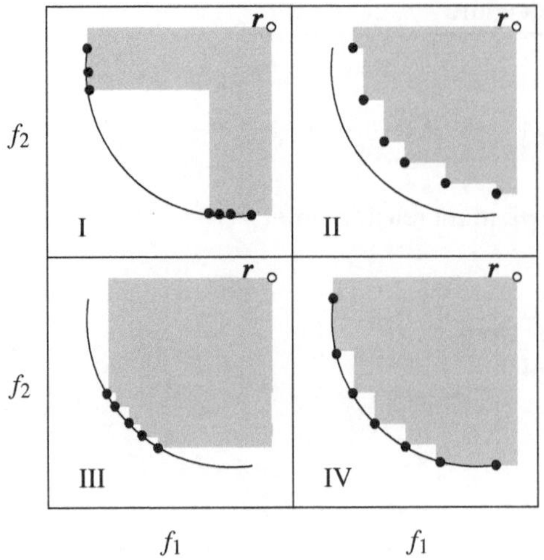

Figure 5.7: Desired features of approximations to the Pareto front. It is also shown the hypervolume for each set of solutions drawn with respect to the reference point $\boldsymbol{r}$. Case IV is the ideal, it shows good convergence and good diversity.

The *hypervolume* (HV) is the most preferred performance indicator because it is the only known indicator which is Pareto compliant. This means that its maximization guarantees to find the Pareto front. For its calculation it is necessary to supply a reference point, and the value of HV depends of this reference point. It is easy to visualize the HV in two dimensions as shown in Fig. 5.7. The HV considers the goals of convergence, spread, and distribution, in this order. Therefore, it has some bias for favoring non-linear Pareto fronts with clusters near to the middle point (called the *knee region*). The project in [155] supplies the code to calculate this indicator, as well the other indicators described in this section.

The R2 indicator is used by MOMBI-II algorithm as described in the previous Section 5.5, to rank the individuals in the population instead of the Pareto dominance (see Section 5.2). There are more possible utility functions to be used by this indicator, as the explained weighted sum (5.6), Tchebyshed (5.7), or ASF (5.9) (see [154]). The main drawbacks of this indicator are that the weight vectors should be supplied by the user, and since the values of different utility functions must be summed up, the values of the different objectives should be normalized.

The *generational distance* (GD) indicates how "far" the given approximation set A is from the discretized Pareto optimal front R. In other words, it measures the average distance from each solution $\boldsymbol{a} \in A$ to its closest reference point $\boldsymbol{r} \in R$ and is calculated as

$$\mathrm{GD}(A,R) = \left(\frac{1}{|A|} \sum_{\boldsymbol{a} \in A} d(\boldsymbol{a},R)^p \right), \tag{5.13}$$

where p is set generally equal to 2, and d is the Euclidean distance from the solution $\boldsymbol{a}$ to the nearest point $\boldsymbol{r}$ on the Pareto front R:

$$d(\boldsymbol{a},R) = \min_{r \in R} \|\boldsymbol{a} - \boldsymbol{r}\|. \tag{5.14}$$

$\mathrm{GD}(A,R) = 0$ indicates that $A = R$; any other value indicates a deviation. This performance indicator assesses convergence. This indicator can be used to test benchmark functions when one knows in advance the solution to the problem (the Pareto front).

The *inverted generational distance* (IGD) indicates how "far" the discretized Pareto optimal front R is from the approximation set A, i.e., it is the average distance from each reference point in R to its nearest solution in A. It is calculated as:

$$\mathrm{IGD}(A,R) = \left(\frac{1}{|R|} \sum_{r \in R} d(\boldsymbol{r},A)^p \right), \tag{5.15}$$

where usually p is set equal to 2, and d is defined as in (5.14). Similarly to the GD indicator (5.13), the interpretation of the IGD values is as follows: when $\mathrm{IGD}(A,R) = 0$, $A = R$; and for any other value $\mathrm{IGD}(A,R) > 0$ represents a deviation. This performance indicator measures both, convergence and diversity.

The Δ_p indicator can be seen as the average Hausdorff distance between the approximation set and the discretized Pareto optimal front. It is defined as:

$$\Delta_p(A,R) = \max\left[\, \mathrm{GD}(A,R), \mathrm{IGD}(A,R) \,\right], \tag{5.16}$$

where GD is the generational distance (5.13), and IGD the inverted generational distance (5.15). Δ_p simultaneously evaluates proximity to the Pareto optimal front and spread of solutions along it.

The *modified inverted generational distance* (IGD+) [156] is an improved version of IGD. This indicator takes into account the dominance relation between a solution and a reference point when their distance is calculated. If a solution is dominated by a reference point, the Euclidean distance is used for their distance calculation with no modification. However, if they are non-dominated with each other, the minimum distance from the reference point to the dominated region by the solution is calculated. It is defined as:

$$\mathrm{IGD+}(A,R) = \frac{1}{R} \sum_{r \in R} f(\boldsymbol{r},A) \tag{5.17}$$

where function f is defined as:

$$f(\boldsymbol{r},A) = \min_{\boldsymbol{a} \in A} \left(\sum_{i=1}^{m} [\max(a_i - r_i, 0)]^2 \right)^{1/2} \tag{5.18}$$

This indicator should have fewer numbers, having an optimum value equal to zero. IGD+ may achieve all the pursued goals, without having a specific order of importance, since its value depends on the distribution of the reference set.

The *s-energy* indicator is given by:

$$E_s(A) = \sum_{i \neq j} \|\boldsymbol{a}_i - \boldsymbol{a}_j\|^{-s}, \tag{5.19}$$

where $s > 0$ is a fixed parameter. This performance indicator has been used to discretize high-dimensional manifolds since its minimization leads to a uniform distribution of the points in A, if $s \geq m-1$. Therefore, s-energy has been used to assess diversity of approximation sets.

All the described indicators were applied to the four fronts shown in Fig. 5.8. The calculated values using the software in [155] are shown in Table 5.1. For HV the reference point $[3.5, 3.5]$ was used. The Pareto front shown in the upper part of Fig. 5.8 was used to calculate indicators GD, IGD, Δ_p, and IGD+. For GD, IGD, and Δ_p, in (5.13), (5.15), and (5.16), respectively, the value of $p = 2$ was used. To calculate the R2 indicator, the ASF utility function, and weights are generated with $m = 2$ and $h = 20$ (see code in Python in 5.1, in page 115), which generated 21 weight vectors. The fronts were generated by using (5.20).

$$c(t) = \{3\cos(t) + c_x, 3\sin(t) + c_y\}, \tag{5.20}$$

The 18 points in the reference set (which is supposed the Pareto front) were generated with $t_i = 180 + i5$, for $i = \{0, 1, \ldots, 18\}$, in (5.20). Set 1 was generated with $t_i = 180 + i10$, for $i = \{0, 1, \ldots, 9\}$, and $c_x = c_y = 3$. Set 2 was calculated as the set 1 but the values of $c_x = 3.3$ and $c_y = 3.3$ were used in (5.20). Set 3 was calculated with $t_i = 200 + i7.1428$, for $i = \{0, 1, \ldots, 7\}$ and the points $[0, 3]$ and $[3, 0]$ were added to this set; also $c_x = c_y = 3$ was used. Set 4 was calculated with $t_i = 190 + i7.777$, for $i = \{0, 1, \ldots, 9\}$, and also $c_x = c_y = 3$.

As set 2 in Fig. 5.8 is just a shifted version of set 1, their s-energy values shown in Table 5.1, are equal. Also it is possible to observe in Table 5.1 that HV prefers a solution in the middle of the front: the best front according to this indicator is the set 4, and not the set 1 as the rest of indicators values mark.

Table 5.1: Value of the indicators of the approximation sets shown in Fig. 5.8. The best results are in boldface.

	HV	R2	GD	IGD	Δ_p	IGD+	*s*-energy
set 1	9.89	**2.19**	**0.00**	**0.18**	**0.18**	**0.051**	**74**
set 2	7.88	28574.29	0.40	0.42	0.42	0.412	**74**
set 3	9.84	2.22	0.06	0.22	0.22	0.074	91
set 4	**9.94**	4345.55	0.07	0.22	0.22	0.055	95

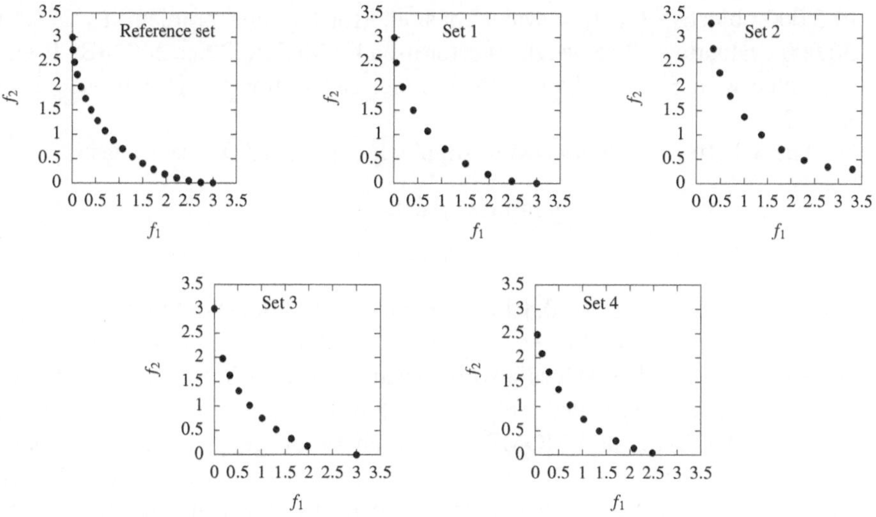

Figure 5.8: The reference Pareto front and 4 set of solutions used to calculate their different indicator values in Table 5.1.

5.7 Examples

The multi-objective optimization algorithms will be applied in the following chapters to maximize LE+, D_{KY} and entropy. In this chapter, NSGA-II (described in Section 5.3), MOEAD (see Section 5.4), and MOMBI-II (in Section 5.5), are applied to solve three MOOPs that are defined with the following mathematical expressions:

1. The Schaffer's problem. A very simple problem with 1 variable, 2 objectives, and without constraints.

2. The Tanaka's problem in which the Pareto front and the Pareto set are equal. This problem has 2 variables, 2 objectives, and 2 constraints.

3. The Osykzka and Kundu problem [157], that has 6 variables, 2 objectives, and 6 constraints. This a much more difficult problem than the two previous ones.

To solve the three MOOPs, the software given in [155], was used to perform the evaluations. For all the three problems, 100 individuals were used. The three algorithms used the simulated binary crossover (SBX) operator and polynomial mutation, thus $\eta_c = 20$ and $\eta_m = 20$ were used for the first two problems; $\eta_c = 5$ and $\eta_m = 5$ were used for the Osykzka and Kundu problem. The crossover probability was set to 0.9, and the mutation probability is equal to $1/n$, where n is the number of variables. The stop condition for the first two problems was set to 5,000 objective function evaluations, and for the third problem it was set to 50,000 evaluations. The weights vector used by MOEA-D and MOMBI-II were generated with Simplex-Lattice Design (see Subsection 5.4.1) with $m = 2$ and $h = 97$.

The Schaffer's problem is very simple, it consists of two parabolas:

$$\begin{aligned} \text{minimize: } f_1(x) &= x^2, \\ f_2(x) &= (x-2)^2, \end{aligned} \tag{5.21}$$

with a search space $x \in [-10, 10]$. The parabolas are shown in Fig. 5.9, where the thick line between coordinates $(0,0)$ and $(2,0)$ form the Pareto set. The Pareto front is the parabola's values also between those coordinates values. Because this is a very simple problem only the result of one single execution is shown in Fig. 5.10. MOEA-D and MOMBI-II use the set of weights vector to produce more points around their Pareto front knees, as one can see in Fig. 5.10. The hypervolume indicator for the three Pareto fronts shown in Fig. 5.10, by using the reference point $(4.1, 4.1)$, were equal to: NSGA-II 14.072532, MOEA-D 14.060906, and MOMBI-II 14.060730. Then NSGA-II obtained the best solution.

The Tanaka's problem is defined as:

$$\begin{aligned} \text{minimize: } & f_1(x_1) = x_1, \\ & f_2(x_2) = x_2, \\ \text{subject to: } & g_1(x_1,x_2) \geq x_1^2 + x_2^2 - 1 - 0.1\cos\left(16 \arctan\frac{x_2}{x_1}\right), \\ & g_2(x_1,x_2) \geq 0.5 - (x_1-0.5)^2 - (x_2-0.5)^2, \end{aligned} \tag{5.22}$$

with search space $x_1, x_2 \in [10^{-6}, \pi]$. One can see in (5.22) that the objective function values are equal to the values of the two variables, and therefore, its Pareto

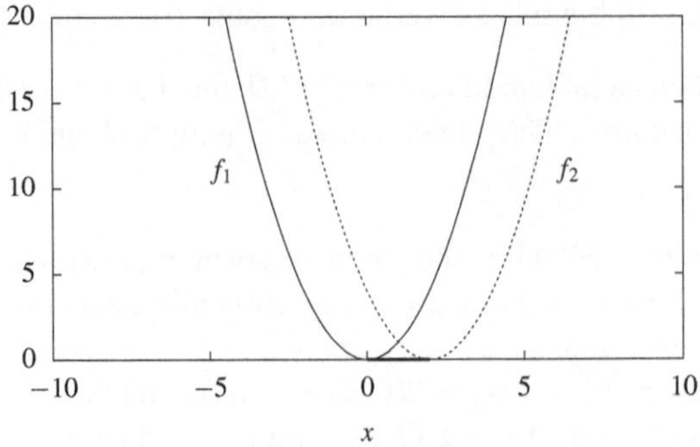

Figure 5.9: Graph of the Schaffer's function (5.21). It has 1 variable, 2 objectives, and without constraints. The thick line between coordinates $(0,0)$ and $(2,0)$ is the Pareto set.

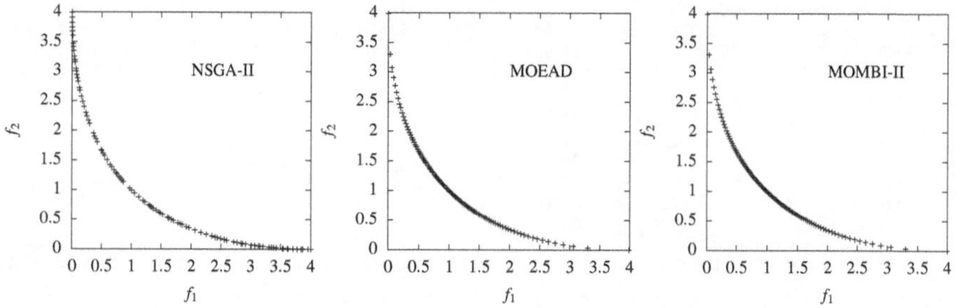

Figure 5.10: Pareto fronts resulted of the execution of NSGA-II, MOEA-D, and MOMBI-II on the Schaffer's problem (5.21).

front is equal to its Pareto set. The first constraint forms an undulated curve when $g_1(x_1,x_2)=0$ (see Fig. 5.11), and the second constraint when $g_2(x_1,x_2)=0$ forms a circle. The feasible space is between the two curves. The obtained results are shown in Fig. 5.12. The hypervolume indicator for the three results shown in Fig. 5.12 is equal to: 0.641094 (NSGA-II), 0.634420 (MOEA-D), and 0.629660 (MOMBI-II). Here again NSGA-II obtained the best value (the greatest one).

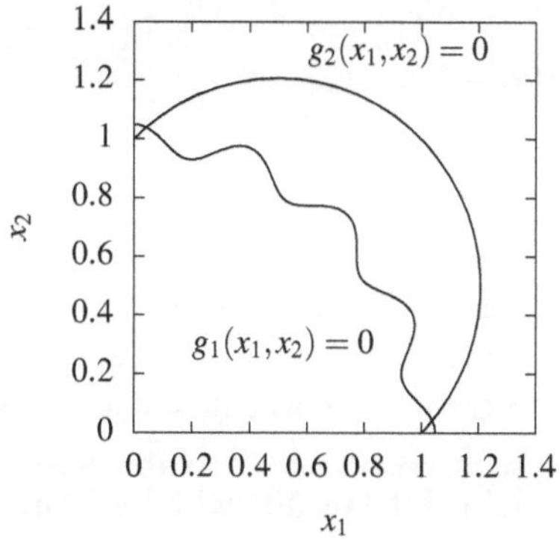

Figure 5.11: Graph of the Tanaka's function (5.21). It has 2 variables, 2 objectives, and 2 constraints. The feasible space is between the two shown curves. The Pareto set is on the border of the g_1 constraint and within the circle $g_2(x_1,x_2)=0$. The Pareto front is equal to the Pareto set and is disconnected.

The Osykzka and Kundu problem [157], which depends on six variables and is subject to four linear inequality constraints and two nonlinear inequality con-

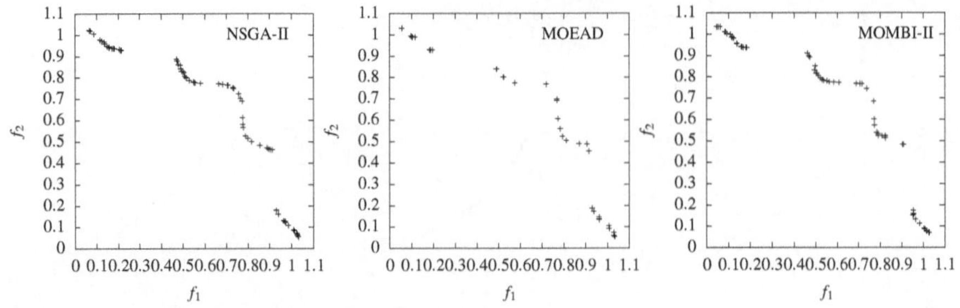

Figure 5.12: Graphs of the Pareto front (or sets) as result of the execution of NSGA-II, MOEA-D, and MOMBI-II to solve the Tanaka problem.

straints is defined as:

$$
\begin{aligned}
\text{minimize: } & f_1(\mathbf{x}) = -25(x_1-2)^2-(x_2-2)^2-(x_3-1)^2-(x_4-4)^2-(x_5-1)^2, \\
& f_2(\mathbf{x}) = x_1^2+x_2^2+x_3^2+x_4^2+x_5^2+x_6^2, \\
\text{subject to: } & g_1(\mathbf{x}) \geq x_1+x_2-2, \\
& g_2(\mathbf{x}) \geq 6-x_1-x_2, \\
& g_3(\mathbf{x}) \geq 2+x_1-x_2, \\
& g_4(\mathbf{x}) \geq 2-x_1+3x_2, \\
& g_5(\mathbf{x}) \geq 4-(x_3-3)^2-x_4, \\
& g_6(\mathbf{x}) \geq (x_5-3)^2+x_6-4,
\end{aligned}
\tag{5.23}
$$

with search spaces defined as $x_1, x_2, x_6 \in [0,10]$, $x_3, x_5 \in [1,5]$, and $x_4 \in [0,6]$. This is a difficult problem, then 30 executions of each algorithm were performed. For each execution its hypervolume was calculated by using the reference point $(-30,80)$. The statistics of the obtained hypervolume values are shown in Table 5.2. MOMBI-II obtained the results with less variation and greatest value. The Pareto front and Pareto set of the best execution, according the greatest hypervolume value, are shown in Fig. 5.13. From the hypervolume values obtained for each algorithm, the Wilcoxon statistical test was performed and the results are shown in Table 5.3. From this table the MOMBI-II algorithm obtained the best results: MOMBI-II hypervolume values are better than MOEA-D and NSGA-II. And also, MOEA-D obtained better results than NSGA-II.

Table 5.2: Hypervolume statistics on 30 execution of each algorithm in Osykzka and Kundu problem.

	Mean	Std. dev.	Min	Max
NSGA-II	13707	2714	0	14420
MOEAD	13323	1854	9000	14383
MOMBI-II	14310	109	14170	14433

Table 5.3: Wilcoxon statistical test applied to the results of the hypervolume indicator for each algorithm. A '+' indicates that the algorithm in the row is statistically better than the algorithm in the column.

	NSGA-II	MOEA-D	MOMBI-II
NSGA-II		−	−
MOEAD	+		−
MOMBI-II	+	+	

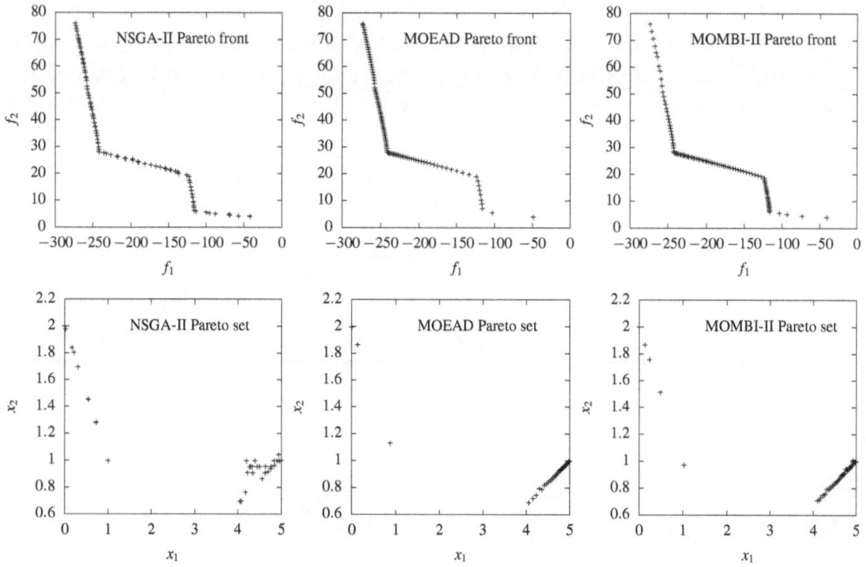

Figure 5.13: Graphs of the best results for the Osykzka and Kundu problem for the three tested algorithms. The top row shows the Pareto front, and in the bottom row the respective Pareto set.

5.8 Optimization of the fractional-order Lorenz chaotic oscillator by applying NSGA-II

As an example of implementing a multi-objective optimization algorithm for the optimization of fractional-order chaotic oscillators, this section shows the results of NSGA-II programmed in C language. The case study is the commensurable fractional-order Lorenz chaotic oscillator, which is optimized by maximizing the maximum Lyapunov exponent (MLE) and at the same time maximizing the Kaplan-Yorke dimension D_{KY}. The optimization process is performed by varying the design parameters γ, ρ, and β, and considering the following values: $q = 0.99$, $h = 0.001$, and initial conditions $(x(0), y(0), z(0)) = (5, 8, 15)$. The initial integration time was set equal to 2 s, for a simulation time of 300 s, and the time to measure the Lyapunov exponents was equal to 0.5 s. In NSGA-II, 40 individuals and 100 generations were used. All other NSGA-II parameters were set to the default values. The obtained Pareto front is shown in Fig. 5.14(a), which results are given in Table 5.4, where one can see the values of the Lyapunov exponents of the original and optimized systems, as also shown in Figs. 5.14(b) and (c), respectively. The running time was 63 minutes (2.6GHz Inter Core i5). This is the time to perform $100 \times 40 = 4,000$ evaluations of the function that calculates the Lyapunov spectrum. During the optimization process, some of the trial oscillators are not functional, i.e., their chaotic behavior decay or vanishes, or oscillates in a periodic way, and these cases take less time to be evaluated.

Table 5.4: Parameters γ, ρ, and β values and Lyapunov exponents of the fractional-order Lorenz chaotic oscillator without and with optimization.

Parameters	Lyapunov Exponents	D_{KY}
10, 28, 2.6666	0.570665, -1.183154, -13.054178	1.482325
5.2251, 109.2780, 1.9573	9.986146, -0.760250, -17.408338	2.529970

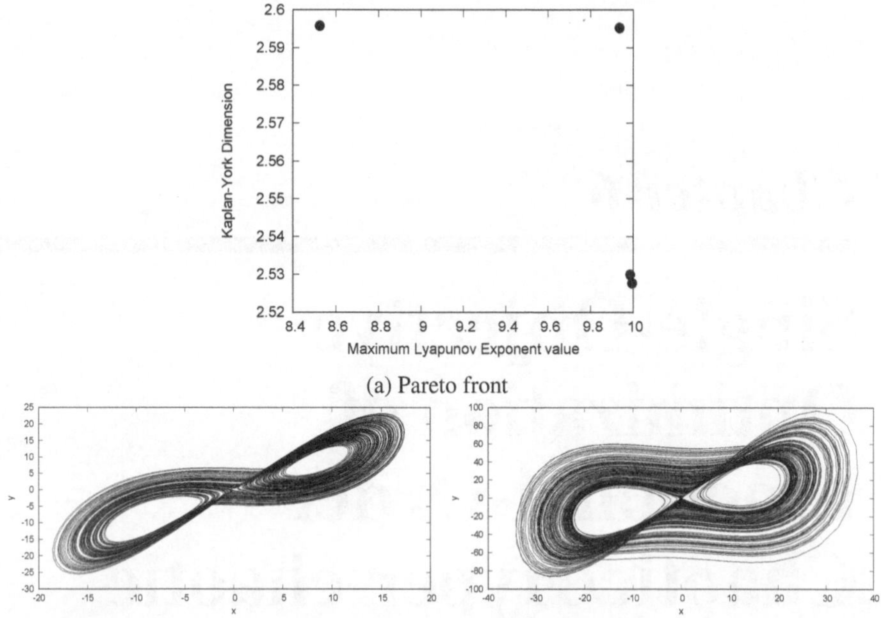

(a) Pareto front

(b) Phase x-y plane of the original system

(c) Phase x-y plane of the optimized system

Figure 5.14: (a) Pareto front of the optimization result of Lorenz system. (b) The original phase portrait, and in (c) the phase portrait of one of the optimized system.

Chapter 6

Single-Objective Optimization of Fractional-Order Chaotic/Hyper-chaotic Oscillators and their FPAA-based Implementation

The single-objective optimization algorithms described in Chapter 4 are adapted herein to optimize different chaotic and hyper-chaotic fractional-order oscillators. The main characteristics as the Lyapunov exponents and Kaplan-Yoke dimension D_{KY} are revisited to be optimized. Some implementations by using fractors are given in [4]. In this chapter the optimized fractional-order chaotic attractors are implemented by using operational amplifiers and the field-programmable gate array (FPAA). In particular, the implementation of the fractional-order derivatives by first order active filter blocks, and synthesized on the FPAA as introduced in [34], is detailed. Simulation and experimental results are provided to show the chaotic behavior of the oscillators.

6.1 Fractional-Order Chaotic/Hyper-chaotic Oscillators

Five fractional-order chaotic oscillators (FOCOs) are described herein. The first FOCO is the Ruan one introduced in [104], and given in (6.1), which has a single design parameter and it generates chaotic behavior by setting $a = 0.5$, with the initial conditions $x_0, y_0, z_0 = (0.1, 0.2, 0.3)$. From the analyses described in Chapter 3 the minimum commensurate fractional-order is equal to $q \geq 0.97$. The numerical simulation by applying FDE12 is shown in Fig. 6.1.

$$\begin{aligned} D_t^{q_1} x(t) &= y, \\ D_t^{q_2} y(t) &= -\frac{1}{3}x + a(1 - z^2)y, \\ D_t^{q_3} z(t) &= -y - 0.6z + yz \end{aligned} \tag{6.1}$$

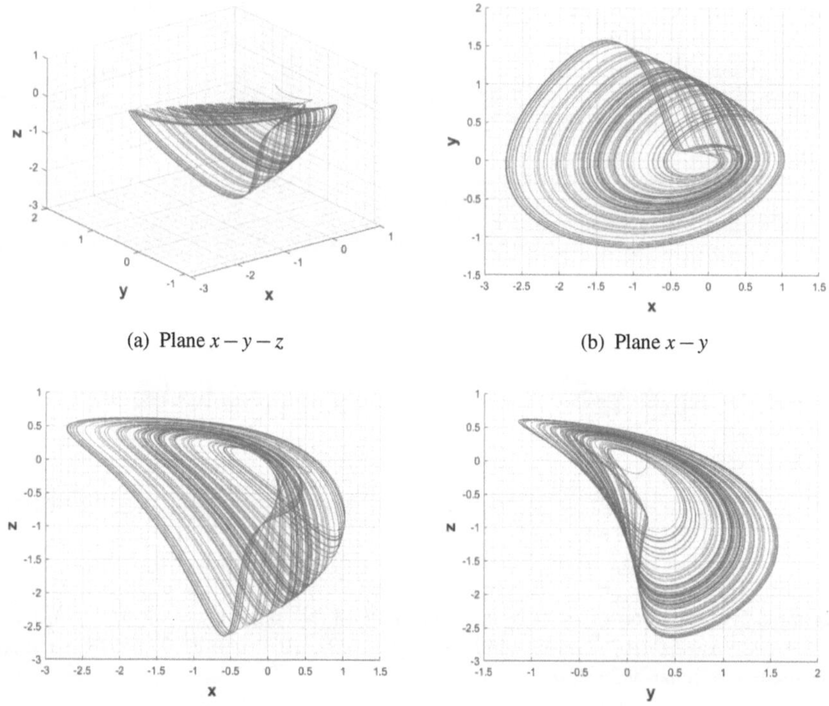

(a) Plane $x - y - z$

(b) Plane $x - y$

(c) Plane $x - z$

(d) Plane $y - z$

Figure 6.1: Simulation of (6.1) by applying FDE12, by setting $a = 0.5$, $T_{sim} = 200s$, $h = 5 \times 10^{-3}$, initial conditions $(x(0), y(0), z(0)) = (0.1, 0.2, 0.3)$, and fractional-order of $q_1 = q_2 = q_3 = 0.97$.

The second FOCO was introduced in [105, 106], and given in (6.2). The Niu FOCO has three design parameters $a = 35, b = 8/3, c = 55$, and it generates chaotic behavior by using the initial conditions $(x(0), y(0), z(0)) = (0.01, 0.01, 0.01)$, and the minimum commensurate fractional-order of $q \geq 0.96$. Fig. 6.2 shows the simulation results of this FOCO by applying FDE12.

$$\begin{aligned} D_t^{q_1} x(t) &= a(y-x) + yz \\ D_t^{q_2} y(t) &= cx - y - xz \\ D_t^{q_3} z(t) &= xy - bz \end{aligned} \tag{6.2}$$

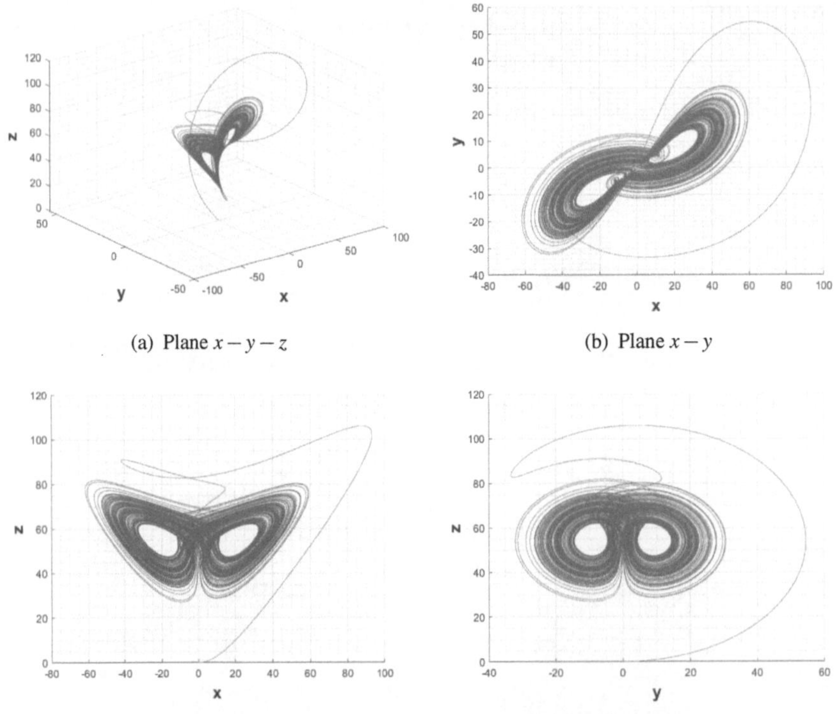

(a) Plane $x-y-z$ (b) Plane $x-y$

(c) Plane $x-z$ (d) Plane $y-z$

Figure 6.2: Simulation of (6.2) by applying FDE12, by setting $a = 35, b = 8/3, c = 55$, $T_{sim} = 100s$, $h = 5 \times 10^{-3}$, initial conditions $(x(0), y(0), z(0)) = (0.01, 0.01, 0.01)$, and commensurate fractional-order of $q_1 = q_2 = q_3 = 0.96$.

The third FOCO is an hyper-chaotic one introduced in [98], and described by (6.3). This FOCO has four state variables x, y, z, w, and the design parameters a, b, c, d, g and k are positive constants.

$$\begin{aligned} D_t^{q_1}x &= a(y-x)+byz+w \\ D_t^{q_2}y &= -cx-dxz^2+gy \\ D_t^{q_3}z &= y^2-kz \\ D_t^{q_4}w &= by-w \end{aligned} \tag{6.3}$$

The FOCO (6.3) generates chaotic behavior by setting the initial conditions $(x_0, y_0, z_0, w_0) = (0.1, 0.1, 2.1, 0.1)$, and the design parameters to (a, b, c, d, g, k) = (35, 2.5, 7, 4, 28, 1/3). This hyper-chaotic system has seven equilibrium points. The minimum commensurate fractional-order is $q > 0.8915$. By applying FDE12, the simulation results are shown in Fig. 6.3.

The fourth FOCO is based on saturated nonlinear function (SNLF) series, as introduced in [100], and its fractional-order description is given in (6.4), where the SNLF is approximated by a piecewise-linear (PWL) function $f_0(x,k)$, and it takes the form given in (6.5) to generate a two-scrolls attractor, where $a = b = c = d_1 = 0.7$ and $k = 1$. The minimum fractional-order is $q \geq 0.873$, and by applying FDE12, the simulation results are shown in Fig. 6.4.

$$\begin{aligned} D_t^{q_1}x &= y, \\ D_t^{q_2}y &= z, \\ D_t^{q_3}z &= -ax-by+d_1 f_0(x,k). \end{aligned} \tag{6.4}$$

$$f_0(x,k) = \begin{cases} k, & \text{if } \ x > 1, \\ kx, & \text{if } \ |x| \leq 1, \\ -k, & \text{if } \ x < 1. \end{cases} \tag{6.5}$$

The fifth case of study is the fractional-order Chen chaotic oscillator introduced in [101], and given in (6.6). This FOCO generates chaotic behavior by setting $(a,b,c) = (35,3,28)$, and the minimum commensurate fractional-order $q \geq 0.8244$.

$$\begin{aligned} D_t^{q_1}x &= a(y-x), \\ D_t^{q_2}y &= (c-a)x-xz+cy, \\ D_t^{q_3}z &= xy-bz \end{aligned} \tag{6.6}$$

6.2 Lyapunov exponents and D_{KY} of fractional-order chaotic/hyper-chaotic oscillators

The analyses of the FOCOs described in the previous section considered commensurate fractional-orders. For incommensurate fractional-orders the design variables increase as well the search space when performing the optimization of the FOCOs. In this case, recall that for self-excited attractors, one can evaluate the equilibrium points and eigenvalues to estimate the minimum fractional-order. But such fractional-order is the sum of all the fractional-orders q_n that can

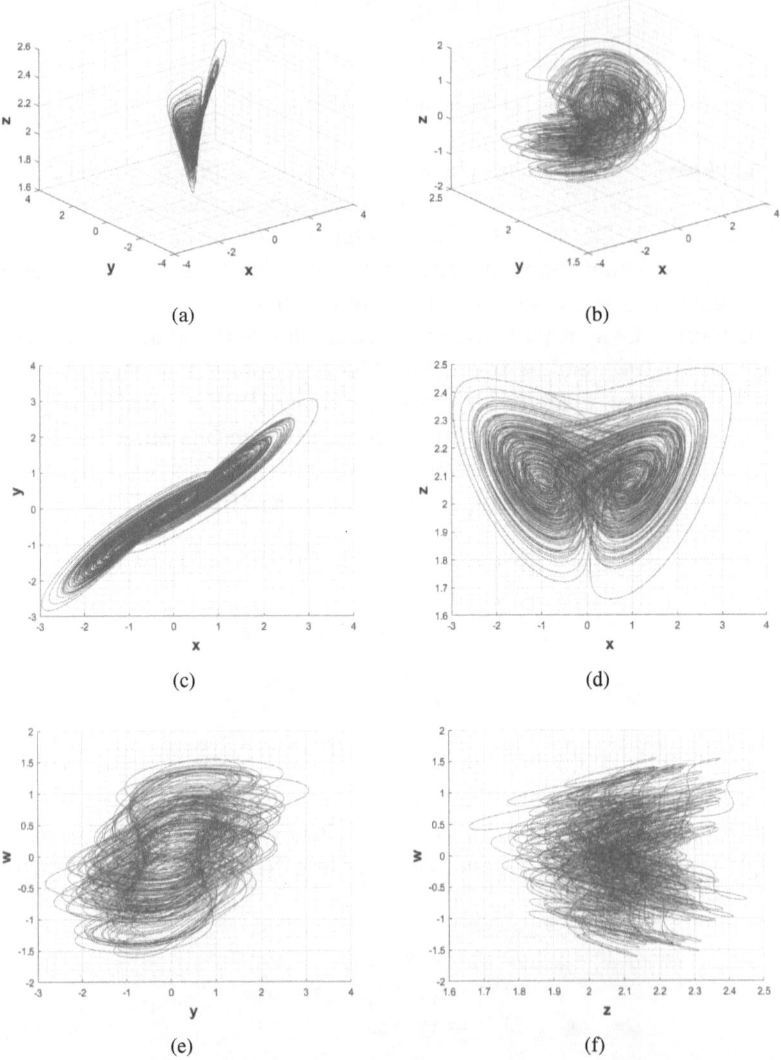

Figure 6.3: Simulation results of the fractional-order hyper-chaotic oscillator (6.3) by applying FDE12 and by setting $(a,b,c,d,g,k) = (35, 2.5, 7, 4, 28, 1/3)$, $T_{sim} = 200s$, $h = 5 \times 10^{-3}$, initial conditions $(x(0),y(0),z(0),w(0)) = (0.1,0.1,2.1,0.1)$ and $q_1 = q_2 = q_3 = q_4 = 0.9$.

be different. In any case, for commensurate and incommensurate FOCOs one can evaluate the dynamical characteristics such as the Lyapunov exponents and D_{KY}.

As mentioned by the majority of authors, fractional calculus is an emerging research area in mathematics and it can have multiple science and engineering applications, some of them reported in [23, 158, 159, 160]. The side of electronic

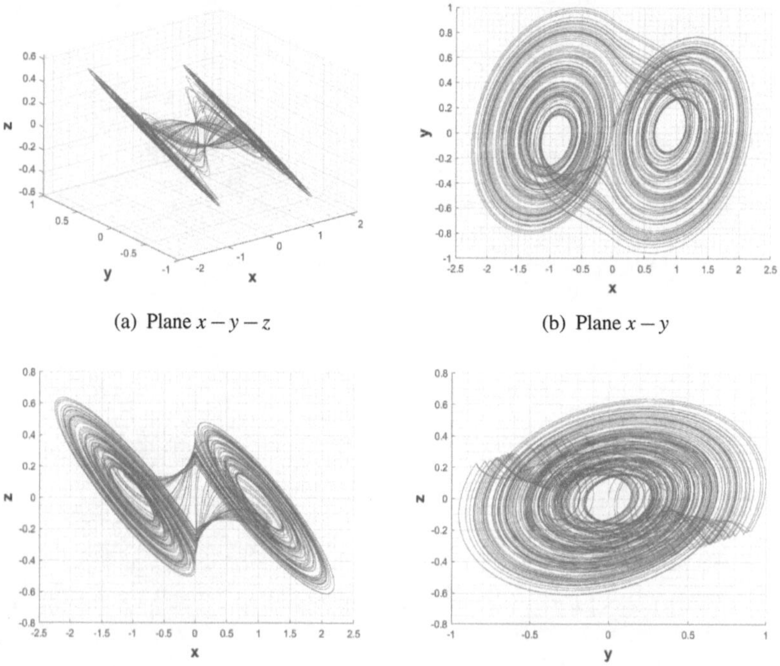

(a) Plane $x-y-z$ (b) Plane $x-y$

(c) Plane $x-z$ (d) Plane $y-z$

Figure 6.4: Simulation of (6.4) by applying FDE12, and by setting $a=b=c=d_1=0.7, k=1$, $T_{sim}=500s, h=2\times10^{-5}$, initial conditions $(x(0),y(0),z(0))=(0.2,0,0)$, and commensurate fractional-order $q_1=q_2=q_3=0.9$.

implementations, as analog or digital designs can be found in [4], but the FOCOs are implemented without optimizing their design parameters, as done herein. In fact, it is well-known that the Lyapunov exponents quantify the unpredictability of the chaotic behavior, and they are considered in the evaluation of D_{KY}, which has already been optimized for integer-order chaotic oscillators, as shown in [135, 136].

As mentioned in Chapter 3, practically all integer-order chaotic oscillators can be transformed to a FOCO, and they can be of commensurate or incommensurate fractional-order. Chapter 1 shows the simulation of the fractional order derivative of x with fractional-order of 0.5 and 0.9 and by applying three numerical methods, namely: Grünwald-Letnikov [36], the predictor-corrector Adams-Bashforth-Moulton [39, 161], and FDE12 [47]. Recall that there are other approximations to fractional-order derivatives such as the Riemmann-Liouville [35] or Caputo definitions [162], but all of them produce some numerical error and therefore the dynamical characteristics such as the Lyapunov exponents and D_{KY} may slightly differ.

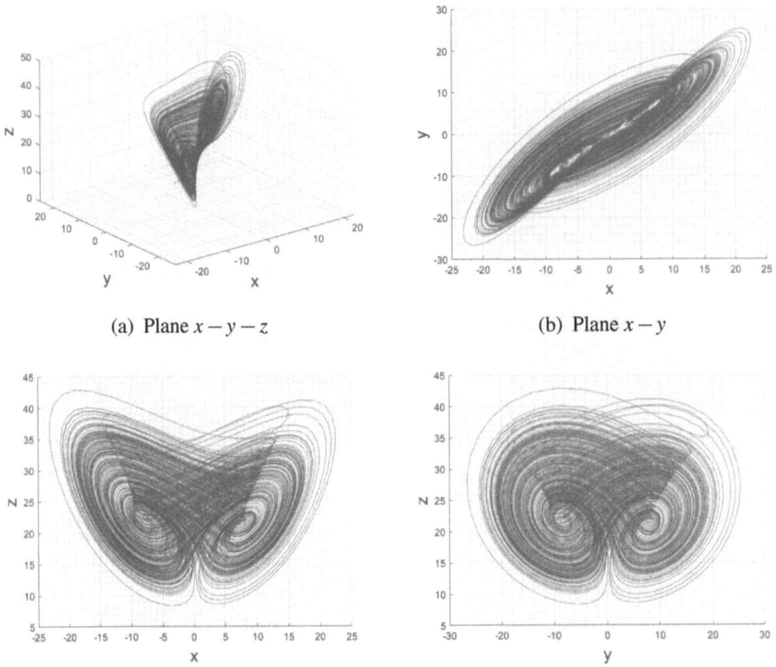

(a) Plane $x-y-z$ (b) Plane $x-y$

(c) Plane $x-z$ (d) Plane $y-z$

Figure 6.5: Simulation of the Fractional-order Chen chaotic oscillator given in (6.6) by applying FDE12, and by setting $(a=35, b=3, c=28)$, $T_{sim}=100s$, $h=5\times10^{-3}$, initial conditions $(x(0), y(0), z(0)) = (-9, -5, 14)$, and commensurate fractional-order $q_1 = q_2 = q_3 = 0.9$.

Before simulating a FOCO by applying any approximation, one must verify if at least one eigenvalue (λ) is in the unstable region, so that the fractional-order q of the derivative, will be the minimum for a commensurate FOCO [163, 164, 53, 165], i.e., that all the n fractional-orders have the same value $q_1 = q_2 = \ldots = q_n$, and is evaluated as given in Conmensurate, where $\mathrm{Im}(\lambda)$ and $\mathrm{Re}(\lambda)$ denote the real and imaginary parts of λ, respectively [45].

$$q > \frac{2}{\pi}\arctan\frac{|\mathrm{Im}(\lambda)|}{|\mathrm{Re}(\lambda)|}, \tag{6.7}$$

For the case of incommensurate FOCOs, where the fractional-orders are different, as $q_1 \neq q_2 \neq \ldots \neq q_n$, the necessary condition to estimate the minimum fractional-order is mathematically equivalent to (6.8), where $\mathrm{Re}(\lambda)$ and $\mathrm{Im}(\lambda)$ denote the real and imaginary parts of λ, respectively [45].

$$\left(\frac{\pi}{2M}\right) - \min_i\{|\arg(\lambda_i)|\} > 0 \tag{6.8}$$

In (6.8), λ_i denotes the eigenvalues from $\det(\mathrm{diag}([\lambda^{Mq_1} \quad \lambda^{Mq_2} \quad \ldots \quad \lambda^{Mq_n}]) - \partial f/\partial x\,|_{x=x^*}) = 0, \forall x^* \in \Omega$ and M is the minimum common multiple of the fractional orders, thus ensuring chaotic behavior [20, 166].

By applying the definitions given in [39, 47, 161, 162] and considering (6.7) for commensurate FOCOs, and (6.8) for incommensurate FOCOs, one can simulate any FOCO to generate a chaotic time series that can be used to estimate the Lyapunov exponents and D_{KY}. The D_{KY} is given in (6.9) and the Lyapunov exponents are ordered from the most positive to the lesser negative [167]. The parameter k is an integer number so that the sum of the ordered Lyapunov exponents (λ_i) is positive. If the chaotic behavior is held for a fractional-order system of three state variables, and accomplishes the constraint imposed by (6.7) or (6.8), then $k = 2$, so that λ_{k+1} is the third Lyapunov exponent and therefore $D_{KY} > 2$. A high value indicates that the FOCO possesses higher unpredictability.

$$D_{KY} = k + \frac{\sum_{i=1}^{k} \lambda_i}{|\lambda_{k+1}|} \tag{6.9}$$

To increase the value of D_{KY} one can apply metaheuristic approaches, as the ones for single objective optimization described in Chapter 4, or for multi-objective optimization algorithms described in Chapter 5. In this chapter we show the application of single-objective optimization algorithms. It is worth mentioning that the maximization of D_{KY} is quite difficult to solve by analytical methods so that they are well suited for a metaheuristic algorithm [168, 169, 170]. On this direction, the evolutive computer algorithms are considered as generic optimization tools that can solve very complex problems and they are characterized by a large or very huge search space. In the case of the FOCOs, as the ones described at the beginning of this chapter, they have large search space values in both, the design parameters and the fractional-order derivatives. The main advantage of the evolutionary computation algorithms is that they can reduce the effective search space through the use of effective search strategies as differential evolution (DE) [171] and particle swarm optimization (PSO) [172].

In the state of the art of FOCOs, the majority of them have fractional orders in the range [0.700,0.999], however, they do not have optimized values of their dynamical characteristics as the D_{KY}, and both the coefficient or design parameters values and the fractional-order derivatives can change during the optimization process.

6.3 Optimizing D_{KY} by DE and PSO

The five FOCOs described in Section 6.1, and given in (6.1), (6.2), (6.3), (6.4) and (6.6), all of them can be optimized to increase D_{KY}. Both DE and PSO can be adapted to optimize this dynamical characteristic to increase the unpredictability. For instance, as described in Section 6.2, the evaluation of D_{KY} requires the computation of the Lyapunov exponents, and to guarantee chaotic behavior one can perform the stability analysis by evaluating the equilibrium points and eigenvalues, from which one can estimate the minimum fractional-order of the derivatives. The simulation of FOCOs can be performed by applying the different numerical methods discussed in Chapter 1. In such a case the chaotic time series of each state variable can be computed and used to evaluate the Lyapunov exponents by applying TISEAN [173]. Afterwards, the evaluation of D_{KY} can be directly performed. This simulation process and evaluation of the dynamical characteristics are performed within the optimization loop in both DE and PSO. The reference values of D_{KY} before performing the optimization process, are given in Table 6.1, where one can see the values of the design parameters or coefficients of the mathematical models of the five FOCOs, the corresponding initial conditions, and minimum fractional-order of the commensurate FOCOs. By using these values, the equilibrium points (*EP*) are given in Table 6.2, and the corresponding eigenvalues are given in Table 6.3.

Table 6.1: Values of the design parameters, initial conditions, minimum fractional-order (FO) and D_{KY} of the FOCOs given in (6.1), (6.2), (6.3), (6.4), and (6.6).

Oscillator	Design Parameters	Initial Conditions	FO	D_{KY}
Ruan [104]	$a = 0.5$	(0.1,0.2,0.3)	0.970	2.0502
Niu [105]	$a = 35, b = 8/3, c = 55$	(0.01,0.01,0.01)	0.960	2.0420
Digy [98]	$a = 35, b = 2.5, c = 7, d = 4, g = 28, k = 1/3$	(0.1,0.1,2.1,0.1)	0.893	3.2176
Liu [100]	$a = b = c = d = 0.7$	(0.1,0.2,0.01)	0.887	2.0247
Chen [101]	$a = 35, b = 3, c = 28$	(0.01,0.01,0.01)	0.874	2.1711

6.3.1 Differential Evolution Algorithm

As described in Chapter 4, DE is a population based algorithm of direct and simple search, and it can find a global optimum in multimodal functions, non-differentiable and nonlinear, as for the case of the FOCOs. DE algorithm establishes the initial point of any problem by performing a sampling in the objective function in multiple points that are randomly selected. The individuals in the population are then perturbed with scaled differences of different members of the same population. The main steps in DE algorithm can be summarized as follows: initialization of the population in an stochastic way over the limits of the problem being optimized, differential mutation of the individuals if a given criterium is accomplished, crossover among the mutated and non-mutated individuals to increase diversity according to a given criterium, and a selection process commonly done by a Greedy criterium [123].

Table 6.2: Equilibrium Points (EP) associated to the FOCOs given in (6.1), (6.2), (6.3), (6.4), and (6.6), as reported in the literature.

Oscillator	Equilibrium Points
Ruan [104]	$EP_1(0,0,0)$ $EP_2(9.7972,-9.7972,16)$ $EP_3(-9.7972,-9.7972,16)$
Niu [105]	$EP_1(-19.3091,-7.5418,54.6094)$ $EP_2(19.3091,7.5418,54.6094)$ $EP_3(0,0,0)$
Digy [98]	$EP_0(0,0,0,0)$ $EP_1(-2.0430,-0.9925,0.81719,2.0034)$ $EP_2(2.0430,0.9925,0.81719,2.0034)$ $EP_3(-j2.2653,-j0.81142,-j0.90611,-2.4631)$ $EP_4(j2.2653,j0.81142,j0.90611,2.4631)$ $EP_5(-j5.5038,-j0.072293,-j2.2015,-14.540)$ $EP_6(j5.5038,j0.072293,j2.2015,-14.540)$
Liu [100]	$EP_1=(k_n d_1/a,0,0)$ $EP_2=(0,0,0)$ $EP_3=(-k_n d_1/a,0,0)$
Chen [101]	$EP_1(0,0,0)$ $EP_2(7.9373,7.9373,21)$ $EP_3(-7.9373,-7.9373,21)$

The DE algorithm begins by generating random vectors of D-dimensions as the population: $x_{i,G}$ $s.t.$ $i=\{1,2,3,...,N_p\}$; where G is the maximum number of generations and N_p the number of vectors in the population. As the generations evolve, new vectors are generated by performing new mutations (6.10) and crossover (6.11) vectors, where a,b,c are randomly selected vectors; g is the current generation; $F\in[0,2]$ is a mutation constant; $j=\{1,2,...,D\}$; $randb(j)\in[0,1]$ is the j-th evaluation of a generated random number; $CR\in[0,1]$ is a crossover coefficient selected by the user; and $rnbr(i)\in[0,1]$ is an index randomly generated.

$$v_i^{g+1}\leftarrow x_c^g+F(x_a^g-x_b^g) \tag{6.10}$$

$$u_{ij}^{g+1}\leftarrow\begin{cases} v_{ij}^{g+1} & \text{if } [randb(j)\leq CR] \text{ or } rnbr(i)=j \\ x_{ij}^{g+1} & \text{if } [randb(j)>CR] \text{ and } rnbr(i)\neq j \end{cases} \tag{6.11}$$

The evaluation and selection processes are executed to evaluate the new vector u_{ij}^{g+1}. If this is better than the previous vector (x_{ij}^{g+1}), the new one replaces

Table 6.3: Eigenvalues associated to the FOCOs given in (6.1), (6.2), (6.3), (6.4), and (6.6), from the equilibrium points listed in Table 6.2.

Oscillator	Eigenvalues
Ruan [104]	$EP_0: \lambda_1 = 0.75, \lambda_2 = 0.6667 - j0.6667, \lambda_3 = 0.6667 + j0.6667$ $EP_1: \lambda_1 = -0.589,\ \lambda_2 = 0.504 + j1.20,\ \lambda_3 = 0.504 - j1.20$ $EP_2: \lambda_1 = 0.589,\ \lambda_2 = 0.504 - j1.20,\ \lambda_3 = 0.504 + j1.20$
Niu [105]	$EP_0: \lambda_1 = -43.0978,\ \lambda_2 = 2.2156 + j24.4545,\ \lambda_3 = 2.2156 - j24.4545$ $EP_1: \lambda_1 = 43.0978,\ \lambda_2 = 2.2156 - j24.4545,\ \lambda_3 = 2.2156 + j24.4545$ $EP_2: \lambda_1 = 0.0487,\ \lambda_2 = 0.0088,\ \lambda_3 = 0.1541$
Diyi [98]	$EP_0: \lambda_1 = -30.847, \lambda_2 = 23.823, \lambda_3 = -0.9763, \lambda_4 = -0.3333$ $EP_1: \lambda_1 = -29.546, \lambda_2 = 21.738, \lambda_3 = 0.4585, \lambda_4 = -0.9829$ $EP_2: \lambda_1 = -28.786, \lambda_2 = 20.924, \lambda_3 = 0.5114, \lambda_4 = -0.9824$ $EP_3: \lambda_1 = -32.849 - j0.1443, \lambda_2 = 26.323 + j0.1397,$ $\lambda_3 = -0.8676 + j0.007746, \lambda_4 = -0.9504 - j0.0031$ $EP_4: \lambda_1 = -32.849 + j0.1443, \lambda_2 = 26.323 - j0.1397,$ $\lambda_3 = -0.8676 - j0.0077, \lambda_4 = -0.9504 + j0.0031$ $EP_5: \lambda_1 = -41.265 + j0.9034, \lambda_2 = 34.349 - j0.8994,$ $\lambda_3 = -1.0238 - j0.0012, \lambda_4 = -0.3930 - j0.0027$ $EP_6: \lambda_1 = -41.265 - j0.9034, \lambda_2 = 34.349 + j0.8994,$ $\lambda_3 = -1.0238 + j0.0012, \lambda_4 = -0.3930 + j0.0027$
Liu [100]	$EP_1: \lambda_1 = 0.848,\ \lambda_2 = 0.074 - j0.9055,\ \lambda_3 = 0.074 + j0.9055$ $EP_0: \lambda_1 = 1.9309,\ \lambda_2 = -1.1154 + j1.6944,\ \lambda_3 = -1.1154 + j1.6944$ $EP_2: \lambda_1 = -0.848,\ \lambda_2 = 0.074 + j0.9055,\ \lambda_3 = 0.074 - j0.9055$
Chen [101]	$EP_0: \lambda_1 = -3, \lambda_2 = 23.8359, \lambda_3 = -30.8359$ $EP_1: \lambda_1 = -18.4280, \lambda_{2,3} = 4.2140 \pm j14.8846$ $EP_2: \lambda_1 = -18.4280, \lambda_{2,3} = 4.2140 \pm j14.8846$

it and becomes part of the new population at generation $(g+1)$. Otherwise, the new vector is discarded. The DE algorithm stops when it reaches the maximum number of generation or a stop criterium, and provides feasible values of D_{KY}. Algorithm 16 shows the pseudocode of DE adapted to maximize D_{KY} for FOCOs. The required parameters are: D = random vectors, G = number of generations, CR = crossover coefficient, F = mutation constant and $func(\bullet)$ assigns the FOCO that is given by (6.1), (6.2), (6.3), (6.4) or (6.6), to optimize its D_{KY}.

6.3.2 Particle Swarm Optimization Algorithm

Different to the DE algorithm, the optimization by particle swarm optimization (PSO) does not perform crossover or mutation among the particles in the population. The PSO operators allow exploring the search space and at the same time as

Algorithm 16 Maximizing D_{KY} by Applying Differential Evolution Algorithm

Require: D, G, N_p, CR, F, and $func(\bullet)$
Initialize the population randomly (**x**)
Evaluate particle's position of the FOCO in $func(x)$.
if $q_1 = q_2 = q_3$ and accomplishes (6.7) **then**
Compute D_{KY} by using (6.9) of the commensurate FOCO
else
if $q_1 \neq q_2 \neq q_3$ and accomplishes (6.8) **then**
Compute D_{KY} by using (6.9) of the incommensurate FOCO
end if
end if
Save the results of D_{KY} in $score$
for $(counter = 1; counter \leq G; counter{+}{+})$ **do**
for $(i = 1; i \leq N_p; i{+}{+})$ **do**
Select three different indexes randomly (a, b, and c in (6.10))
for $(j = 1; j \leq D; j{+}{+})$ **do**
if $U(0,1) < CR \;||\; j = D$ **then**
$trial_j \leftarrow x_{aj} + F(x_{bj} - x_{cj})$
else
$trial_j \leftarrow x_{ij}$
end if
end for
$fx \leftarrow func(trial)$
if fx is better than $score_i$ **then**
$score_i \leftarrow fx$
$x_i \leftarrow trial$
end if
end for
end forreturn x and $score$

the regions with high probability of finding a global optimum. PSO considers the candidate solutions as particles that in its initial form are randomly distributed in a search space, thus forming the initial population. The quality of the population is determined by evaluating the individuals in the objective function, and also determining the best element. The new positions and velocities of the particles are computed considering the best global element and the current value of each particle together with a random number. Each time the particles are displaced, they are evaluated again in the objective function. Only the best current and global particles are updated when the algorithm finds better values than the ones saved. The PSO algorithm starts with a random initialization of the vectors $x_i \in \mathbb{R}^J$, which are seen as particles in the space. The behavior of these vectors is defined by two equations to measure: velocity (6.12) and position (6.13), where

i denotes the particle index, j the dimension, p_i the best position found in i, and p_g the best position found during the optimization. $\alpha \in \mathbb{R}$ is the inertial weight, $\beta \in \mathbb{R}$ is the acceleration constant and $U(\bullet)$ is a random number generator and uniformly distributed.

$$v_{ij}^{t+1} \leftarrow \alpha v_{ij}^{t} + U(0,\beta)\left(p_{ij} - x_{ij}^{t}\right) + U(0,\beta)\left(g_j - x_{ij}^{t}\right) \tag{6.12}$$

$$x_{ij}^{t+1} \leftarrow x_{ij}^{t} + v_{ij}^{t+1} \tag{6.13}$$

The optimization strategy is based on evaluating $f(x_i)$ that is associated with D_{KY}. If the last result is better than the $i-th$ registered, it will be the new vector p_i. The best global value (p_g) is also compared, and if the last values are better, it becomes replaced. This process is performed until reaching the stop criterium or the maximum number of generations G. The main characteristic of PSO is that each particle is moving around a centroid region determined by p_i and p_g. Therefore, the particles search for new positions to find the best solution of D_{KY}, as described in the pseudo-code given in Algorithm 17, where the required parameters are: D=random vectors, G=generations, α=inertial weight, β=acceleration constant, and $func(\bullet)$ to assign the FOCO that can be (6.1), (6.2), (6.3), (6.4) or (6.6).

6.4 Search spaces, design variables and optimization results

The simulation of the FOCOs to evaluate D_{KY} is executed by using the initial conditions given in Table 6.1, which are close to the attractor basin according to the unstable region. Both DE and PSO were run with the same conditions and for the five FOCOs described in this chapter, i.e., they were executed with the same number of populations (P) and maximum generations (G). The simulations were performed for 10,000 iterations, and eliminating the first 1000 transient iterations for the five FOCOs. The parameters of PSO are: $\alpha = 0.7298$ and $\beta = 1.4961$, according to [172]. In all the cases, the design variables were varied by using one digit for the integer part and three digits for the fractional part. The fractional-order derivatives were also encoded by three digits. In this manner, the optimization is performed for the maximization of D_{KY} and in both DE and PSO: $G = 100$, and $P = 40$.

From Table 6.4, one can observe that the first optimization case is to commensurate FOCOs in which the fractional-order derivatives can be fixed to search for feasible solutions to maximize D_{KY} of each FOCO.

In a second optimization case, the five FOCOs are considered as incommensurate systems, so that $q_1 \neq q_2 \neq q_3$. This implies that each computation requires more time than for commensurate FOCOs, because each fractional-order of the

Algorithm 17 Maximizing D_{KY} by Applying Particle Swarm Optimization

Require: D, G, N_p, α, β, and $func(\bullet)$.
Initialize the population randomly (**x**)
Initialize particle's velocities (v)
Evaluate particle's positions of the FOCO in $func(x)$.
if $q_1 = q_2 = q_3$ and accomplishes (6.7) **then**
    Compute D_{KY} by using (6.9) of the commensurate FOCO
else
    if $q_1 \neq q_2 \neq q_3$ and accomplishes (6.8) **then**
        Compute D_{KY} by using (6.9) of the incommensurate FOCO
    end if
end if
Save the results of D_{KY} in *score* and $p \leftarrow x$
Find the best value from p and save it in g
for $(counter = 1; counter \leq G; counter{+}{+})$ **do**
    for $(i = 1; i \leq N_p; i{+}{+})$ **do**
        for $(j = 1; j \leq D; j{+}{+})$ **do**
            $v_{ij} \leftarrow \alpha v_{ij} + U(0,\beta)(p_{ij} - x_{ij}) + U(0,\beta)(g_j - x_{ij})$
This evaluates the new velocity by using (6.12)
            $x_{ij} \leftarrow x_{ij} + v_{ij}$
This evaluates the new position by using (6.13)
        end for
        $f_x \leftarrow func(x_i)$
        if f_x is better than $score_i$ **then**
            $score_i \leftarrow f_x$
            $p_i \leftarrow x_i$
            if p_i is better than g **then**
                $g \leftarrow p_i$
            end if
        end if
    end for
end forreturn x, p, g, and *score*

derivatives are varied considering three digits. The optimization of D_{KY} of the five FOCOs is performed under the same search spaces for the design variables, and for incommensurate systems each fractional-order derivative adds another design variable. The search space of the fractional derivatives is the same for the five FOCOs and is given by the range: $0.600 \leq q_1, q_2, q_3 \leq 0.999$.

Table 6.4: Search spaces of the design variables of the FOCOs given in (6.1), (6.2), (6.3), (6.4), and (6.6).

Oscillator	Search Spaces	Fractional-order
Ruan [104]	$0.01 \leq a \leq 5$	$q_1 = q_2 = q_3 = 0.970$
Niu [105]	$30 \leq a \leq 50, 0.01 \leq b \leq 5, 40 \leq c \leq 70$	$q_1 = q_2 = q_3 = 0.960$
Digy [98]	$30 \leq a \leq 50, 0.01 \leq b \leq 5, 0.01 \leq c \leq 20$	$q_1 = q_2 = q_3 = q_4 = 0.900$
Liu [100]	$0.01 \leq a, b, c, d \leq 2$	$q_1 = q_2 = q_3 = 0.900$
Chen [101]	$30 \leq a \leq 50, 0.01 \leq b \leq 5, 10 \leq c \leq 40$	$q_1 = q_2 = q_3 = 0.900$

The optimization results for commensurate FOCOs by applying DE and PSO are summarized in Tables 6.5 and 6.6, respectively. The five best feasible solutions of $D_{KY} > 2$ and their design variables are listed, where the maximum variation is ± 0.05, thus demonstrating the usefulness of DE and PSO to perform single-objective optimization in FOCOs.

On the side of incommensurate FOCOs, the best feasible solutions of D_{KY} are summarized in Tables 6.7 and 6.8, where the design variables and the fractional-orders of the derivatives are also given for the DE and PSO algorithms, respectively.

In both cases of commensurate and incommensurate FOCOs, the application of DE and PSO provided best values of D_{KY}, compared to the ones reported in the current literature.

6.5 Implementation of optimized fractional-order chaotic/hyper-chaotic oscillators by using amplifiers

The optimal solutions for the five FOCOs provided by DE and PSO algorithms, and summarized in Tables 6.5, 6.6, 6.7 and 6.8, can be implemented by electronic devices, as already done in [4] for some FOCOs by using amplifiers, FPAAs, and FPGAs.

Lets us consider the implementation of the fractional-order hyper–chaotic oscillator given in (6.3) [98]. In this Chapter the fractional-order derivatives are approximated by a ratio of polynomials in the Laplace domain, as done in [34], and for the case of a fractional-order of 0.9 one can use (3.10), which can be designed by using first-order active filter blocks. The new design parameters of (6.3) who provides higher values of D_{KY} are given in Table 6.6. Those values can be replaced in (6.14), which generates chaotic behavior by setting the initial conditions to $(x_0, y_0, z_0, w_0) = (0.1, 0.1, 2.1, 0.1)$, and the design variables to

Table 6.5: Design variables of the best feasible solutions of D_{KY} by applying DE for the commensurate FOCOs listed in Table 6.4.

Oscillator	DE	
	Design Variables	D_{KY}
Ruan [104]	$a = 0.4278$	2.0896
	$a = 0.5711$	2.0789
	$a = 0.4381$	2.0828
	$a = 0.5266$	2.0919
	$a = 0.4924$	2.1089
Niu [105]	$a = 25.5482$ $b = 2.6654$ $c = 53.3262$	2.1778
	$a = 30.5482$ $b = 2.6321$ $c = 66.5631$	2.1889
	$a = 42.5482$ $b = 3.6915$ $c = 65.1323$	2.1514
	$a = 36.9121$ $b = 1.5895$ $c = 44.9173$	2.1468
	$a = 47.8765$ $b = 3.1589$ $c = 61.4625$	2.1678
Digy [98]	$a = 33.1493$ $b = 2.8546$ $c = 9.2312$	3.3142
	$a = 38.2132$ $b = 4.2154$ $c = 16.1945$	3.3239
	$a = 41.1567$ $b = 3.2532$ $c = 14.8531$	3.3191
	$a = 40.4212$ $b = 2.7154$ $c = 11.6248$	3.3239
	$a = 37.8925$ $b = 3.1447$ $c = 9.1625$	3.3274
Liu [103]	$a = 0.3987$ $b = 0.3457$ $c = 0.3321$ $d = 0.4321$	2.1856
	$a = 0.4863$ $b = 0.3256$ $c = 0.4853$ $d = 0.4112$	2.1912
	$a = 0.9273$ $b = 0.4345$ $c = 0.7853$ $d = 0.6122$	2.1825
	$a = 0.8123$ $b = 0.5349$ $c = 0.4823$ $d = 1.1230$	2.1938
	$a = 0.9564$ $b = 0.5184$ $c = 0.7211$ $d = 1.2690$	2.1816
Chen [174]	$a = 39.1502$ $b = 3.0215$ $c = 28.4562$	2.1821
	$a = 41.2119$ $b = 3.3512$ $c = 29.6231$	2.1789
	$a = 40.2512$ $b = 3.7266$ $c = 33.1478$	2.1947
	$a = 42.5222$ $b = 4.6544$ $c = 32.5646$	2.2013
	$a = 41.3461$ $b = 4.1289$ $c = 31.2937$	2.1988

$(a,b,c,d,g,k) = (35.3453, 3.3136, 12.2812, 4, 28.0000, 0.3333)$. The fractional-order hyper-chaotic oscillator has seven equilibrium points, as shown in Table 6.9. The Jacobian of this system is given in (6.15), from which the computed eigenvalues are given in Table 6.10.

$$\begin{aligned}
D_t^{q_1}x &= a(y-x)+byz+w \\
D_t^{q_2}y &= -cx-dxz^2+gy \\
D_t^{q_3}z &= y^2-kz \\
D_t^{q_4}w &= by-w
\end{aligned} \tag{6.14}$$

Table 6.6: Design variables of the best feasible solutions of D_{KY} by applying PSO for the commensurate FOCOs listed in Table 6.4.

Oscillator	PSO	
	Design Variables	D_{KY}
Ruan [104]	$a = 0.4871$	2.0790
	$a = 0.4922$	2.0889
	$a = 0.5126$	2.0826
	$a = 0.4935$	2.0978
	$a = 0.5236$	2.0891
Niu [105]	$a = 32.0256$ $b = 2.8526$ $c = 47.2535$	2.1767
	$a = 36.4351$ $b = 4.1643$ $c = 42.2535$	2.1969
	$a = 41.9431$ $b = 4.3261$ $c = 62.2257$	2.1836
	$a = 39.2878$ $b = 2.5648$ $c = 59.8622$	2.1847
	$a = 43.3341$ $b = 4.8473$ $c = 60.8941$	2.1953
Digy [98]	$a = 35.3453$ $b = 3.3136$ $c = 12.2812$	3.3323
	$a = 37.3232$ $b = 4.4277$ $c = 12.1355$	3.3226
	$a = 40.1217$ $b = 3.5818$ $c = 14.6431$	3.3158
	$a = 41.6412$ $b = 4.1162$ $c = 13.8648$	3.3179
	$a = 39.9965$ $b = 3.4733$ $c = 14.1762$	3.3117
Liu [103]	$a = 1.2561$ $b = 0.7557$ $c = 0.5486$ $d = 1.1101$	2.1470
	$a = 1.1523$ $b = 0.7442$ $c = 0.5324$ $d = 1.2310$	2.1769
	$a = 1.2112$ $b = 0.7578$ $c = 0.5732$ $d = 1.1789$	2.2165
	$a = 1.1335$ $b = 0.8743$ $c = 0.2821$ $d = 1.3487$	2.1735
	$a = 1.1299$ $b = 0.8957$ $c = 0.2821$ $d = 1.3267$	2.1833
Chen [174]	$a = 38.2232$ $b = 3.2371$ $c = 28.4667$	2.1933
	$a = 40.7489$ $b = 4.1593$ $c = 29.6383$	2.1812
	$a = 41.2512$ $b = 4.7646$ $c = 28.1576$	2.1911
	$a = 41.3622$ $b = 4.9634$ $c = 30.5693$	2.1989
	$a = 40.4961$ $b = 3.5889$ $c = 29.2937$	2.1937

Table 6.7: Best values of D_{KY} considering incommensurate FOCOs by applying DE.

Oscillator	Design Variables	q_1	q_2	q_3	D_{KY}
Ruan [104]	$a = 0.5298$	0.973	0.984	0.976	2.1056
Niu [105]	$a = 42.9719$ $b = 3.4402$ $c = 61.2546$	0.962	0.981	0.962	2.2520
Digy [98]	$a = 38.2654$ $b = 4.2591$ $c = 11.2213$	0.959	0.922	0.965	3.3221
Liu [103]	$a = 1.1294$ $b = 0.7798$ $c = 0.4987$ $d = 1.0546$	0.926	0.957	0.941	2.1865
Chen [174]	$a = 39.1964$ $b = 3.8872$ $c = 30.4928$	0.956	0.962	0.953	2.2097

Table 6.8: Best values of D_{KY} considering incommensurate FOCOs by applying PSO.

Oscillator	Design Variables	q_1	q_2	q_3	D_{KY}
Ruan [104]	$a = 0.5112$	0.964	0.978	0.986	2.1328
Niu [105]	$a = 39.8523$ $b = 4.1124$ $c = 62.1365$	0.978	0.967	0.982	2.2124
Digy [98]	$a = 38.2654$ $b = 4.2591$ $c = 11.2213$	0.963	0.941	0.958	3.3196
Liu [103]	$a = 1.1322$ $b = 0.7133$ $c = 0.4710$ $d = 1.122$	0.931	0.946	0.952	2.1932
Chen [174]	$a = 40.4421$ $b = 3.1672$ $c = 30.1126$	0.944	0.963	0.946	2.2116

Table 6.9: Equilibrium points of the fractional-order hyper-chaotic oscillator given in (6.14).

$EP_0(0,0,0,0)$
$EP_1(-2.0430,-0.9925,0.81719,2.0034)$
$EP_2(2.0430,0.9925,0.81719,2.0034)$
$EP_3(-j2.2653,-j0.81142,-j0.90611,-2.4631)$
$EP_4(j2.2653,j0.81142,j0.90611,2.4631)$
$EP_5(-j5.5038,-j0.072293,-j2.2015,-14.540)$
$EP_6(j5.5038,j0.072293,j2.2015,-14.540)$

Table 6.10: Eigenvalues associated to the equilibrium points given in Table 6.9.

$EP_0:$	$\lambda_1 = -30.847, \lambda_2 = 23.823, \lambda_3 = -0.97638, \lambda_4 = -0.33333$
$EP_1:$	$\lambda_1 = -29.546, \lambda_2 = 21.738, \lambda_3 = 0.4585, \lambda_4 = -0.98297$
$EP_2:$	$\lambda_1 = -28.786, \lambda_2 = 20.924, \lambda_3 = 0.51142, \lambda_4 = -0.98249$
$EP_3:$	$\lambda_1 = -32.849 - j0.1443, \lambda_2 = 26.323 + j0.1397,$
	$\lambda_3 = -0.8676 + j0.007746, \lambda_4 = -0.95049 - 0.003157i$
$EP_4:$	$\lambda_1 = -32.849 + j0.1443, \lambda_2 = 26.323 - j0.1397,$
	$\lambda_3 = -0.8676 - j0.007746, \lambda_4 = -0.95049 + j0.003157$
$EP_5:$	$\lambda_1 = -41.265 + j0.9034, \lambda_2 = 34.349 - j0.8994,$
	$\lambda_3 = -1.0238 - j0.001251, \lambda_4 = -0.39305 - j0.002759$
$EP_6:$	$\lambda_1 = -41.265 - j0.9034, \lambda_2 = 34.349 + j0.8994,$
	$\lambda_3 = -1.0238 + j0.001251, \lambda_4 = -0.39305 + j0.002759$

$$J_{(x^*,y^*,z^*,w^*)} = \begin{bmatrix} -a & a+bx & by & 1 \\ -c-d & g & 2z & 0 \\ 0 & 2y & -k & 0 \\ 0 & b & 0 & -1 \end{bmatrix} \tag{6.15}$$

The eigenvalues given in Table 6.10 are used to determine the minimum fractional-order, and for the fractional-order hyper-chaotic oscillator given in

(6.14), the commensurate order is computed by (6.16) and it becomes $q > 0.8915$.

$$
\begin{aligned}
q &\geq \frac{2}{\pi} \arctan\frac{|\mathrm{Im}(\lambda)|}{|\mathrm{Re}(\lambda)|} \\
q &\geq \frac{2}{\pi} \arctan\frac{|0.13977|}{|26.323|}
\end{aligned}
\tag{6.16}
$$

The portraits of the attractor of the fractional-order hyper-chaotic oscillator given in (6.14), is shown in Fig. 6.6. The design parameters are set to: $(a,b,c,d,g,k) = (35.3453, 3.3136, 12.2812, 4, 28, 0.3333)$, and the fractional-order derivatives to $q_1 = q_2 = q_3 = q_4 = 0.9$.

According to the design process introduced in [34], the equations from (6.14) can be transformed to the Laplace domain and represented by the block diagram shown in Fig. 6.7, where $H(s)$ corresponds to the fractional-order approximated in (3.10). By using operational amplifiers, the Laplace approximation of the fractional-order of 0.9 is implemented by first-order active filter blocks, as shown in Fig. 3.14.

The electronic implementation of the fractional-order hyper-chaotic oscillator given in (6.14) , is shown in Fig. 6.8. The operational amplifiers can be AD712 / LM324 / TL082, the bias power supply can be established as $\pm$ 12V, the multipliers can be implemented by AD633, and $H(s)$ in Fig. 6.7 corresponds to the cascade topology introduced in Fig. 3.14. The values of the resistances are set to: R_1=70 kΩ, R_2=R_5=2 kΩ, R_3=R_{10}=7 kΩ, R_6=28 kΩ, R_{13}=R_{19}=1 kΩ, R_9=17.5 kΩ, R_{11}=2.5 kΩ, R_{30}=10 kΩ, R_{21}=4 kΩ, R_4=R_7=R_{12}=R_{14}=R_{15}=R_{18}=R_{20}=R_2=R_{24}=10 kΩ.

The simulation of the electronic implementation of the fractional-order hyper-chaotic oscillators is shown in Fig. 6.9. As one sees, they are quite similar to the results provided by the time simulation shown in Fig. 6.6, and the ranges of the state variables are also within $\pm$ 3V. In this manner, this design is suitable for an FPAA-based implementation.

Another example is the electronic implementation of the fractional-order chaotic oscillator based on SNLF or PWL functions given in (6.4), and by using the optimized parameters given in Table 6.6. The generation of a two-scrolls attractor is based on the PWL function, as discussed in Chapter 3. The design parameters are set to: a = 1.2112 b = 0.7578 c = 0.5732 d = 1.1789 and $k = 1$. The block diagram of the description of (6.4), in the Laplace domain, is shown in Fig. 6.10, where $H(s)$ again corresponds to the approximation of the fractional-order derivative of 0.9 given in (3.10).

The synthesis of the blocks are shown in Fig. 6.11, which is based on operational amplifiers that again they can be AD712/LM324/TL082, to allow a power supply of $\pm$ 12V. The fractional-order of 0.9 can be implemented as already

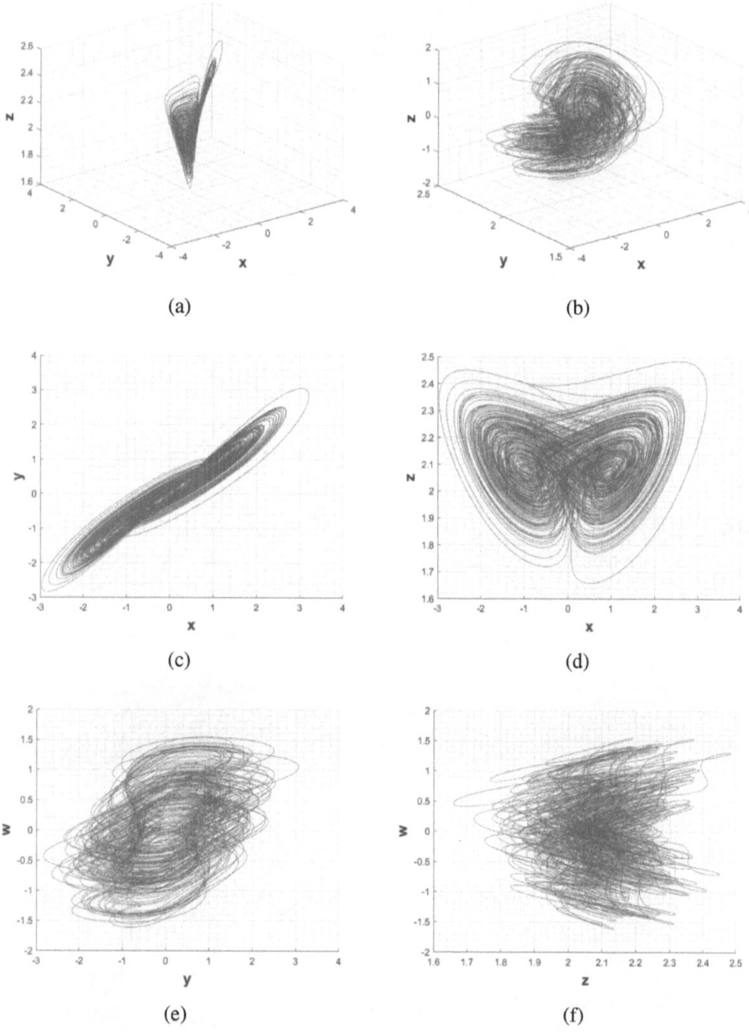

Figure 6.6: Portraits of the attractor from (6.14), and by setting: $(a,b,c,d,g,k) = (35.3453, 3.3136, 12.2812, 4, 28, 0.3333)$, $T_{sim} = 200s$, $h = 5 \times 10^{-3}$, initial conditions $(x(0), y(0), z(0), w(0)) = (0.1, 0.1, 2.1, 0.1)$, and fractional-order derivatives of $q_1 = q_2 = q_3 = q_4 = 0.9$.

shown in Fig. 3.14, and the resistances are set to: R_1=R_2=R_{14}=R_{15}=R_{32}=R_{33}=100 kΩ, R_{27}=369.2 kΩ, R_{28}=R_{31}=400 kΩ, R_{29}=40 kΩ, R_{30}=950 kΩ. The simulation results are shown in Fig. 6.12 for the attractor in the planes: (a) $x-y$, (b) $x-z$ and (c) $y-z$.

One can use the optimized design parameters given in Table 6.6, to generate more scrolls from the FOCO given in (6.4). In this case, the attractor is a multi-

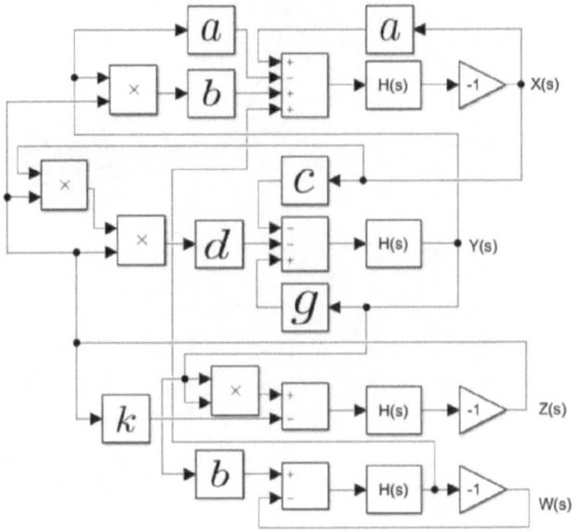

Figure 6.7: Block description of the fractional-order hyper-chaotic oscillator given in (6.14) for a fractional-order of 0.9.

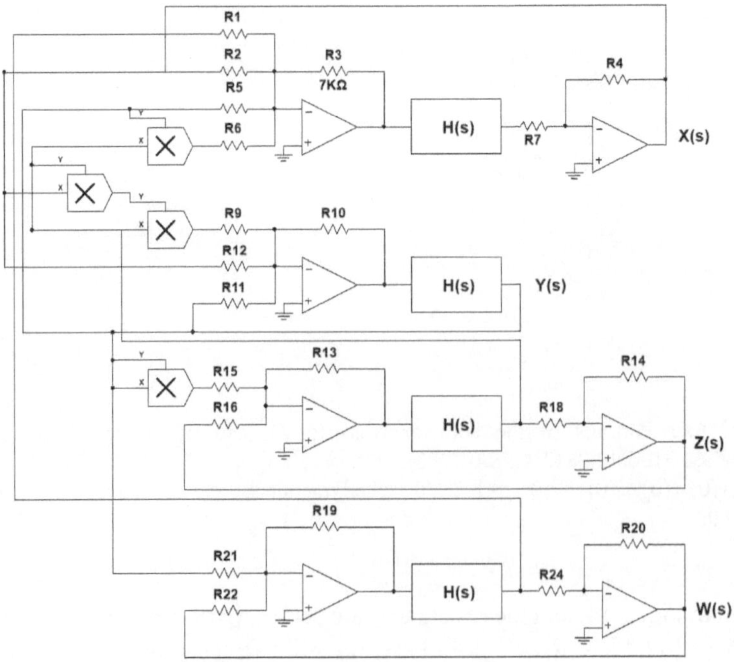

Figure 6.8: Electronic implementation of (6.14) for the commensurate fractional-order of 0.9.

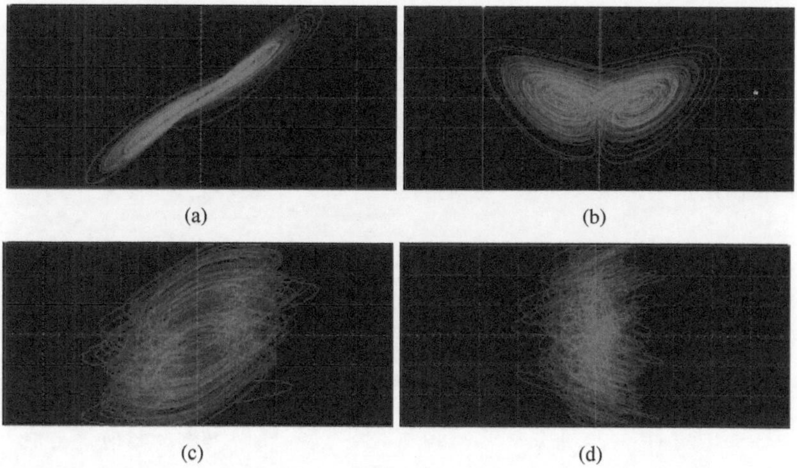

Figure 6.9: Simulation of the electronic implementation of (6.14): (a) plane $x-y$, (b) plane $x-z$, (c) plane $x-w$, and (d) plane $z-w$, with scale of 500mV/Div.

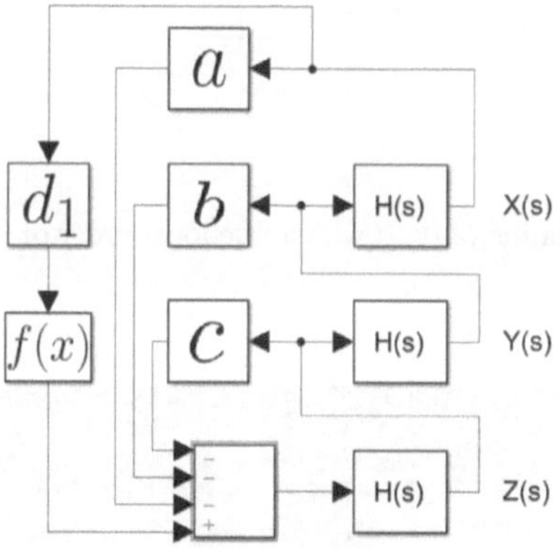

Figure 6.10: Block description of the FOCO given in (6.4), where $H(s)$ is approximated by (3.10).

scroll in one direction. Basically, the PWL function is increased and the number of saturation levels is equal to the number of scrolls being generated. For example, to generate 3-scrolls the PWL function is extended as shown in Fig. (6.13), where it can be seen at three saturation levels of amplitude -2k, 0 and 2k. Equation (6.17) shows this description considering the ranges in the horizontal axis. The 3-scrolls attractor is shown in Fig. 6.14, in the portraits: (a) plane $x-y$, (b) plane $x-z$, and (c) plane $y-z$.

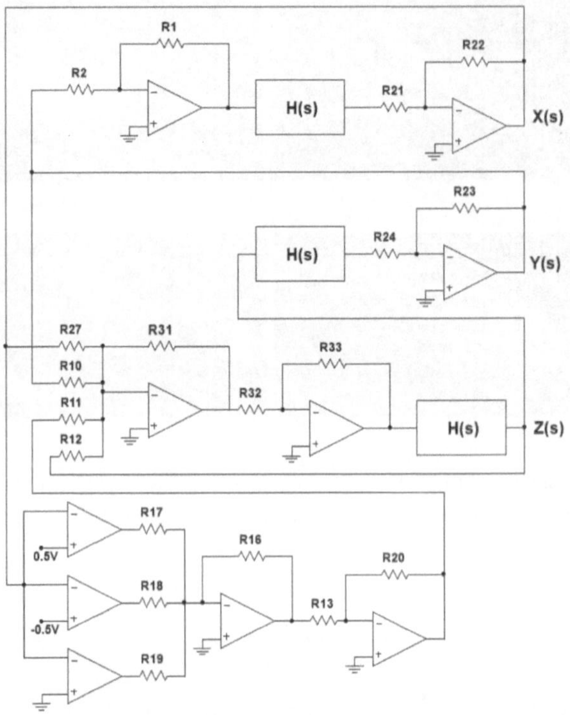

Figure 6.11: Implementation of (6.4) with a fractional-order of 0.9 by using operational amplifiers.

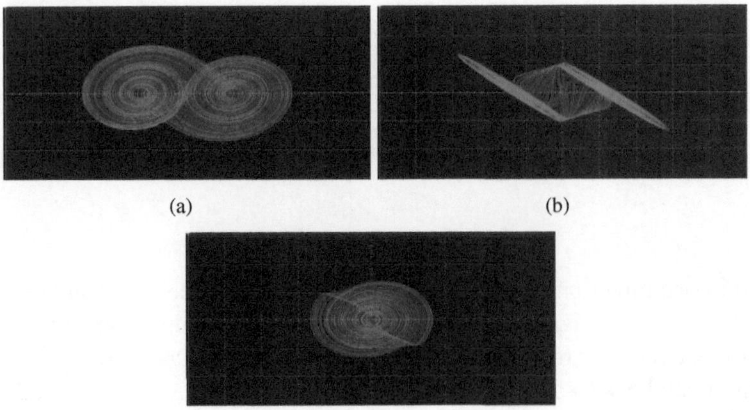

(a) (b)

(c)

Figure 6.12: Simulation of the electronic implementation of (6.4): (a) plane $x-y$, (b) plane $x-z$, and (c) plane $y-z$, with scale of 500mV/Div.

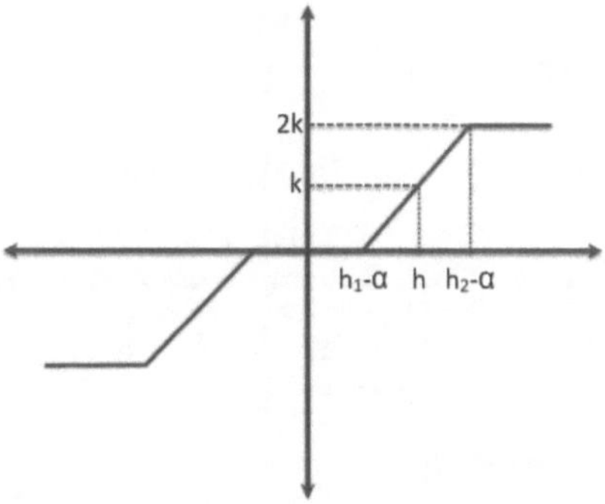

Figure 6.13: PWL function to generate a 3-scrolls attractor from (6.4).

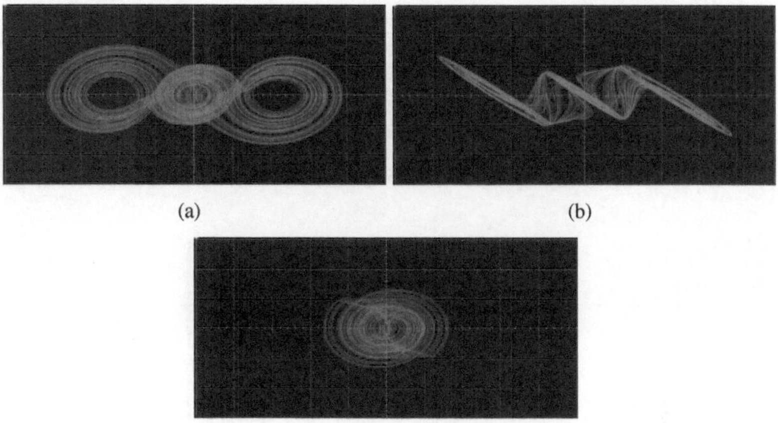

(a) (b)

(c)

Figure 6.14: Simulation of the electronic implementation of (6.4) to generate a 3-scrolls attractor: (a) plane $x-y$, (b) plane $x-z$, and (c) plane $y-z$, with scale of 500 mV/Div.

$$f_0(x,k) = \begin{cases} (2q)k, & \text{if} \quad x > qh+\alpha \\ s(x-h)+k, & \text{if} \quad h-\alpha \leq x \leq h+\alpha \\ s(x+h)-k, & \text{if} \quad -h-\alpha \leq x \leq -h+\alpha \\ 0, & \text{if} \quad -h-\alpha \leq x \leq h-\alpha \\ -(2q)k, & \text{if} \quad x > -ph-\alpha, \end{cases} \tag{6.17}$$

The same process is performed to increase the generation of more scrolls. Equation (6.18) shows the description of the PWL function to generate a 4-scrolls attractor. It has now four saturation levels as multiples of k, as sketched in Fig. 6.15. The simulation of the 4-scrolls attractor is shown in Fig. 6.16, for the planes: (a) $x-y$, (b) $x-z$, and (c) $y-z$.

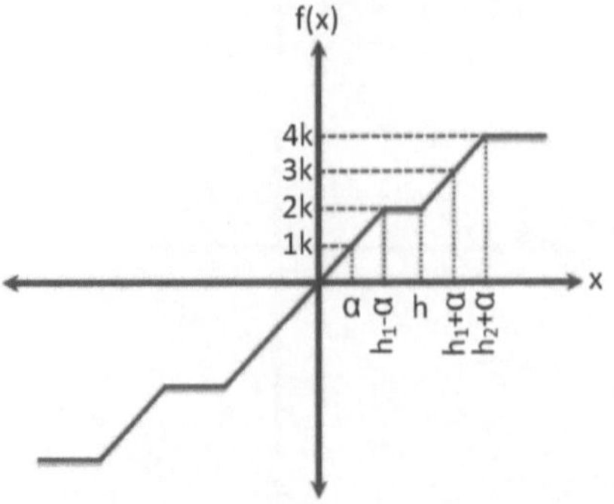

Figure 6.15: Sketching the PWL function to generate a 4-scrolls attractor.

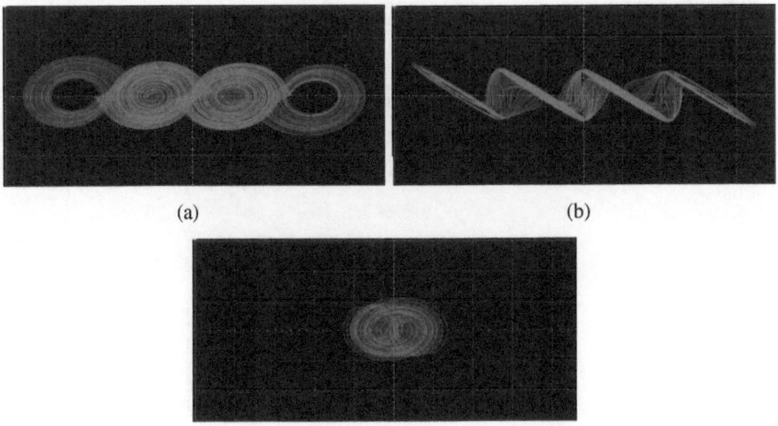

(a) (b) (c)

Figure 6.16: Simulation of the electronic implementation of (6.4) to generate a 4-scrolls attractor: (a) plane $x-y$, (b) plane $x-z$, and (c) plane $y-z$, with scale of 500mV/Div.

$$f_0(x,k) = \begin{cases} (2q+1)k, & \text{if} \quad x > qh+\alpha \\ s(x+h)-2k, & \text{if} \quad -h-\alpha \leq x \leq -h+\alpha \\ sx, & \text{if} \quad -\alpha \leq x \leq \alpha \\ s(x-h)+2k, & \text{if} \quad h-\alpha \leq x \leq h+\alpha \\ -k, & \text{if} \quad -h-\alpha \leq x \leq h-\alpha \\ k, & \text{if} \quad \alpha < x < h-\alpha \\ -(2q+1)k, & \text{if} \quad x < ph-\alpha \end{cases} \tag{6.18}$$

The fractional-order chaotic oscillator given in (6.4) can generate multi-scrolls in one direction (1D), two directions (2D) and three directions (3D). The previous designs generate 1D multi-scrolls because the mathematical model adds only one PWL function as $f_0(x,k)$. To generate multi-scrolls in 2D, one can add another PWL function, to have one for the state variable x and another for the state variable y, as given in (6.19), where k and h are associated to the PWL function, as the one shown in Fig. 6.15. The other parameters of the PWL functions p and q, represent the slope and distance between the saturated slopes, that can be varied as shown in [175], where the functions are irregular. Varying the PWL functions associated to $f(x)$ and $f(y)$ in (6.19), one can generate attractors of $n \times m$ scrolls. For instance, (6.20) shows the PWL function with the parameters k, h, p, q, and more details on the synthesis of a determined number of linear segments can be found in [77, 176].

$$
\begin{aligned}
D_t^{q_1} x &= y - \frac{d_2}{b} f(y; k_2, h_2, p_2, q_2) \\
D_t^{q_2} y &= z \\
D_t^{q_3} z &= -ax - by - cz + d_1 f(x; k_1, h_1 . p_1, q_1) + d_2 f(y; k_2, h_2, p_2, q_2)
\end{aligned}
\tag{6.19}
$$

$$
f_0(x,k) = \begin{cases} (2q+1)k, & \text{if } \; x > qh + \alpha \\ k/\alpha(x+ih) + 2ik, & \text{if } \; |x+ih| \leq \alpha, -p \leq i \leq q \\ (2i+1)k, & \text{if } \; ih + \alpha < x < (i+1)h - \alpha, -p \leq i \leq q-1 \\ -(2p+1)k, & \text{if } \; x < ph - \alpha \end{cases}
\tag{6.20}
$$

The fractional-order chaotic oscillator given in (6.19) can be transformed to the Laplace domain by performing the process detailed in [34], and it can be represented by the block diagram shown in Fig 6.17. The Laplace function $H(s)$ corresponds to the synthesis of the fractional-order of 0.9 and given in (3.10). In Chapter 3 one can see its electronic implementation by using operational amplifiers, as shown in Fig. 3.14.

The other blocks given in Fig. 6.17 can also be implemented by using operational amplifiers, and the PWL functions expressed as $f(x)$ and $f(y)$ can systematically be designed as already shown in [177]. The design includes the use of parameters k, α, s, h, that are combined to the values of the PWL functions designed with amplifiers that are biased by a voltage level E_i, and saturated by using voltages V_{sat} and currents I_{sat}. These voltages and currents are calibrated by the use or resistances R, as given in (6.21). In this manner, the topologies of the PWL functions are shown in Fig. 6.18 for the $f(x)$ and in Fig. 6.19 for $f(y)$, respectively.

$$
k = R_{ix} I_{sat} \quad I_{sat} = \frac{V_{sat}}{R_c} \quad \alpha = \frac{R_i |V_{sat}|}{R_f} \quad s = \frac{k}{\alpha} \quad h = \frac{E_i}{1 + \frac{R_i}{R_f}}
\tag{6.21}
$$

As shown in [177], one can choose a saturation voltage to find the other parameters. In this case $V_{sat} = \pm 2.5V$, and the resistors are set to: $R_{ix} = 10K\Omega$,

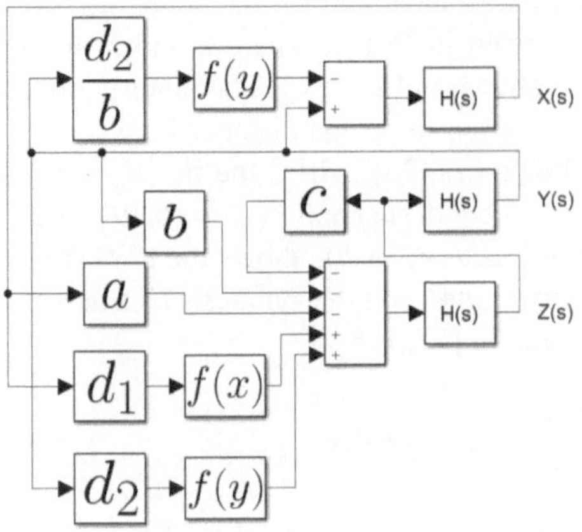

Figure 6.17: Block diagram description of the fractional-order chaotic oscillator given in (6.19) to generate multi-scrolls in 2D.

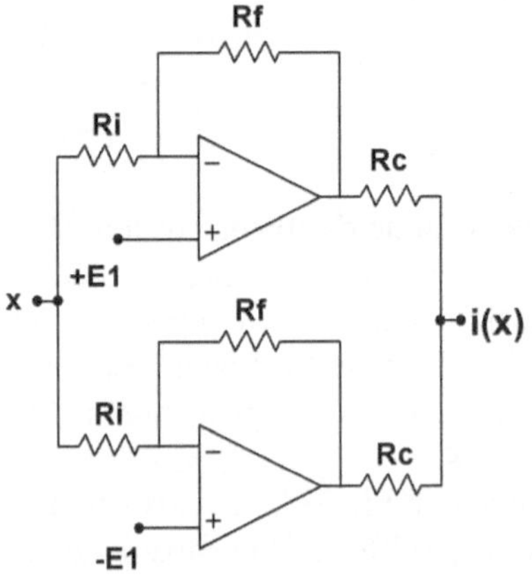

Figure 6.18: Synthesis of the PWL function $f(x)$ from (6.20) to generate a 2D attractor of 3×3-scrolls.

$R_f = 1M\Omega$, $R_c = 100K\Omega$, so that by using (6.21), one gets: k=200mV, $I_{sat} = 20\mu$A, α=2mV, s=100 and $h = E_1$=200mV, to generate a 2D 2×2-scroll attractor.

A 2D 2×2-scroll attractor can be generated from (6.19) by setting $a = b = c = d_1 = d_2 = 0.7$, $k = 250 \times 10^{-3}$, $\alpha = 2.5 \times 10^{-3}$, $h = 250 \times 10^{-3}$, and $p1 = q1 = p2 = q2 = 1$ to evaluate (6.19) and (6.20). In this case, the electronic design of the FOCO given in (6.19) is shown in Fig. 6.20, which synthesizes the blocks given in Fig. 6.17. By choosing $V_{sat} = \pm 2.5V$, then k=250mV, $I_{sat} = 25\mu$A, α=2.5mV, s=100, $h = E_1$ =250mV, and the resistors are set to: R1=R3=R22=R23=7KΩ, R2=R4=R7=R10=R11=R13=R15=R19=10KΩ,

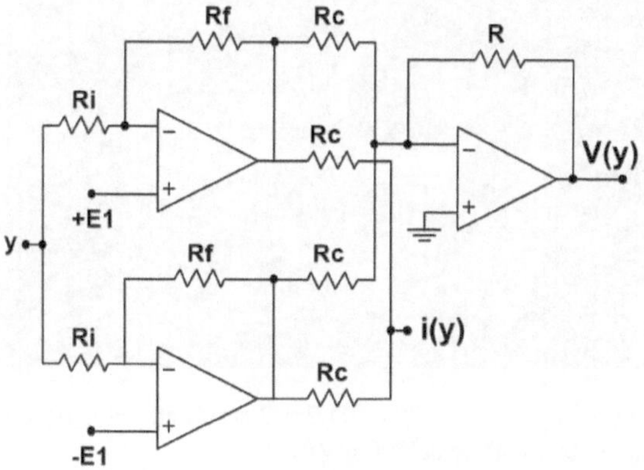

Figure 6.19: Synthesis of the PWL function $f(y)$ from (6.20) to generate a 2D attractor of 3×3-scrolls.

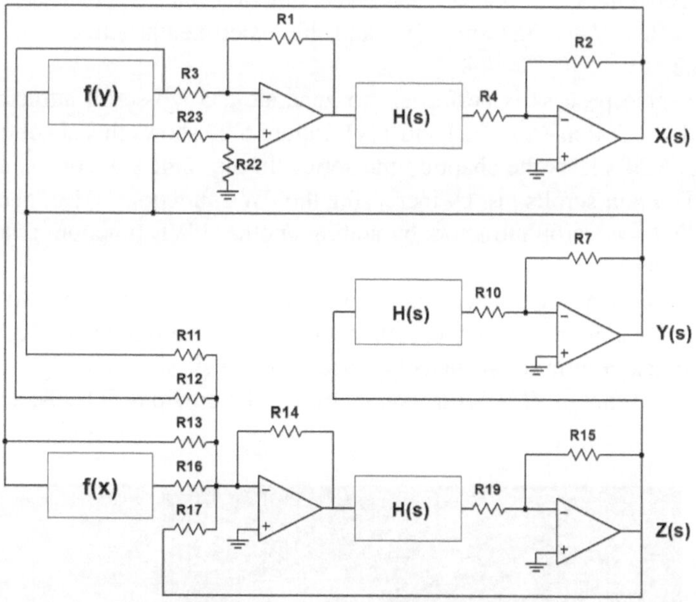

Figure 6.20: Electronic implementation of the FOCO given in (6.19) to generate 2D multi-scroll attractors.

and R12=R14=R6=100KΩ. The experimental result of the 2D 2×2-scroll attractor is shown in Fig. 6.21.

As one can infer, increasing the PWL functions one can generate different number of $n \times m$-scrolls. The synthesis of $f(x)$ and $f(y)$ can also be designed

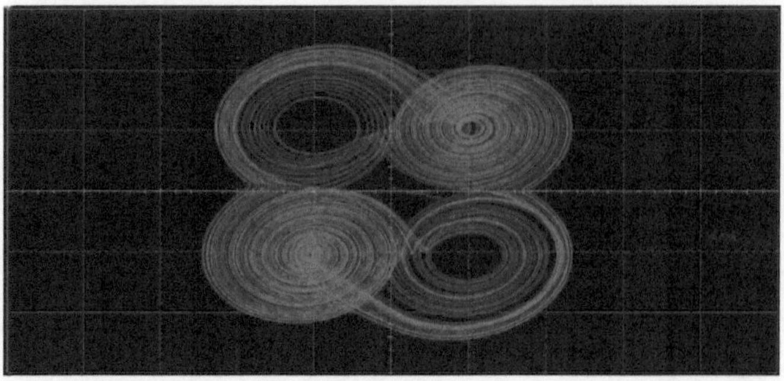

Figure 6.21: Experimental result of the 2D 2×2-scroll attractor from (6.19), of fractional-order $q_1 = q_2 = q_3 = 0.9$, with scale 500 mV/Div.

by increasing the number of amplifiers in a systematic way from Fig. 6.18 and Fig. 6.19, and by h and the reference voltages E_i, as shown in [177]. For example: to generate a 3×2-scrolls the PWL functions can be designed by setting E1 = ± 250 mV in $f(x)$ and $f(y)$, and the experimental attractor is shown in Fig. 6.22.

The same process is performed to generate a 3×3-scroll attractor. In this case: E1 = ± 500 mV for $f(x)$ and $f(y)$. Figure 6.23 shows this chaotic attractor, and Fig. 6.24 shows the chaotic time series. In the same way, one can generate more 2D $n \times m$-scrolls just by increasing the PWL functions. Also, one can generate 3D $n \times m$-scroll attractors by adding another PWL function, associated to $f(z)$, as detailed in [177].

The fractional-order Chen chaotic oscillator given in (6.6), can also be implemented by using operational amplifiers and setting the fractional-order equal to 0.9. Its block description from its Laplace transform is shown in Fig. 6.25, where the transfer function $H(s)$, is the same used in the previous FOCOs, and again it

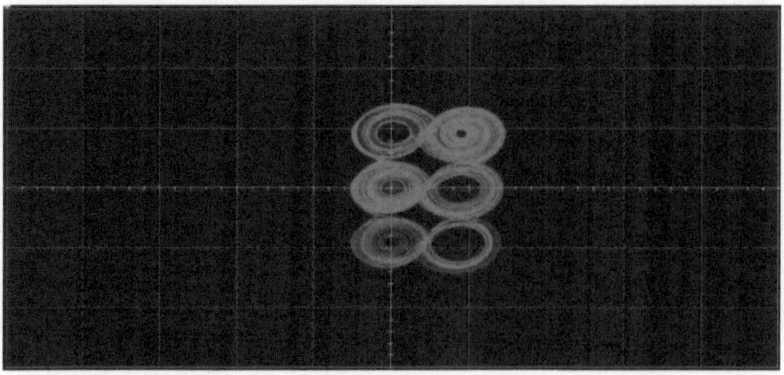

Figure 6.22: Experimental result of the 2D 3×2-scroll attractor from (6.19), of fractional-order $q_1 = q_2 = q_3 = 0.9$, with scale 500 mV/Div.

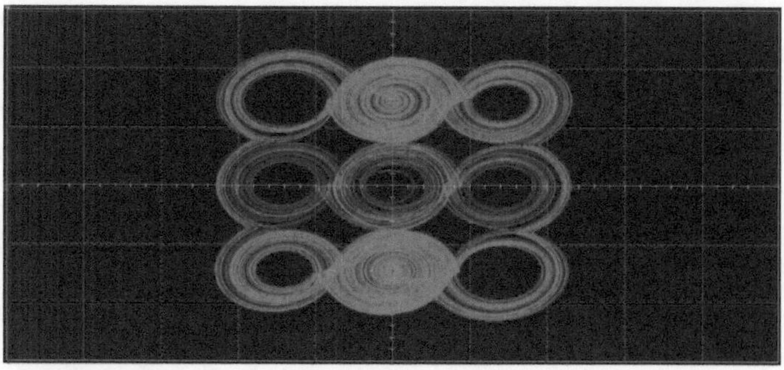

Figure 6.23: 2D 3 × 3-scroll attractor from (6.19) with a fractional-order $q_1 = q_2 = q_3 = 0.9$, with scale 500 mV/Div.

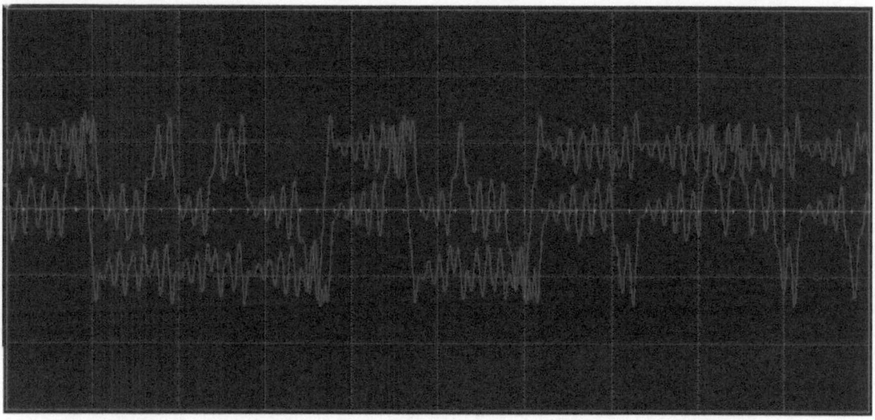

Figure 6.24: Chaotic time series of the 2D 3 × 3-scroll attractor from (6.19) with a fractional-order $q_1 = q_2 = q_3 = 0.9$, with scale 500 mV/Div.

is approximated in the Laplace domain by (3.10), whose electronic implementation is already given in Chapter 3 in Fig. 3.14.

The blocks shown in Fig. 6.25 can be synthesized as shown in Fig. 6.26, where the operational amplifiers can be of type AD712/LM324/TL082, the multipliers can be synthesized by the commercially available AD633, with output coefficient of 0.1, and the whole design can be supplied by ± 12V. The fractional-order of 0.9 can be approximated by(3.14). The resistors can be set to: R_3=R_4=R_5=R_8=R_9=R_{16}=R_{24}=R_{25}=R_{26}=10 kΩ, R_6=R_7=2 kΩ, R_{22}=7.5 kΩ, R_{15}=R_{23}=1 kΩ, R_{14}=20 kΩ, and R_{17}=715Ω. The attractor is shown in Fig. 6.27.

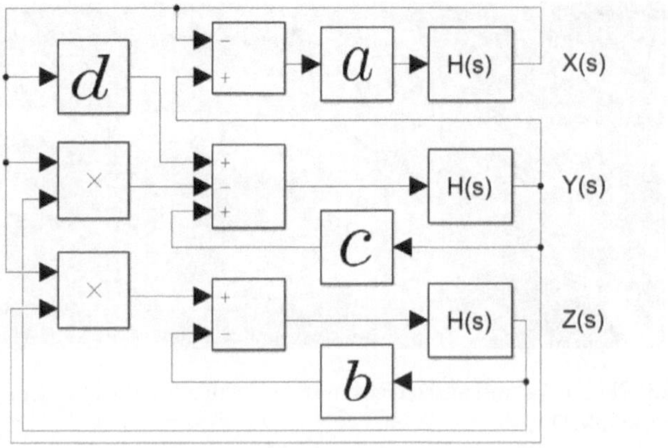

Figure 6.25: Block description of the fractional-order Chen chaotic oscillator given in (6.6), whose $H(s)$ is approximated by (3.10) for the fractional-order of 0.9.

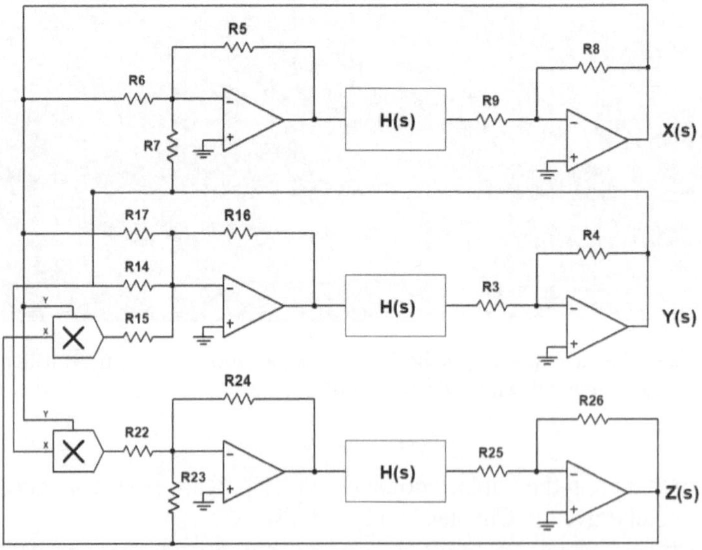

Figure 6.26: Electronic implementation of the fractional-order Chen chaotic oscillator given in (6.6).

6.6 Implementation of optimized fractional-order chaotic/hyper-chaotic oscillators by FPAA

The electronic implementations of fractional-order chaotic oscillators by using commercially available operational amplifiers, as shown in the previous section, can generate errors due to the fact that not all the electronic devices and pas-

Figure 6.27: Fractional-order chaotic attractor of (6.6) with $q_1 = q_2 = q_3 = 0.9$, and with scale 500 mV/Div.

sive circuit elements such as resistors and capacitors, can have the same characteristics, they can vary due to fabrication errors. To mitigate such mismatch errors, one can use field-programmable analog arrays (FPAAs), which embed amplifiers with characteristics quite similar and with the advantage of allow re-programming of the block functions, such as: amplifier gain, differentiators, integrators, multipliers and active filters of first and second orders. In this manner, this section shows the FPAA-based of the FOCOs given in (6.3), (6.4), (6.6) and (6.19). These FOCOs have the fractional-order of (3.10) and can be synthesized as shown in [34]. The FPAA that is used herein is from Anadigm QuadApex Development Board, which embeds four AN231E04 chips that are re-programmed by using the software tool AnadigmDesigner®2 (AD2). The design begins by approximating the fractional-order derivatives of 0.9, which is designed by cascading three first-order active filters, and it can be implemented in the FPAA by using two configurable analog modules (CAMs). As shown in Chapter 3, the fractional-order of 0.9 can be implemented by cascading three first-order transfer functions $H_1(s), H_2(s)$ and $H_3(s)$ that are expanded from (3.10). Taking into account (3.11), those Laplace transfer functions can be implemented by using a CAM Low Pass Bilinear Filter (6.22)to synthesize $H_1(s)$ and a CMA Pole and Zero Bilinear Filter (6.23) to synthesize $H_2(s)$ and $H_3(s)$. The corresponding transfer functions embedded into the FPAA are given as (6.22) and (6.23) [74].

$$T(s) = \frac{w_o G}{(s + w_o)} \tag{6.22}$$

$$T(s) = \frac{G_H(s + w_z)}{(s + w_p)} \tag{6.23}$$

In those blocks the integration constants are described by: $K = 1/RC$. In AnadigmDesigner®2 (AD2) these integration constants are in $1/\mu s$, not in $1/s$, so that when one computes K, required by the CAM in the FPAA, one must consider that $K = 1x10^{-6}/RC$. In the FPAA the IOCELL pins are configured as Bypass type. Combining and doing the cascade connection of 2 Bilinear Filters and one low-pass Bilinear filter, one gets the Laplace transfer function that corresponds to the one shown in Fig. 3.14. This trasfer function from the FPAA is given in (6.24), and it should be equated to the desired transfer function to approximate the fractional-order integrator given in (3.10).

$$T(s) = \frac{G_{H1}(w_{z1}+s)}{(w_{p1}+s)} \cdot \frac{G_{H2}(w_{z2}+s)}{(w_{p2}+s)} \cdot \frac{w_o G}{w_o+s} \tag{6.24}$$

In (6.24): $w = 2\pi f$, so that it can be associated to the time constants given in (6.25), which determine the frequencies of the poles that are evaluated in (6.26), and the zeroes frequencies are evaluated in (6.27).

$$w_{p1} = \frac{1}{\tau_1}, \qquad w_{p2} = \frac{1}{\tau_2}, \qquad w_o = \frac{1}{\tau_3} \tag{6.25}$$

$$\begin{aligned} w_{p1} &= \frac{1}{\tau_1} = 2\pi f_{p1} \Rightarrow f_{p1} = 0.00206\text{Hz} \\ w_{p2} &= \frac{1}{\tau_2} = 2\pi f_{p2} \Rightarrow f_{p2} = 0.346\text{Hz} \\ w_o &= \frac{1}{\tau_3} = 2\pi f_o \Rightarrow f_{po} = 57.87\text{Hz} \end{aligned} \tag{6.26}$$

$$\begin{aligned} w_{z1} &= 2\pi f_{z1} = 1.3061 \Rightarrow f_{z1} = 0.208\text{Hz} \\ w_{z2} &= 2\pi f_{z2} = 218.075 \Rightarrow f_{z2} = 37.71\text{Hz} \end{aligned} \tag{6.27}$$

Within the FPAA, f_{p1} is programmed into the first Pole and Zero Filter, f_{p2} in the second Pole and Zero Filter, and f_o is programmed to the Low Pass Filter. The cascade connection of these three first-order active filters within AnadigmDesigner2 is sketched in Fig. 6.28. This figure is used as a block to solve the fractional-order derivative of 0.9 of the FOCOs given in (6.3), (6.4), (6.19), and (6.6), which have been optimized to have higher values of D_{KY}.

The FPAA-based implementation of the fractional-order hyper-chaotic oscillator given in (6.3), begins by considering that this FOCO does not require scaling. In fact, the ranges of the amplitudes of its state variables are within the ranges required by the AN231E04 chip of ± 3V, as already observed in Fig. 6.6. Therefore, and according to the block diagram shown in Fig. 6.7, and its corresponding electronic circuit shown in Fig. 6.8, one can use 27 CAMs into the FPAA. The multiplication operations require four CAMs to synthesize yz, xz^2 and y^2, the four state variables require 1 CAM each to perform the SumDiff operation, 3 CAMs are required to invert the sign and 9 CAMs are required to

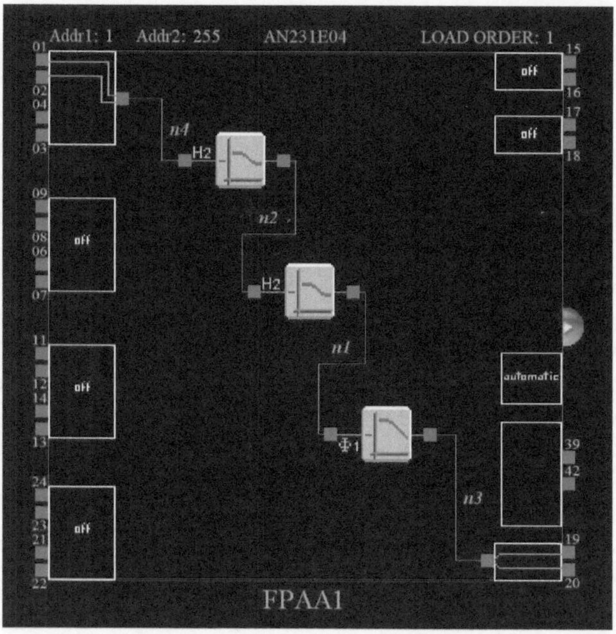

Figure 6.28: Synthesis of the fractional-order integrator $\frac{1}{s^{0.9}}$ by the cascade connection of three first-order active filters $H_1(s)$, $H_2(s)$ and $H_3(s)$ to approximate (3.10).

implement the three fractional-order integrators of 0.9 that use (6.26) and (6.27). As the hardware resources of the QuadApex board have four AN231E04 chips, they are enough to synthesize (6.3). Every chip embeds eight CAMs, and the implementation is performed as follows:

- Select the CAM taking into account the Laplace transfer functions given in (3.11)
- Evaluate the parameters of the constant coefficients in the blocks Sum/Differentiator, inverter, bilinear Pole-Zero and Low-Pass filters, multipliers and Clock frequencies A and B, to program the CAMs.
- From the block diagrams, e.g., Fig. 6.7, and the electronic circuit shown in Fig. (6.8), select the input as inverter or noninverter in the CAMs (Sum/Difference).
- As the FOCO given in (6.3) requieres four multipliers, the CAM needs two Clocks, A and B, in which the relation is 16 times Clock B with respect to Clock A.
- The inversion of the outputs in the state variables x, z and w is performed by using a CAM of Inverting gain stage type, and with unity gain.

- The fractional-order integrators of 0.9 are synthesized by using the CAMs sketched in Fig. 6.28.

The hyper-chaotic oscillator given in chaotic (6.3) is implemented into the FPAA as sketched in Fig. 6.29, but of integer order, which was synthesized from the block diagram shown in Fig. 6.7, where the integrator is based on an RC array (R=10kΩ and C= 20nF) that is implemented with the CAM Integrator having $1/RC$ [$1/\mu$s] = $1/(1 \times 10^6 * 1 \times 10^4 * 20 \times 10^{-9})$ = 0.005. It is quite easy to observe that its fractional-order implementation, shown in Fig. 6.30, differs only by the design of the fractional-order integrators. In the FPAA-based implementations, one can reprogram the design variables and therefore, one can test all the optimal solutions for D_{KY}, provided by DE and PSO algorithms shown in this chapter. For instance, Fig. 6.31 shows the chaotic time series of (6.3) in its integer-order version. Figure 6.30 shows the fractional-order version of (6.3) with a commensurate fractional-order of 0.9. It is easily to identify that the main difference is just the implementation of the fractional-order integrator by using the CAM Low Pass Bilinear Filter given in (6.22) to synthesize $H_1(s)$ and the Pole and Zero Bilinear Filter (6.23) to synthesize $H_2(s)$ and $H_3(s)$.

To show the differences between the integer and fractional-order versions of the hyper-chaotic oscillator, the chaotic time series of the integer order is shown in Fig. 6.32, and its fractional-order of 0.9 is shown in Fig. (6.3). In both cases the simulation parameters are set to: simulation time = 100 ms, $h = 31.25ns$, and the ranges of the voltages of the state variables x, y, z, w are marked with the cursor at time 5.703 ms.

The FPAA-based implementation of the FOCO based on PWL functions given in (6.4) provides chaotic time series with ranges below $\pm$ 3V, so that its implementation on an FPAA does require scaling of the amplitudes. It requires three AN231E04 chips of the Quadpex FPAA, and although it does not require multipliers, the hardware resources for the synthesis of 1D multi-scroll attractors is of 16 CAMs, and they increase as the number of generated scrolls increase. Figure 6.33 shows the FPAA-based implementation and its corresponding chaotic time series for generating a 1D 2-scroll attractor is shown in Fig. 6.34.

The generation of more than 2-scroll attractors in 1D, 2D or 3D can be performed by increasing the PWL functions. In the case of generating a 1D 3-scroll attractor, it can be performed by using (6.17) and for 4-scrolls by using (6.18). As one sees, the FOCO given in (6.4) can be used to generate 2D $n \times m$-scroll attractors by adding two PWL functions as shown in (6.19). The FPAA-based implementation of (6.19) to generate 2D 2×2-scrolls and 3×3-scroll attractors requiere four AN231E04 chips, with a total of 30 CAMs and four to set the reference voltages V_{sat}. The design of this FOCO is shown in Fig. 6.35, which uses 95% of the hardware resources. The block labeled FPAA1 performs the PWL function implemented as shown in Fig. 6.18 to synthesize $F(x)$, and FPAA2 performs the PWL function shown in Fig. 6.19 to synthesize $F(y)$. The simulation of

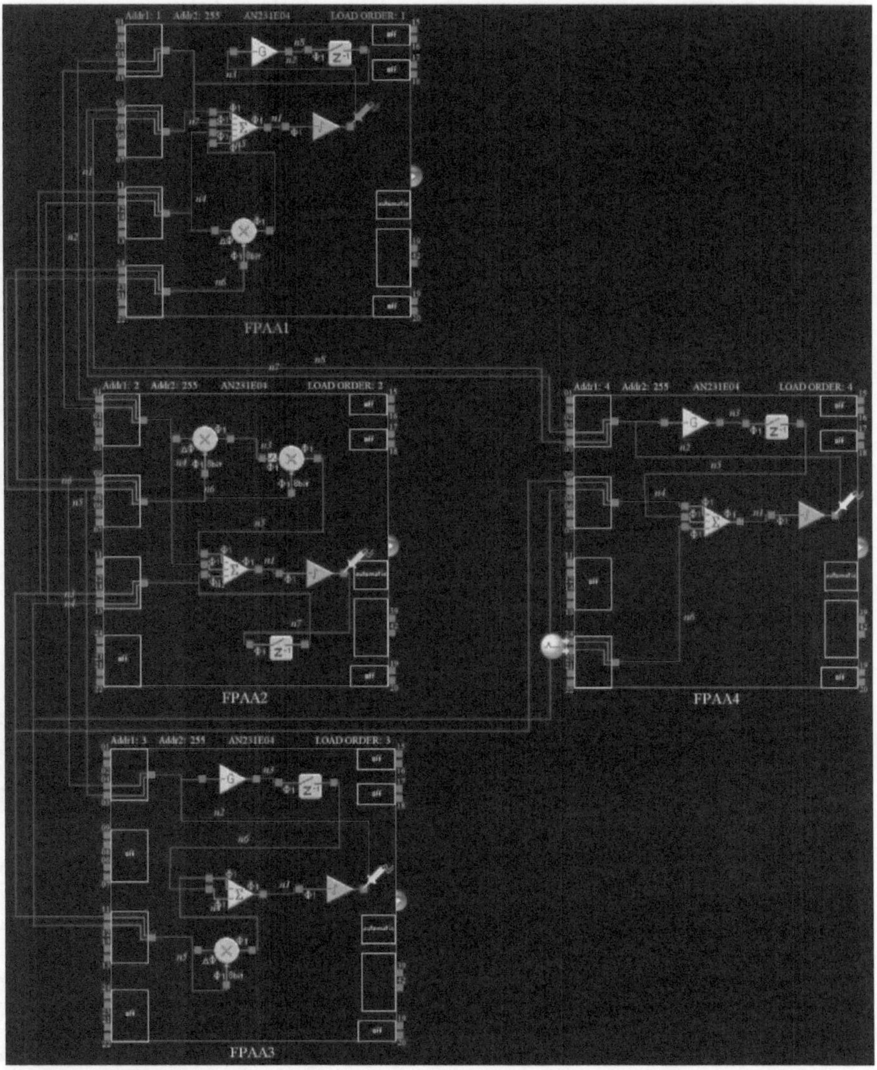

Figure 6.29: FPAA-based implementation of the hyper-chaotic oscillator given in (6.3).

of this FOCO generates the chaotic time series shown in Fig. 6.36, it is associated to (6.19) with a fractional-order of 0.9.

Finally, the fractional-order Chen chaotic oscillator given in (6.6) requires analyzing the ranges of the state variables x, y, z, to perform its FPAA-based design by using the optimized design parameters provided by DE and/or PSO algorithms. In this manner, for the optimization of the commensurate FOCO and observing Table 6.5, the design parameters are: a = 42.5222, b = 4.6544, and c = 32.5646, which provide the highest D_{KY}. In this case, the ranges of the state

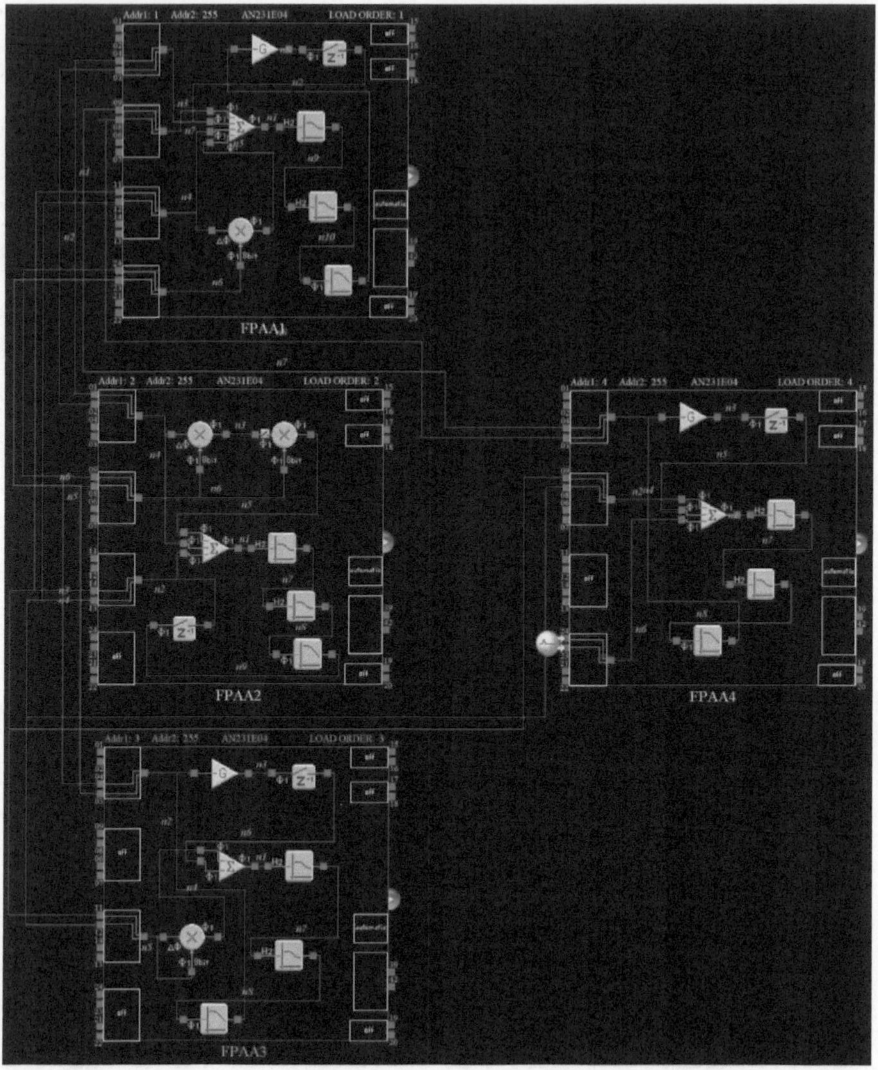

Figure 6.30: FPAA-based implementation of the fractional-order hyper-chaotic oscillator given in (6.3) by designing the fractional-order integrator approximated by (3.10).

variables have amplitudes within the ranges: $x = (-25, 25)$, $y = (-30, 29)$, and $z = (5, 40)$, so that one must perform a scaling process, as shown in [34].

The ranges must be down scaled to be within $\pm 3V$, to allow their design within the FPAA AN231E04. The scaling process is performed by adding the scaling factor to (6.6), as shown in (6.28), where $k_1 = k_2 = k_3 = \frac{1}{18}$. The updated equations from (6.6), take the form given in (6.29).

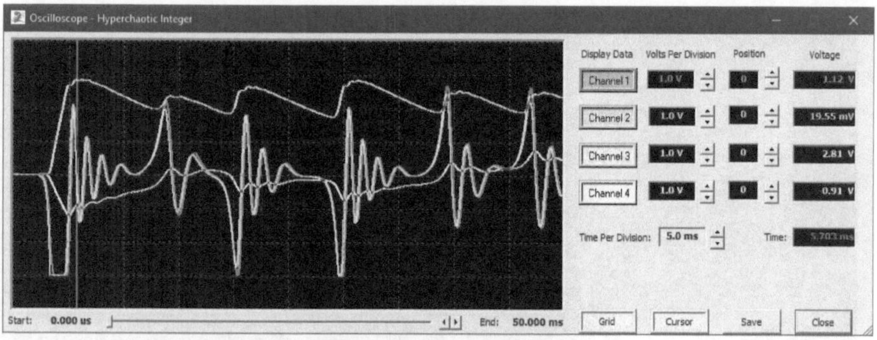

Figure 6.31: Chaotic time series of (6.3) of integer order.

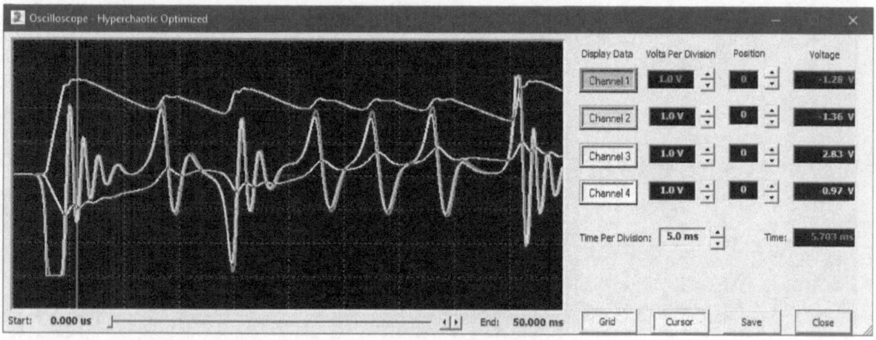

Figure 6.32: Chaotic time series of (6.3) of fractional-order of 0.9.

$$X = k_1 x \qquad Y = k_2 y \qquad Z = k_3 z \tag{6.28}$$

$$\begin{aligned} {}_0D_t^{q_1} X &= a(Y - X), \\ {}_0D_t^{q_2} Y &= (c - a)X - 18XZ + cY, \\ {}_0D_t^{q_3} Z &= 18XY - bZ \end{aligned} \tag{6.29}$$

The FPAA-based implementation of the chaotic oscillator given in (6.29) with a commensurate fractional-order of 0.9, requieres the use of 17 CAMs: 2 to perform the multiplication operations of the state variables xz and xy, 3 CAMs to perform the Sum/Diff, 3 CAMs to invert the variables with unity-gain, and 9 bilinear filters (6 Bilinear pole-zero and 3 Bilinear low-pass filters). The required hardware resources to synthesize (6.29) into the FPAA Anadigm QuadApex Development Board, requieres three AN231E04 chips, and the circuit is sketched in Fig. (6.37). The corresponding chaotic time series is shown in Fig. 6.38.

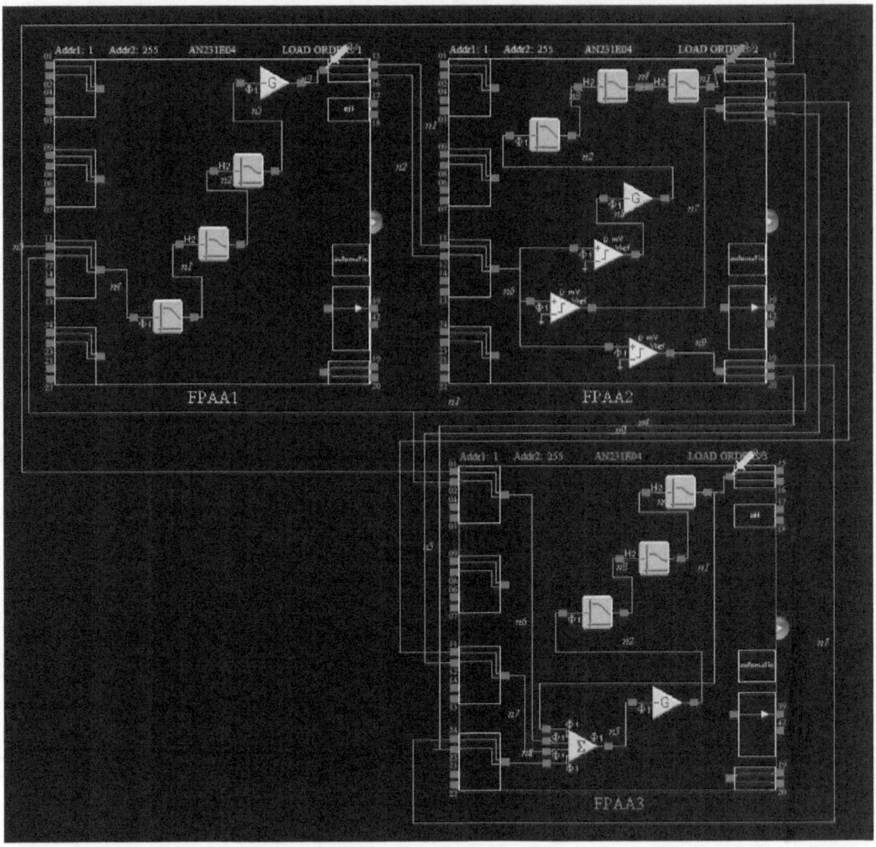

Figure 6.33: FPAA-based implementation of the fractional-order chaotic oscillator given in (3.17) whose fractional-order integrator is approximated by (3.10).

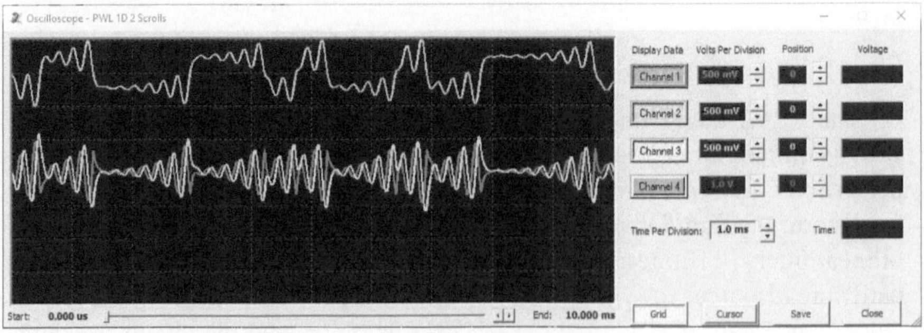

Figure 6.34: Chaotic time series generated from the FPAA-based implementation shown in Fig. 6.33.

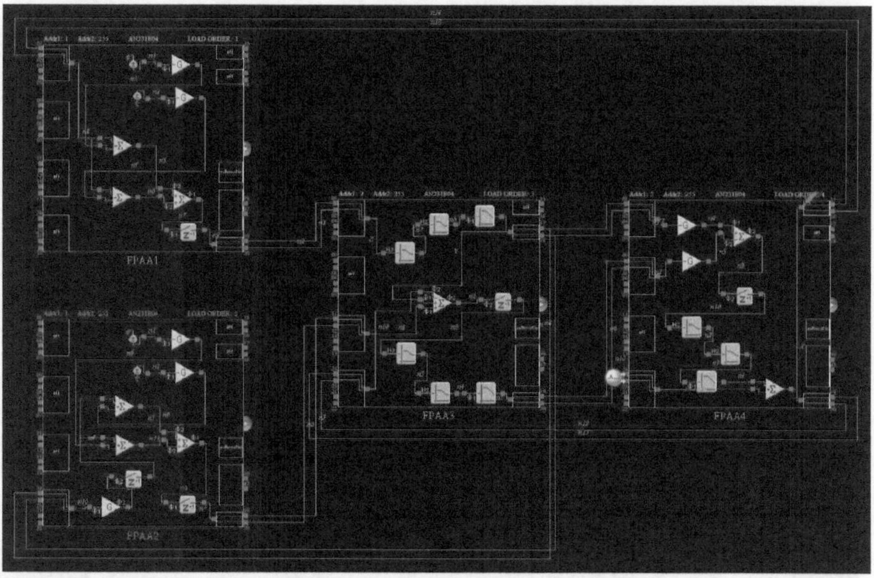

Figure 6.35: FPAA-based implementation of the FOCO given in (3.17) of fractional-order 0.9 approximated by (3.10).

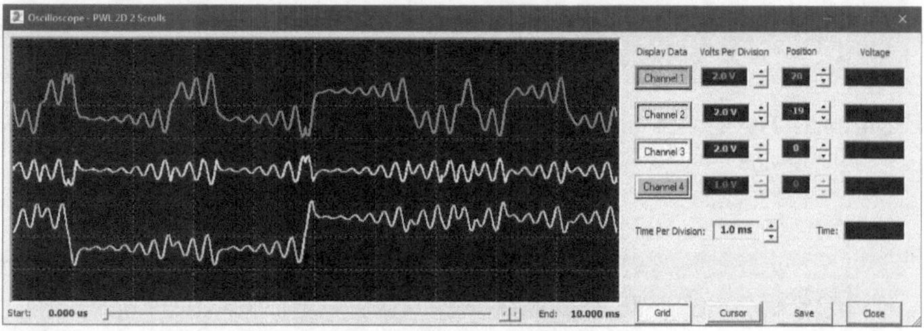

Figure 6.36: Chaotic time series generated from the FPAA-based implementation of (6.4), shown in Fig. 6.35 for a 2×2-scroll attractor.

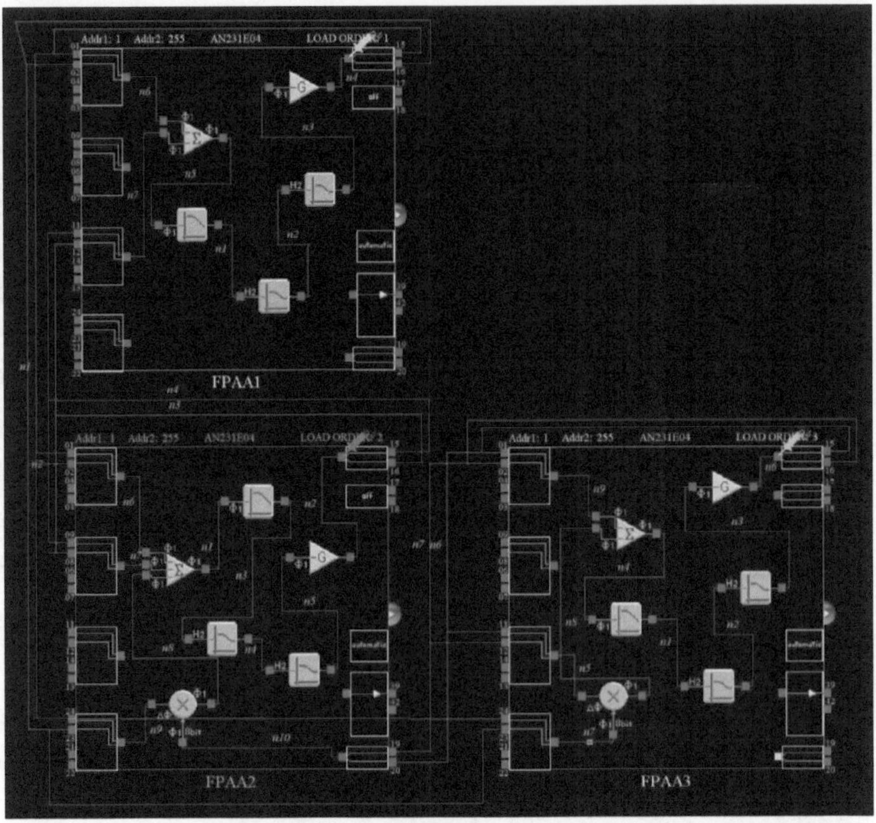

Figure 6.37: FPAA-based implementation of the FOCO given in (6.29) with a fractional-order of 0.9 that is approximated in the Laplace domain by (3.10).

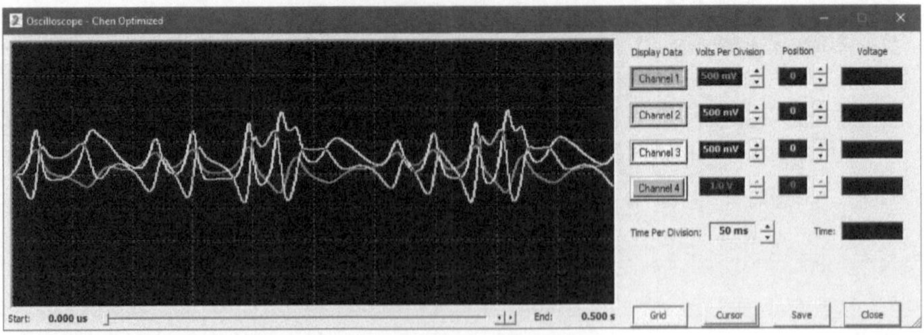

Figure 6.38: Chaotic time series of the FOCO shown in Fig. 6.37, with scale 500mV/Div.

Chapter 7

Multi-Objective Optimization of Fractional-Order Chaotic/Hyper-chaotic Oscillators and their FPGA-based Implementation

As mentioned in Chapter 5, multi-objective optimization is associated to optimize more than one objective simultaneously, with multiple criterions, and where the objectives are in conflict [178]. The metaheuristics that perform multi-objective optimization try to find a set of feasible solutions that represent the optimal values for two or more objectives.

The optimization of fractional-order chaotic oscillators (FOCOs), takes into account dynamical characteristics such as the value of the Lyapunov exponents (LEs), the Kaplan-Yorke dimension D_{KY}, and Kolmogorov-Sinai entropy $Entropy_{KS}$. These characteristics quantify the unpredictability of chaotic behavior and it is desired to find their highest values. Since there is not an analytical equation to maximize those dynamical characteristics, then metaheuristic algorithms are a good option to find the trade-offs among LEs , D_{KY}, and $Entropy_{KS}$. In the multi-objective optimization process, basically, the coefficients and nonlinear functions of the mathematical models of the FOCOs, are varied, and the search spaces of the design variables can be very huge [68].

This chapter shows the multi-objective optimization process considering three FOCOs as cases of study. As some authors have shown that $Entropy_{KS}$ for chaotic systems is equal or lower than the sum of the positive LEs, then the objective functions to be optimized are the maximum or positive LE (LE+) and D_{KY}.

7.1 Multi-objective optimization of fractional-order chaotic/hyper-chaotic oscillators

The heuristics that allow performing the multi-objective optimization of engineering problems were discussed in Chapter 5, and among them, one of the most used is the non-dominated sorting genetic algorithm (NSGA-II). This metaheuristic is applied herein to maximize LE+ and D_{KY}. One can find the code of this algorithm in different programming languages and also for MATLAB® users, which can be found in this link `https://www.mathworks.com/matlabcentral/fileexchange/10429-nsga-ii-a-multi-objective-optimization-algorithm`. This MATLAB code is focused on minimizing objective functions. In this manner, as we are interested in maximizing the dynamical characteristics of FOCOs, then the optimization problem can be formulated considering that the objective functions be negative, i.e., one can minimize -LE+ and -D_{KY}. This is equivalent to maximising both objective functions. This chapter also discusses the use of the C-programming language to perform the multi-objective optimization of FOCOs, so that one can see that one can reduce computing time when programming in C-language instead of using MATLAB tools.

The multi-objective optimization by applying NSGA-II is first performed for the fractional-order Chen chaotic oscillator [101], which is given in Fig. 7.1, and whose design parameters that generate chaotic behavior are set to: $a = 35, b = 3$ and $c = 28$. In Chapter 3, this FOCO was simulated by applying the Grünwald-Letnikov method, by setting the fractional-order as $q_1 = q_2 = q_3 = 0.9, h = 0.001$, Memory length $L_m = 32$, and initial conditions $x_0 = y_0 = z_0 = 0.1$.

$$\begin{aligned} D_t^{q_1} x &= a(y-x) \\ D_t^{q_2} y &= (c-a)x - xz + cy \\ D_t^{q_3} z &= xy - bz \end{aligned} \tag{7.1}$$

The LEs spectrum can be evaluated by applying Benettin-Wolf 's algorithm for FOCOs that are modeled by Caputo's derivatives, as shown in [179]. This algorithm is already programmed for MATLAB users and it can be linked to NSGA-II, so that one can evaluate the first objective associated to LE+. The second objective function associated to D_{KY} given in (7.2), can also be evaluated from the evaluation of the LEs, which are ordered from the highest to the lowest (in a descending order) and labeled as $\lambda_1 \geq \lambda_2 \geq \cdots \geq \lambda_n$, and k is an index that for integer-order chaotic oscillators is set to the number of state variables or number of LEs minus one. For FOCOs, it should be a correct integer and positive number to accomplish $\sum_{i=1}^{k} \lambda_i \geqslant 0$ and $\sum_{i=1}^{k+1} \lambda_i < 0$.

$$D_{KY} = k + \frac{\sum_{i=1}^{k} \lambda_i}{|\lambda_{k+1}|} \tag{7.2}$$

As the two objective functions have been established to be maximized, this process is done by varying the design parameters in the mathematical models. In the case of the fractional-order Chen chaotic oscillator given in (7.1), the design parameters are varied in the following ranges: a between [25.000000, 45.000000], b between [1.000000, 15.000000], and c between [15.000000, 65.00000]. The nonlinear functions of this FOCO are associated to the multiplication of two state variables, namely: xz and xy. They have the original values listed in the first row in Table 7.1, which provide the values of the objective functions LE+=3.5579 and D_{KY}=2.1706. These values are optimized by applying NSGA-II and the best 10 feasible solutions are given in Table 7.1. This metaheuristic was run with a population of 120 individuals during 100 generations. One can see the new values of the design parameters that include six digits to represent the fractional part for a,b and c. As one sees, in the 10 new values of the design parameters, both objective functions LE+ and D_{KY} provided better values than the original ones given in the first row in Table 7.1. The optimal values of the objective functions provided by NSGA-II are shown in Fig. 7.1. As observed from Table 7.1, the feasible solutions of both objective functions increase considerably with respect to the original values of $LE+ = 3.5579$ and $D_{KY} = 2.1706$.

The chaotic time series of the original FOCO are shown in Fig. 7.2(a), and the corresponding chaotic time series for one feasible solution from Table 7.1, are shown in Fig. 7.2(b). Those data seems similar, however, the one that is optimized has higher values of the objective functions LE+ and D_{KY}. Figure 7.3 shows the fractional-order chaotic attractors of the original design parameters given in the first row of Table 7.1, and the other attractors correspond to the next five values given in rows 2 to 6 in the same Table 7.1.

The second FOCO that is optimized by applying NSGA-II for the same objective functions LE+ and D_{KY} is an hyper-chaotic one. It has four state variables and is known as Dadras [108]. The mathematical equation is given in 7.3, which generates chaotic behavior by setting $a = 8.0, b = 40.0$ and $c = 14.9$, the commensurate fractional-orders to $q_1 = q_2 = q_3 = q_4 = 0.95$, $h = 0.001$, Memory length $L_m = 256$, and initial conditions $x_0 = y_0 = z_0 = w_0 = 0.01$.

Table 7.1: Original values and feasible solutions of the Pareto front generated by applying NSGA-II to Chen's FOCO given in (7.1) for the commensurate fractional-order of $q_1 = q_2 = q_3 = 0.9$.

a	b	c	Objective Function 1 LE+	Objective Function 2 D_{KY}
35.0	3.0	28.0	3.5579	2.1706
34.217571	2.4901145	26.873330	4.5861167	2.2098571
33.633243	2.0457396	26.955938	4.4756555	2.2263003
33.657800	2.0557202	26.962981	4.4574060	2.2242929
34.199548	2.4972190	26.866305	4.5404883	2.2087159
34.220637	2.4963387	26.872236	4.5474488	2.2077899
33.636953	2.0520262	26.947779	4.4292136	2.2243685
33.630652	2.0448927	26.956416	4.4292565	2.2240404
34.219221	2.4901466	26.873318	4.5374559	2.2078722
34.206397	2.4925091	26.867481	4.5316228	2.2079541
33.733437	2.0393436	26.957236	4.4546842	2.2236021

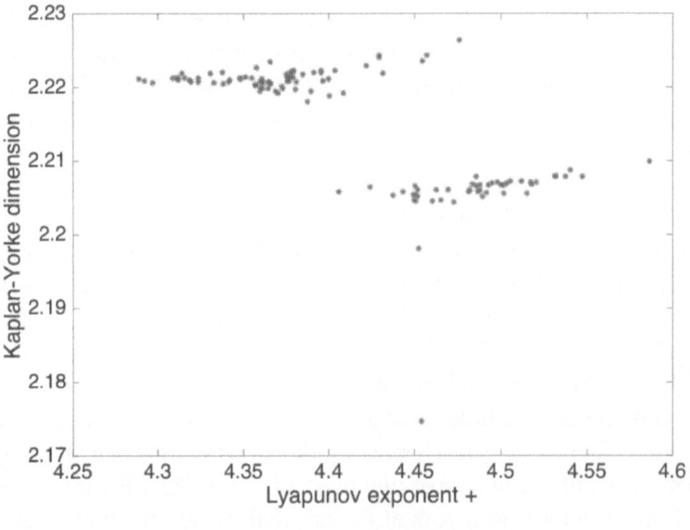

Figure 7.1: Feasible solutions provided by NSGA-II for the optimization of the fractional-order Chen chaotic oscillator given in (7.1) for the commensurate fractional-order of $q_1 = q_2 = q_3 = 0.9$.

$$
\begin{aligned}
D_t^{q_1} x &= ax - yz + w \\
D_t^{q_2} y &= xz - by \\
D_t^{q_3} z &= xy - cz + xw \\
D_t^{q_4} w &= -y
\end{aligned}
\tag{7.3}
$$

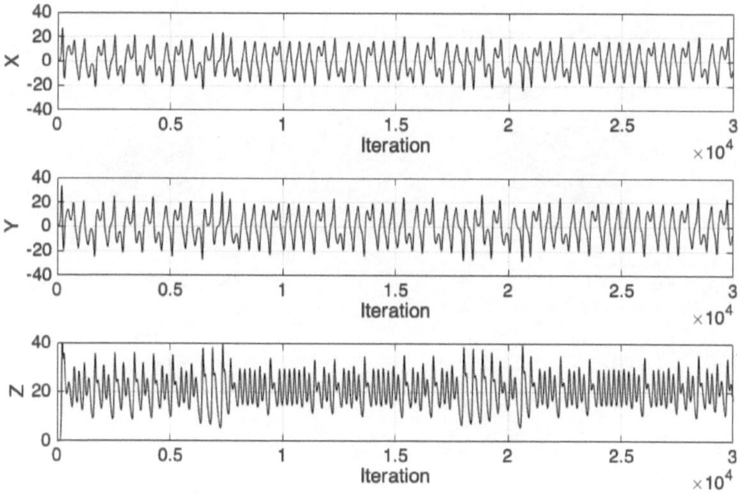

(a)

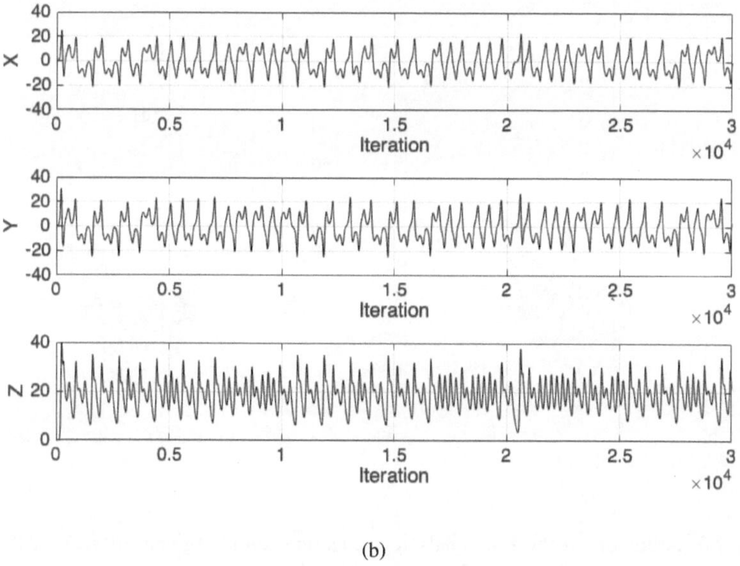

(b)

Figure 7.2: Chaotic time series of (7.1) with: (a) a = 35.0, b = 3.0 and c = 28.0; and (b) a = 34.217571, b = 2.4901145 and c = 26.873330.

The optimization of this FOCO is also performed by varying the design parameters a, b and c. In this case, the original values of the coefficients are given in the first row of Table 7.2, and the next rows show ten feasible solutions provided

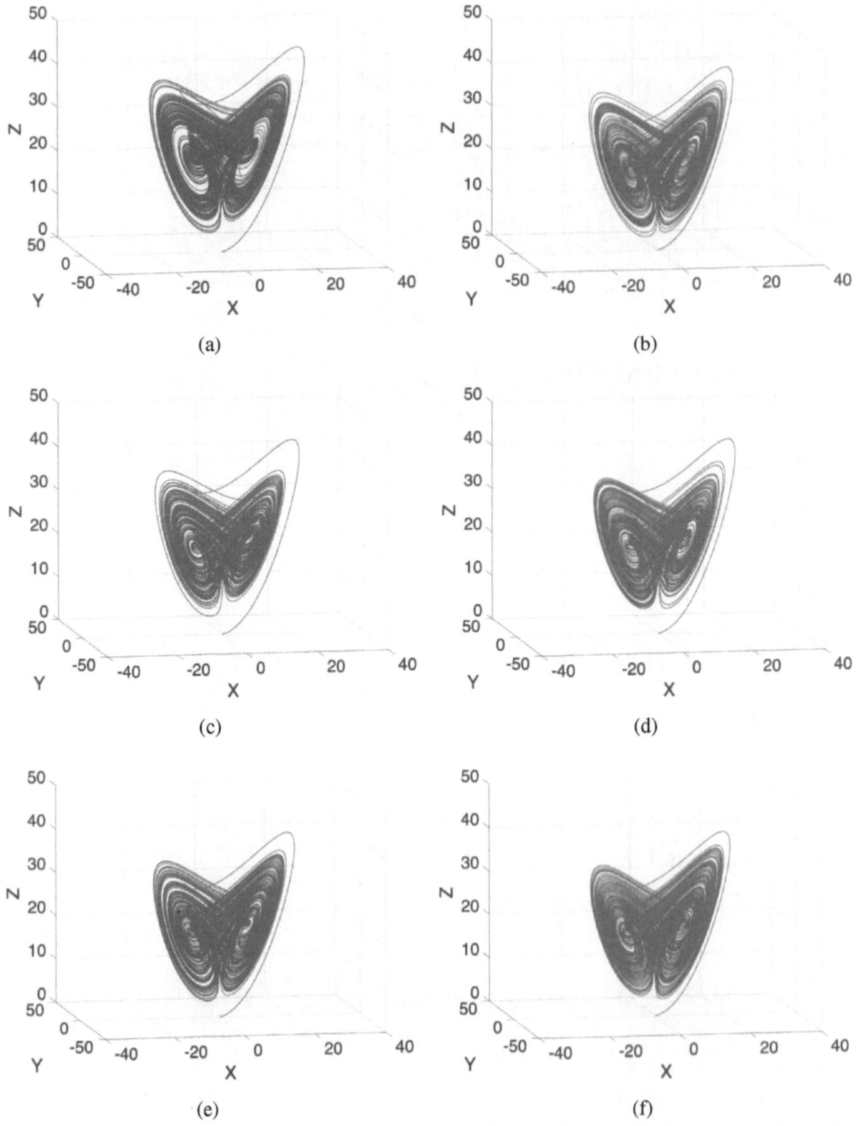

Figure 7.3: Attractors of the fractional-order chaotic oscillator given in (7.1) for the design parameters a, b, c listed in the first six rows in Table 7.1.

by NSGA-II by using a population of 100 individuals evolved during 100 generations. The ranges of the search spaces were set to: a=[1.000000, 20.000000], b=[20.000000, 60.000000], and c=[10.000000, 60.000000]. The feasible solutions are shown in the Pareto front in Fig. 7.4, again, both objective functions have higher values than the original ones that provide $LE+ = 2.3069$, and

Table 7.2: Original values and feasible solutions of the Pareto front generated by applying NSGA-II to Dadras's FOCO given in (7.3), for the commensurate fractional-order of $q_1 = q_2 = q_3 = q_4 = 0.95$.

a	b	c	Objective function 1 LE+	Objective function 2 D_{KY}
8.0	40.0	14.9	2.3069	3.0441
19.999867	32.635312	16.091803	13.121091	3.2335176
19.997078	32.407735	16.156424	13.520973	3.2250168
19.980261	32.566974	16.483646	13.289829	3.2273018
20.000000	32.402967	16.078260	13.180775	3.2327830
19.999955	32.407612	16.068609	13.083356	3.2335770
19.955326	32.461772	16.116914	13.567555	3.2215922
19.998642	32.418668	16.141167	13.404083	3.2265556
19.998299	32.407296	16.134051	13.341318	3.2263055
19.999995	32.411577	16.135208	13.383644	3.2250946
19.977973	32.448042	16.154806	13.164622	3.2298827

$KYD = 3.0441$. All the feasible solutions listed in Table 7.2 overpassed the original value of LE+ from 2.3069 to more than 13.083356. The feasible solutions of the second objective function D_{KY} also provided higher values than 3.0441.

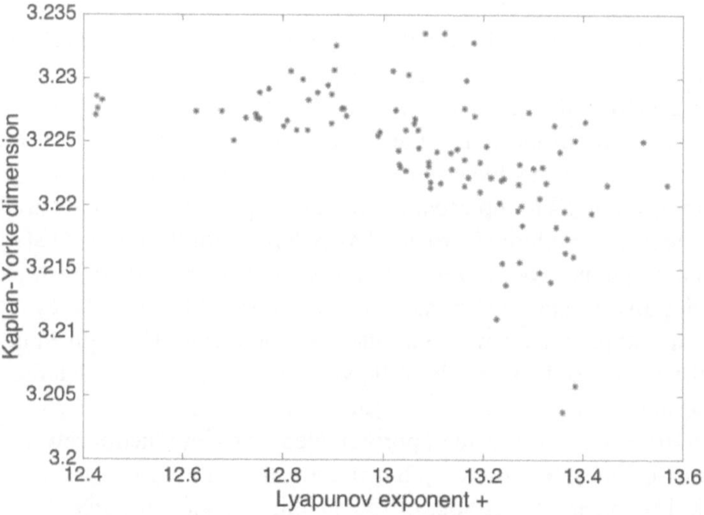

Figure 7.4: Feasible solutions provided by NSGA-II for the optimization of the fractional-order Dadras chaotic oscillator given in (7.3) for the commensurate fractional-order of $q_1 = q_2 = q_3 = q_4 = 0.95$.

The comparison between the chaotic time series generated by using the original design parameters of (7.3) versus a chaotic time series generated by one feasible solution is shown in Fig. 7.5. The corresponding fractional-order chaotic attractors of the original design parameters and the first five feasible solutions given in Table 7.2, are shown in Fig. 7.6.

The third case of study is the optimization of the fractional-order hyper-chaotic oscillator introduced by Zhang [111], and given in (7.4), which has the original parameters: $a = 12, b = 40$ and $c = 3.6$, the commensurate fractional-order derivatives of $q_1 = q_2 = q_3 = q_4 = 0.98$, $h = 0.001$, Memory length $L_m = 256$ and initial conditions $x_0 = y_0 = z_0 = w_0 = 1.0$.

$$\begin{aligned} D_t^{q_1} x &= a(y-x)+w \\ D_t^{q_2} y &= -xz+w \\ D_t^{q_3} z &= xy-b \\ D_t^{q_4} w &= -cx \end{aligned} \tag{7.4}$$

The multi-objective optimization by applying NSGA-II was performed by setting the search space ranges of the design parameters as: a = [1.0000000, 50.000000], b = [1.0000000, 50.0000000], and c = [1.0000000, 50.0000000]. Table 7.3 shows the original values of the two objective functions, LE+ = 0.6598 and D_{KY} = 3.0597, and the remaining ten rows list the 10 feasible solutions provided by NSGA-II executed with a population of 100 individuals during 100 generations. The feasible solutions are shown in the Pareto front shown in Fig. 7.7, again, both objective functions increase with respect to the original values of the design parameters.

Figure 7.8 shown the chaotic time series of the fractional-order Zhang chaotic oscillator given in (7.4), by using the original design parameters and a set of values provided by NSGA-II. The fractional-order attractors associated with the first six feasible solutions listed in Table 7.3, are shown in Fig. 7.9, where they can be appreciated that for this hyper-chaotic system, the optimization was performed by using seven digits to represent the fractional part of the design variables.

All the feasible solutions provided by NSGA-II for the three FOCOs are good options to implement optimized oscillators by using FPAAs. In fact, the embedded device allows reprogramming of the design variables so that they are suitable for fast prototyping. However, only the first optimized FOCO given in (7.1) can be synthesized into the FPAA by following the design process detailed in Chapter 6, because it has the commensurate fractional-order of $q_1 = q_2 = q_3 = 0.9$, and therefore one can use the approximated transfer function given in (3.10), which can be designed by using first-order active filter blocks as already shown in [34]. The other FOCOs optimized in this Chapter, namely the fractional-order Dadras chaotic oscillator given in (7.3), and the fractional-order Zhang chaotic oscillator given in (7.4), they have commensurate fractional-orders of $q_1 = q_2 = q_3 = q_4 = 0.95$ and $q_1 = q_2 = q_3 = q_4 = 0.98$, respectively. However,

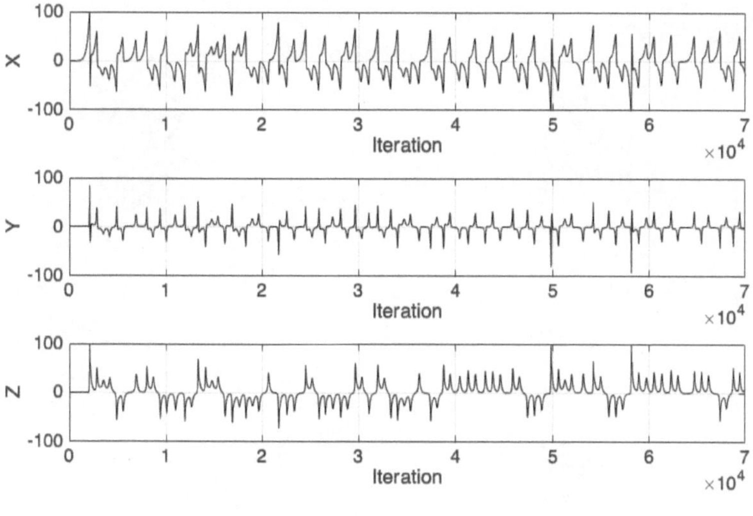

(a)

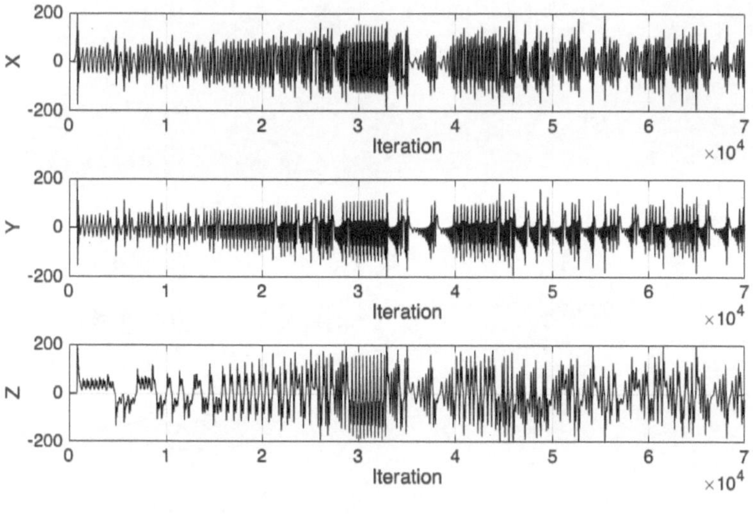

(b)

Figure 7.5: Chaotic time series of (7.3) with: (a) a = 8.0, b = 40.0 and c = 14.9; and (b) a = 19.999867, b = 32.635312 and c = 16.091803.

the Laplace approximation of their fractional-orders can have large polynomials in the frequency domain, making difficult the use of the FPAA due to its limited number of hardware resources. Henceforth, one can implement those FOCOs by

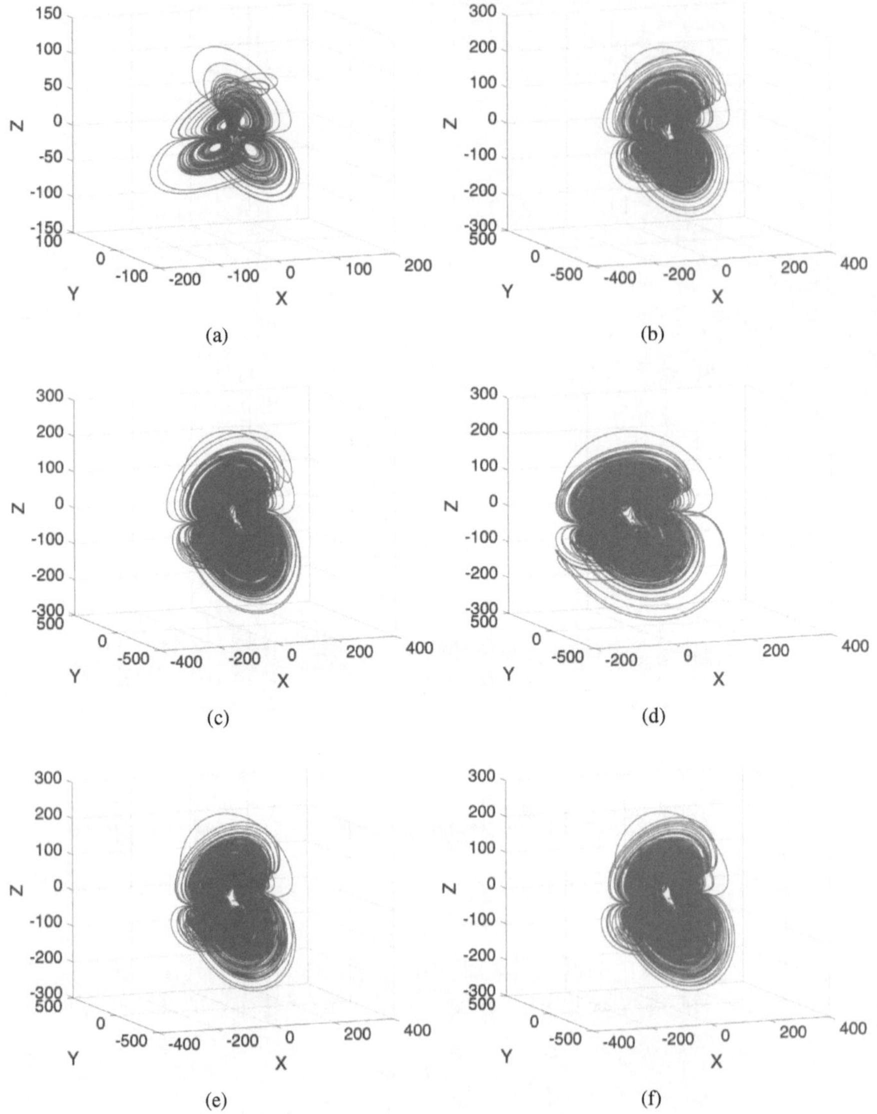

Figure 7.6: 4D attractors of the fractional-order chaotic oscillator given in (7.3) for the design parameters a, b, c listed in the first six rows in Table 7.2.

using field-programmable gate arrays (FPGAs), which are also suitable for fast prototyping, and allows reprogramming of the configurable blocks. This design process is described below.

Table 7.3: Original values and feasible solutions of the Pareto front generated by applying NSGA-II to Zhang's FOCO given in (7.4), for the commensurate fractional-order of $q_1 = q_2 = q_3 = q_4 = 0.98$.

a	b	c	Objective function 1 LE+	Objective function 2 D_{KY}
12.0	40.0	3.6	0.6598	3.0597
5.0336401	33.155891	1.0355292	1.4713409	3.2051412
5.4763167	36.575757	1.0999028	1.4718018	3.1925047
5.0956524	34.645062	1.0405253	1.4038000	3.1940974
5.3790271	36.595382	1.1243777	1.4279497	3.1908182
5.0000000	34.801949	1.0000829	1.3829954	3.1950684
5.3918448	36.564904	1.1264313	1.4632449	3.1899111
5.0000000	34.045174	1.1284280	1.3613612	3.1950447
5.3775357	36.548365	1.1316606	1.4163893	3.1889193
5.4711318	36.497926	1.0975353	1.4236058	3.1867587
5.0033366	33.501179	1.0008479	1.3664501	3.1936711

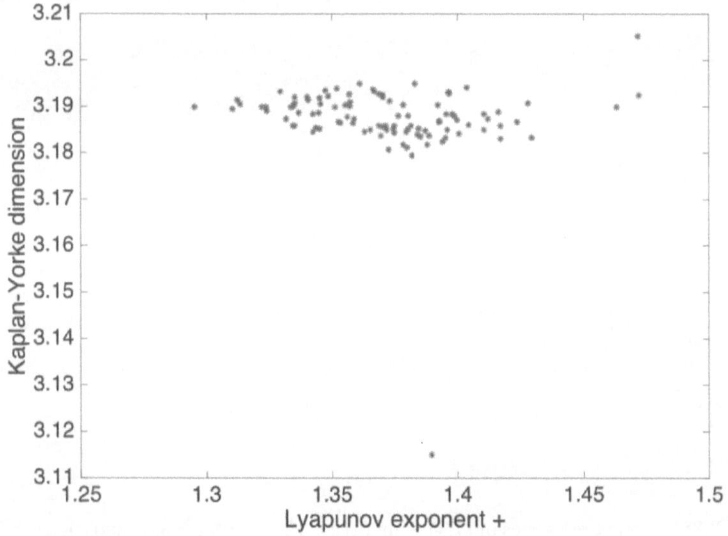

Figure 7.7: Feasible solutions provided by NSGA-II for the optimization of the fractional-order Zhang chaotic oscillator given in (7.4) for the commensurate fractional-order of $q_1 = q_2 = q_3 = q_4 = 0.98$.

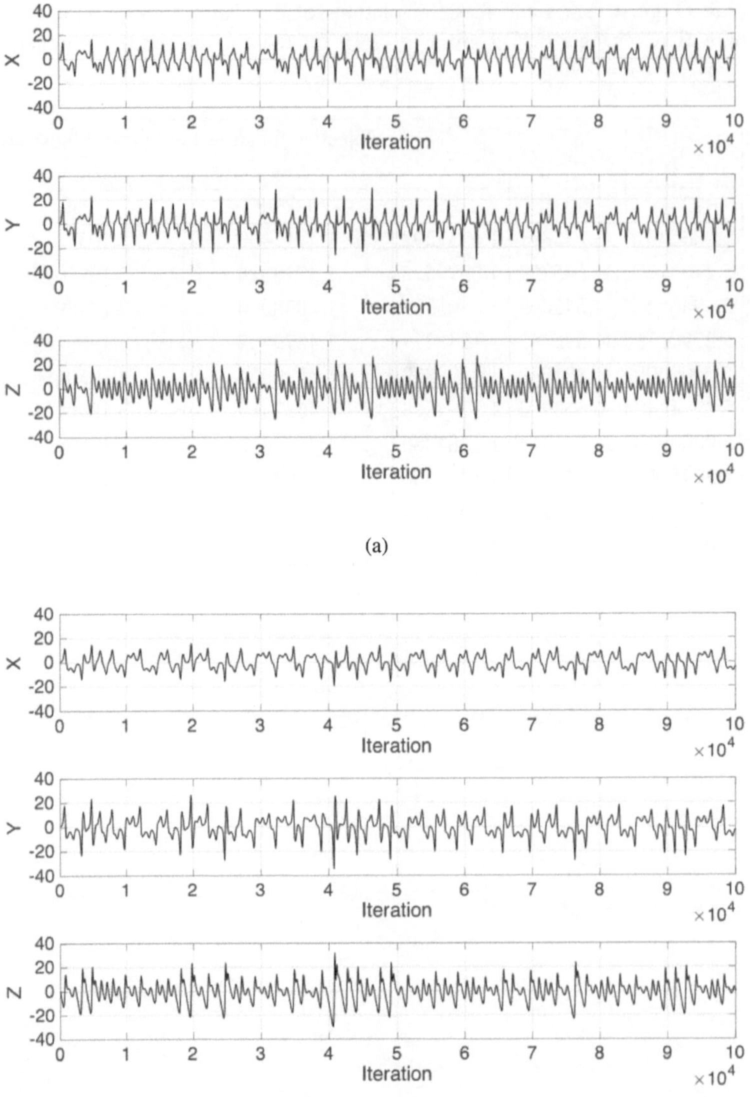

Figure 7.8: Chaotic time series of (7.4) with: (a) a = 12, b = 40, c = 3.6 ; and (b) a = 5.0336401, b = 33.155891 and c = 1.0355292.

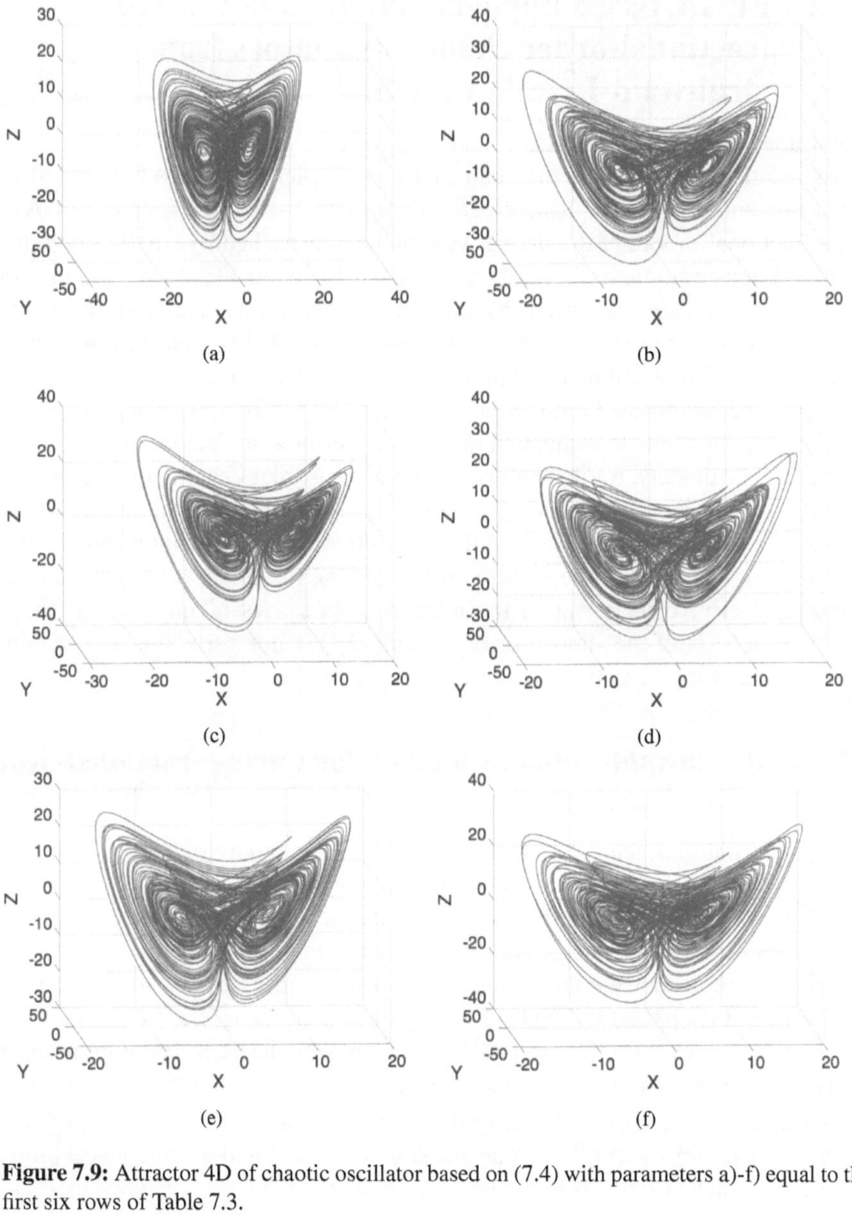

Figure 7.9: Attractor 4D of chaotic oscillator based on (7.4) with parameters a)-f) equal to the first six rows of Table 7.3.

7.2 FPGA-based implementation of optimized fractional-order chaotic oscillators from Grünwald-Letnikov method

Nowadays, fractional calculus is an emerging area and in the case of fractional-order dynamical systems, they can find a good plethora of applications if one can implement them by using electronic devices, as the FPAAs and FPGAs. If the fractional-orders of the derivatives can be approached by small polynomials in the Laplace domain, they can be designed by using FPAAs. Otherwise, one must perform time simulations by applying the definitions discussed in Chapters 1 and 3, in order to synthesize the numerical methods by using FPGAs. One of the most used time simulation approximations for fractional-order chaotic oscillators is the Grünwald-Letnikov method. This method is slow because it requires large computer time to evaluate large sums of terms associated to binomial coefficients. Fortunately, for the majority of fractional-order chaotic oscillators, those memory terms can be reduced by applying the short memory principle, as already shown in [4]. In such a case, one fixes a finite number of memory blocks to be used (L_m), and as the time solution increases (t), the older terms are vanishing and only survive the terms in the interval $[t - L_m]$. As highlighted in Chapter 1, the selection of the short memory length (L_m) is not a trivial task, it requires attention and it is directly related to the step-size (h).

7.2.1 *Grünwald-Letnikov method for solving fractional-order chaotic oscillators*

The electronic implementation of the Grünwald-Letnikov method by using the short memory principle is more efficient when it is synthesized into an embedded system as the FPGA. The first challenge is the selection of the correct h and L_m, which can vary as the design parameters are optimized.

Lets us consider the Grünwald-Letnikov method described in Chapter 1. If this method is applied to a FOCO having three state variables, as the fractional-order Chen chaotic oscillator given in (7.1), then the discretized equations can be described as shown in (7.5), where α is associated to the fractional-order of the derivative and it can be in the range $0 < \alpha < 1$, $c_j^{(\alpha)}$ are the binomial coefficients, which higher terms tend to zero as the sum increases, and v is the lower limit of the sum, which is equal to 1 when $n \leq L_m/h$ or equal to $n - L_m/h$ when $n > L_m/h$. From the experimental examples given in [4], it is worth mentioning that L_m is related to the exactness of the approximated solution, and to the hardware resources required in the FPGA.

$$x(t_n) = f(x(t_n), t_n)h^{\alpha_1} - \sum_{j=v}^{k} c_j^{(\alpha_1)} x(t_{n-j})$$
$$y(t_n) = f(y(t_n), t_n)h^{\alpha_2} - \sum_{j=v}^{k} c_j^{(\alpha_2)} y(t_{n-j}) \quad (7.5)$$
$$z(t_n) = f(z(t_n), t_n)h^{\alpha_3} - \sum_{j=v}^{k} c_j^{(\alpha_3)} z(t_{n-j})$$

7.2.2 Block diagram description of the Grünwald-Letnikov method for FPGA implementation

The analysis of (7.5), leads us to identify building blocks that can be associated with arithmetic and logic operations that can be described under the very-high-speed-integrated-circuit hardware description language (VHDL). Figure 7.10 shows the description of the Grünwald-Letnikov method from (7.5), which consists of three state variables (x, y and z), and it is straightforward to identify digital blocks as subtractors, adders, multipliers, shift registers (to save the history of the terms $y(t_n - j)$), counters (to evaluate the sums of the products of the binomial coefficients and the history of the state variables), and memories. The memories are of read only memory (ROM) and random access memory (RAM) types. One can also use look-up tables (LUTs) when the date is small. All the digital blocks can be designed by adding a clock (CLK) and a reset (RST) pins, and the iterations and CLK cycles are controlled by the block labeled as Control unit of states machine. As one sees, the block description shown in Fig. 7.10 is a generic design that can be used to solve any fractional-order chaotic oscillator that can be synthesized intro the block labeled Chaotic system. In the case of the fractional-order Chen chaotic oscillator given in (7.1), its block description is shown in Fig. 7.11. In this case, one can see the use of single constant multiplier (SCM) blocks, which are quite useful when the multiplication between two operands includes one constant. See for example the operation in the first equation in (7.1) that is: $D_t^{q_1} x = a(y - x)$, where $(y - x)$ is evaluated by using a subtractor block, and the result is multiplied by a. In this case, the SCM saves hardware resources because its design consists of adders and shift-registers, as shown in [180]. The second equation $D_t^{q_2} y = (c - a)x - xz + cy$, requires two SCM blocks to evaluate $(c - a)x$ and cy. The operation xz cannot be performed by using an SCM block, it must be performed by using a multiplier that is available into the FPGA resources. The same is for the third equation $D_t^{q_3} z = xy - bz$, which requires a multiplier to evaluate xy, and an SCM to evaluate bz.

As detailed in [4, 180] the blocks can be designed to have two inputs (A,B) and one output. The blocks adders and subtractors have buses of N bit, and the

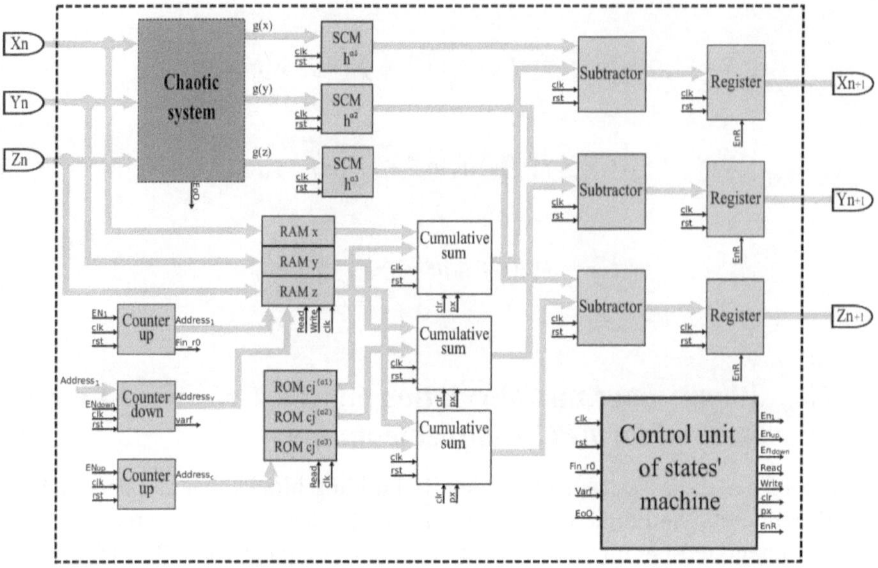

Figure 7.10: Block description of the FPGA-based implementation of the Grünwald-Letnikov method.

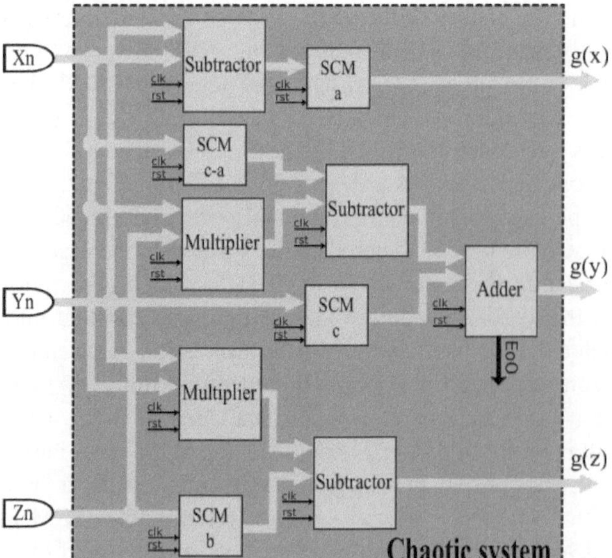

Figure 7.11: Block description of the FOCO given in (7.1), which can be used to synthesize the block labeled Chaotic system in Fig. 7.10.

operations are performed by using two's complements with fixed-point arithmetic. The number of bits are chosen according to the simulate amplitudes of the state variables, and estimating the highest number that can be obtained among the multiplication operations. In this manner, to implement the fractional-order Chen chaotic oscillator given in (7.1) into an FPGA, $N = 32$ divided into the fixed-point format of 12.20 bits, which means that 1 bit is associated to the sign, 11 bits to represent the integer part, and 20 bits to represent the fractional part. The blocks are of synchronous type to include a CLK and a RST pins. The multiplier is a two inputs and one output blocks of N bits. Besides, the multiplication operation between the inputs A and B, may produce as output a number of $2N$ bits, so that in this case a truncation operation takes place to match the N bits that are processed by the other blocks and to accomplish the fixed-point format of 12.20 bits. The challenge here is guaranteeing that the operations do not lose relevant information making the truncation error as small as possible.

When in the mathematical model the multiplication is performed by one state variable and one constant, one can implement a Single-Constant Multiplier (SCM) block that has one input and one output and the constant value is used to synthesize the hardware with a determined number of adders (subtractors) and shift-registers. This saves hardware resources in the FPGA and reduces the execution time. As one sees in Fig. 7.10, the SCM block can also be used to multiply the functions with the fractional-order step-size h^α. In that block diagram the shift registers are described under VHDL and synthesized to work in parallel, i.e., the input passes or it is shifted directly to the output when the pin enable is turned on. They can be synthesized by D-type flip-flops and are also essentials to control the data at each iteration. In Fig. 7.10, the counters control the directions of the ROMs and RAMs to read or write data. The Cumulative sum is a block that is associated with the Grünwald-Letnikov method and performs the sum of the terms (binomial coefficients) stored in the ROM and RAM. The sum is performed at each CLK cycle and when all the terms are accumulated, the signal px is turned on to enable the parallel registers to show the result of the total sum, and the counter restarts its count controlling the signal clear (clr) to compute the next sum. The detail of the blocks within the cumulative sum is sketched in Fig. 7.12, where one can see the blocks multiplier, adder and parallel registers.

The ROM memory is in charge of pre-save the binomial coefficient values $c_j^{(\alpha)}$, and to allow the reading of the saved values, depending on the address pointed by the counter and the "Read" enable signal. In the FPGA the ROM blocks can address different number of bits, depending on the application at hand. For the implementation of the fractional-order Chen chaotic oscillator given in (7.1), the memories have a size of M9K, which have 356 address of 32 bit each. The RAM blocks are in charge of saving the values of the state variables x, y and z when it is enabled by the signal "Write", and the first ascending counter controls the addresses. This RAM provides the saved

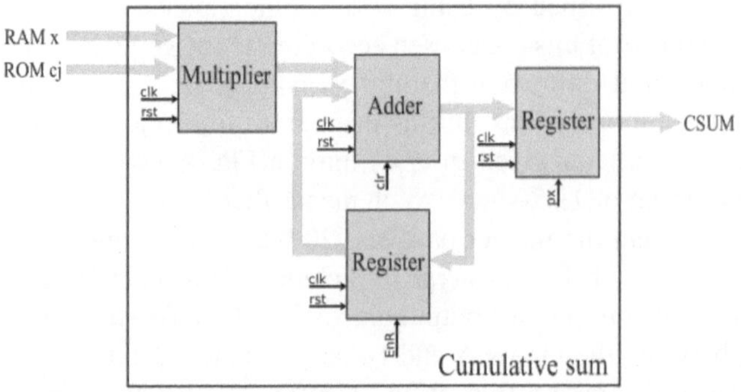

Figure 7.12: Detail of the cumulative sum block to implement the Grünwald-Letnikov method.

values to the "cumulative sum" blocks through the signal "Read" and the second descending counter controls its address.

The oscillator block shown in Fig. 7.11 is in charge to describe the main equations of the FOCOs, and it includes adders, subtractors, multipliers and SCMs. At the last step, this block provides the values of the oscillator functions, as for example: $f(x), f(y)$ and $f(z)$ of the FOCO given in (7.1).

The state machine of the control unit block is a Finite State Machine (FSM) in charge of controlling the iterations of the whole system sketched in Fig. 7.10. It begins by controlling the fill of the RAM by enabling the signal "EN_1" of the first counter "counter up" to control the address to write in the memory limit (L_M) previously fixed, and to send a logic '1' to the "Write" enabled at each CLK cycle. In parallel, the values of the oscillator block are generated, who are enabling the signal "EoO" send a logic '1' when the values $f(x), f(y)$ and $f(z)$ are available. Afterwards they are passed to the SCM block that multiplies by h^{α}. The next step consists of activating the "counter down" block that takes the current counting of the "counter up" block, and the "FSM" enables both the "counter down" and "counter up" by controlling the signals "EN_{Down}" and "EN_{up}", and the values are given one by one through activating the signals "Read" of both the RAM and ROM at each CLK cycle. At the same time the values of the memories are provided or read, the cumulative sum is executed for each state variable, and the "FSM" sends the signal "px" when the sum is finished. The values of the sum are cleared by activating the signal "clr". The data of the "cumulative sum" blocks are subtracted to the values of the SCMs, and the FSM enables the output registers by controlling the signal "EnR" to provide the final values and the data is fed back to the input to begin with the new iteration. The pseudocode of this process is given in Algorithm 18.

Algorithm 18 Pseudocode for the VHDL description of the FPGA-based implementation of Grünwald-Letnikov method for fractional-order chaotic oscillators.

Define the FOCO to synthesize the oscillator block
Inputs:
$x_n, y_n, z_n \leftarrow$ initial conditions
$L_m \leftarrow$ memory size
$h \leftarrow$ step-size
$\alpha_1, \alpha_2, \alpha_3 \leftarrow$ fractional-orders of the derivatives
while $rst = 1$ **do**
    $address_1 \leftarrow address_1 + 1$ *{**Address for** x,y,z}*
    $write = 1$
    $RAM(address_v, write) \leftarrow x_n, y_n, z_n$
    $write = 0$
    $(f(x_n), f(y_n), f(z_n)) \leftarrow oscillator(x_n, y_n, z_n)$
    $SCMs \leftarrow (x_n, y_n, z_n) * (h^{\alpha_1} h^{\alpha_2}, h^{\alpha_3})$
    $clr = 1$
    $address_v \leftarrow address_1$
    $address_c \leftarrow 1$ *{**Address for** $cj^{\alpha_1}, cj^{\alpha_2}, cj^{\alpha_3}$}*
    while $address_c \leq L_m$ **do**
        $read = 1$
        $(x_n, y_n, z_n) \leftarrow RAM(address_v, read)$
        $(cj^{\alpha_1}, cj^{\alpha_2}, cj^{\alpha_3}) \leftarrow ROM(address_c, read)$
        $(cs_1, cs_2, cs_3) \leftarrow cummulativesum(x_n, y_n, z_n, cj^{\alpha_1}, cj^{\alpha_2}, cj^{\alpha_3})$
        $address_v x_{n+1} \leftarrow address_v - 1$
        $address_c \leftarrow address_c + 1$
    end while
    $clr = 0$
    $read = 0$
    $(x_{n+1}, y_{n+1}, z_{n+1}) \leftarrow SCMs - (cs_1, cs_2, cs_3)$
    $Outputs \leftarrow (x_{n+1}, y_{n+1}, z_{n+1})$
end while

7.3 VHDL descriptions of the Adder, Subtractor, Multiplier, SCM, Shift Register, and Accumulator

Depending on the computer arithmetic, the VHDL descriptions of the blocks shown in the previous section can be done as shown herein. In this case, the data bus consist of 32 bit, and each block includes a CLK and RST pines that are synthesized within the FPGA. Recall that the arithmetic operations such as $a+b$ and $a-b$ are performed in binary and therefore the out returns a logic vector. The multiplier having two inputs is described and one can see the difference with respect to a single-constant multiplier (SCM). The evaluation of the Grünwald-Letnikov method requires shift registers and accumulators, whose VDL descriptions are also sketched in the following Listings. One can find more details on the VHDL description styles in [4, 180].

Listing 7.1: VHDL description of the ADDER block using 32 bit.

```
-------------------------------------------------------
--Project: Adder of two inputs of n bits
--Instructions: Assign the inputs a and b
--Inputs: clk - Depends on the FPGA model, e.g., 50 MHz
--        rst - puts a button/switch
--        a,b - logic vectors of n bits with sign
--Output: out1- Returns the arithmetic sum a+b
-------------------------------------------------------
library ieee;
use ieee.std_logic_1164.all;
use ieee.numeric_std.all;

Entity adder2 is
        generic(n: integer:= 32);
        port (
        clk,rst :in std_logic;
        a,b     :in std_logic_vector(n-1 downto 0);
        out1    :out std_logic_vector(n-1 downto 0));
end adder2;

Architecture arch of adder2 is
begin
        sequential: process(rst,clk)
        begin
                if rst = '0' then
                        out1 <= (others => '0');
                elsif (rising_edge(clk)) then
                        out1 <= std_logic_vector(
                        signed(a)+signed(b));
                end if;
        end process;
end arch;
```

Listing 7.2: VHDL description of the SUBTRACTOR using 32 bit.

```
-------------------------------------------------------
--Project: Subtractor of two inputs of n bits
--Instructions: Assign the inputs a and b
--Inputs: clk - Depends on the FPGA model,  e.g., 50 MHz
--        rst - puts a button/switch
--        a,b - logic vectors of n bits with sign
--Output: out1- Returns the arithmetic subtraction a-b
-------------------------------------------------------
```

```
library ieee;
use ieee.std_logic_1164.all;
use ieee.numeric_std.all;

entity subtractor2 is
        generic(n: integer:= 32);
        port (
        clk,rst :in std_logic;
        a,b     :in std_logic_vector(n-1 downto 0);
        out1    :out std_logic_vector(n-1 downto 0));
end subtractor2;

architecture arch of subtractor2 is
begin
        sequential: process(rst,clk)
        begin
                if rst = '0' then
                        out1 <= (others => '0');
                elsif (rising_edge(clk)) then
                        out1 <= std_logic_vector(
                        signed(a)-signed(b));
                end if;
        end process;
end arch;
```

Listing 7.3: VHDL description of the MULTIPLIER using 32 bit.

```
-----------------------------------------------------
--Project: Multiplier of two inputs of n bits
--Instructions: Assign the inputs a and b
--Inputs: clk - Depends on the FPGA model, e.g., 50 MHz
--        rst - puts a button/switch
--        a,b - logic vectors of n bits with sign
--Output: out1- Returns the arithmetic multiplication a*b
-----------------------------------------------------
library ieee;
use ieee.std_logic_1164.all;
use ieee.numeric_std.all;

entity multiplier2 is
        generic(n: integer:= 32);
        port (
        clk,rst :in std_logic;
        a,b     :in std_logic_vector(n-1 downto 0);
        out1    :out std_logic_vector(n-1 downto 0));
end multiplier2;

architecture arch of multiplier2 is
signal aux: signed(n+n-1 downto 0):=(others => '0');
```

```
begin
        sequential: process(rst,clk)
        begin
                if rst ='0' then
                        out1<=(others =>'0');
                elsif (rising_edge(clk)) then
                        aux <= signed(a)*signed(b);
                        out1 <= std_logic_vector(
                        aux(51 downto 20));
                end if;
        end process;
end arch;
```

Listing 7.4: VHDL description of the Single-Constant Multiplier (SCM) using 32 bit.

```
-------------------------------------------------------
---Project: Single Constant Multiplier by h^q=0.001995
---   where h=0.001 and the fractional-order q=0.9
---Instructions: Assign the inputs
---Inputs: clk - Depends on the FPGA model, e.g., 50 MHz
---          rst - puts a button/switch
---          in1 - logic vector of n bits with sign
---Output: out1- Returns the arithmetic multiplication
---          in1*h^q
-------------------------------------------------------
library ieee;
use ieee.std_logic_1164.all;
use ieee.numeric_std.all;

entity scm_hq is
        generic(n: integer:= 32);
        port (
        clk,rst : in std_logic;
        in1     : in std_logic_vector(n-1 downto 0);
        out1    : out std_logic_vector(n-1 downto 0));
end scm_hq;

architecture arch of scm_hq is
-- 1. Creation of signals to use in the mapping
signal w1,w128,w127,w4  : signed (51 downto 0);
signal w123,w1968,w2091: signed (51 downto 0);
begin
-- 2. Adjust the size of the input in1 to 52 bits
--    by applying the resize instruction
    w1 <= resize(signed(in1), w1'length);

-- 3. Evaluate shifts and sums of the signals
--    The symbol "&" is used to concatenate
    w128        <= w1(44 downto 0)&"0000000";
```

```
    w127         <= w128 - w1;
    w4           <= w1(49 downto 0)&"00";
    w123         <= w127 - w4;
    w1968        <= w123(47 downto 0)&"0000";
    w2091        <= w123+ w1968;

-- 4. Sequential part
        Process(clk, rst)
        begin
                if rst ='0'     then
                        out1 <= (others=>'0');
                elsif (rising_edge(clk)) then
-- Select the output bits according to the format
-- for example: fixed-point of 12.20
                        out1 <= std_logic_vector(
                        w2091(51 downto 20));
                end if;
        end process;
end arch;
```

Listing 7.5: VHDL description of the Shift Register using 32 bit.

```
--------------------------------------------------------
---Project: Shift Register of n bits
---Instructions: Assign the inputs
---Inputs: clk - Depends on the FPGA model, e.g., 50 MHz
---           rst - puts a button/switch
---           en - is the input that enables the register
---             d - logic vectors of n bits
--- Output:   q - Returns d when en is enabled
--------------------------------------------------------
library ieee;
use ieee.std_logic_1164.all;

entity register1 is
        generic(n: integer := 32;
                initial : std_logic_vector:=x"00000000");
        port (
        clk,rst,en      :in std_logic;
        d:in std_logic_vector(n-1 downto 0);
        q:out std_logic_vector(n-1 downto 0));
end register1;

architecture arch of register1 is
signal qi: std_logic_vector(n-1 downto 0);
begin
        process(clk,rst,en)
        begin
                if rst = '0' then
```

```
                        qi <= initial;
                elsif (rising_edge(clk)) then
                        if en = '1'     then
                                qi <= d;
                        else
                                qi <= qi;
                        end if;
                end if;
        end process;
q <= qi;
end arch;
```

Listing 7.6: VHDL description of a simple accumulator using 32 bit.

```
------------------------------------------------------------
---Project: Simple Accumulator
---Instructions: Assign the inputs
---Inputs: clk - Depends on the FPGA model, e.g., 50 MHz
---         clear - puts a botton/switch
---         Din - logic vector of n bits with sign
---Output: Q - Returns the cumulative sum of the input
--              Din of n-bits and resets to 0 the out1
--              when the pin clear is high.
------------------------------------------------------------
library ieee;
use ieee.std_logic_1164.all;
use ieee.std_logic_unsigned.all;

entity accumulator is
        generic(n: integer:= 32);
        port(
        clk, clear: in std_logic;
        Din : in std_logic_vector(n-1 downto 0);
        Q : out std_logic_vector(n-1 downto 0));
end accumulator;

architecture arch of accumulator is
signal tmp: std_logic_vector(n-1 downto 0);
begin
        process (clk, reset)
        begin
                if (clear='1') then
                        tmp <= (others=>'0');
                elsif (rising_edge(clk)) then
                        tmp <= tmp + Din;
                end if;
        end process;
Q <= tmp;
end arch;
```

7.4 Results of the FPGA-based implementation of FOCOs

The FPGA-based implementation of the Grünwald-Letnikov method to perform the time simulation of FOCOs, can be described under the VHDL language and in this chapter is it performed by using the FPGA Cyclone IV GX EP4CGX150DF31C7 from ALTERA. Table 7.4 shows the hardware resources for the synthesis of the fractional-order Chen chaotic oscillator given in (7.1). The output data provided by the FPGA-based implementation have their chaotic behavior shown in Fig. 7.13. These chaotic time series of the state variables x, y and z, described under the VHDL code were generated by setting $a = 35, b = 3$ and $c = 28$. This chaotic time series can be compared to the ones shown in Fig. 7.14 by using the optimized design parameters provided by NSGA-II, listed in the second row in Table 7.1, and having the values: $a = 34.217571, b = 2.4901145$ and $c = 26.873330$. Finally, Fig. 7.15 shows the state variables by using the optimized values in the third row in the same Table 7.1, which are $a = 33.633243, b = 2.0457396$ and $c = 26.955938$.

Table 7.4: Resources used in the FPGA Cyclone IV EP4CGX150DF31C7 for the implementation of the fractional-order Chen chaotic oscillator given in (7.1), by applying the Grünwald-Letnikov (GL) method.

Resources	GL Method	Available
Logic Elements	2,182	149,760
Registers	2096	149,760
9*9 Bits multipliers	88	720
Max Freq (MHz)	88.2	50

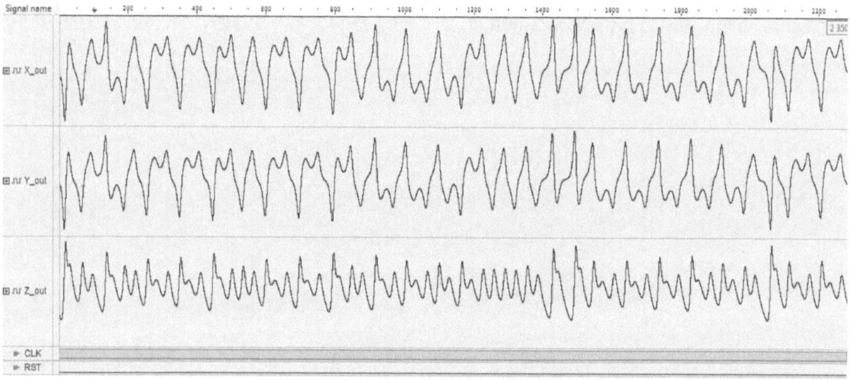

Figure 7.13: Chaotic time series of the state variables of the fractional-order Chen chaotic oscillator given in (7.1), with parameters: $a = 35, b = 3$ and $c = 28$.

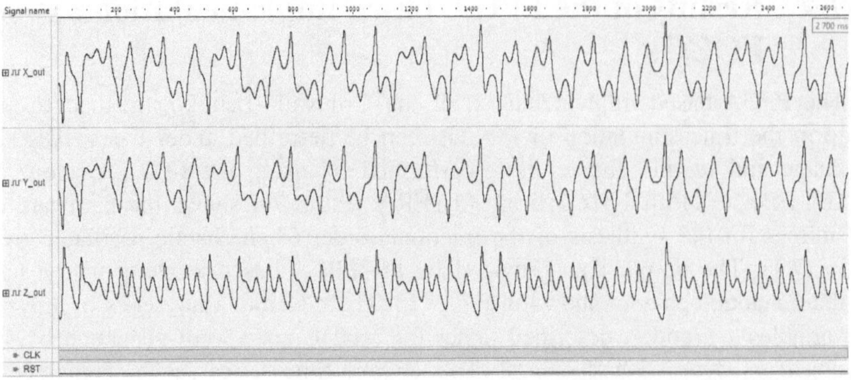

Figure 7.14: Chaotic time series of the state variables of the fractional-order Chen chaotic oscillator given in (7.1), with parameters: $a = 34.217571, b = 2.4901145$ and $c = 26.873330$.

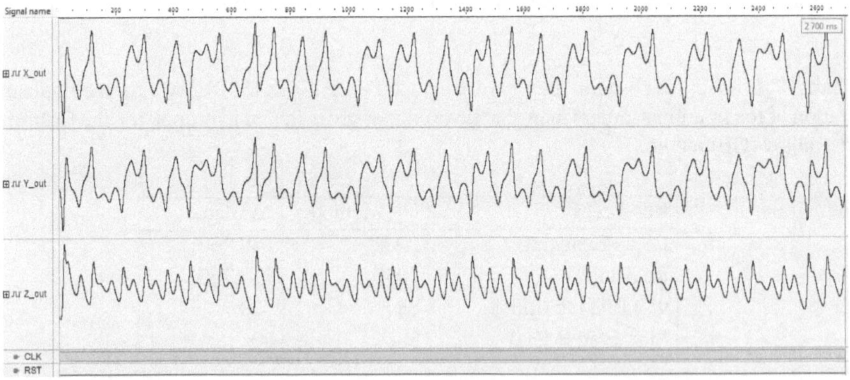

Figure 7.15: Chaotic time series of the state variables of the fractional-order Chen chaotic oscillator given in (7.1), with parameters: $a = 33.633243, b = 2.0457396$ and $c = 26.955938$.

Chapter 8

Applications of Optimized Fractional-Order Chaotic/Hyper-chaotic Oscillators

Two chaotic systems can be synchronized in a master-slave topology to develop applications in chaotic secure communication systems [1]. The synchronization occurs when the trajectories of the state variables of the master and slave systems meet in the same time with a minimum synchronization error, so that they adjust their behavior temporarily due to coupling or forcing [181]. Chaotic secure communication systems have been developed since the introduction of the first synchronization approach between two chaotic oscillators by Pecora and Carroll [182, 183]. Nowadays, several challenges remain open to accomplish and guarantee privacy and high security of the transmitted information [3], so that researchers are searching for the best chaotic oscillator and synchronization approaches. Some synchronization techniques for chaotic dynamical systems can be found in the following references: Hamiltonian forms and observer approach [184], open-plus-closed-loop (OPCL) [185], sliding mode [186, 187, 188], based on Kalman filter [189], adaptive control [190, 191, 192], projection based [193, 194], among others that are suitable for the development of secure communication systems [195, 196, 197, 198].

8.1 Synchronization of FOCOs by applying Hamiltonian forms and observer approach

The fractional-order chaotic oscillators (FOCOs) described in the previous chapters can be synchronized to develop cryptographic applications. This section shows the synchronization of two FOCOs connected in a master-slave topology by applying Hamiltonian forms and observer approach, according to the seminal work given in [184]. In particular, the cases of study are the FOCOs optimized in Chapters 6 and 7.

The Hamiltonian forms technique is based on the representation of a system as an energy conservation one, so that it is suitable for the design of state observers. A chaotic oscillator is described by $\dot{x} = f(x)$, and the Hamiltonian approach can be described by (8.1), where ∂H is the gradient vector of the energy function H, positive definite in R^n. H is a quadratic function defined by $H(x) = \frac{1}{2}X^T Mx$, with M as a symmetrical matrix and positive definite. $J(x)$ and $S(x)$ are matrices representing the conservative and non-conservative parts of the system, respectively, and must satisfy: $J(x)+J^T(x)=0$ and $S(x)=S^T(x)$. There exists the possibility to add a destabilizing vector as $F(x)$, to get the form of a Hamiltonian system, as shown in (8.2). This can consider suppositions to get the form given in (8.1), without $F(x)$.

$$\dot{x} = J(x)\frac{\partial H}{\partial x} + S(x)\frac{\partial H}{\partial x}, \qquad x \; \varepsilon \; R^n \tag{8.1}$$

$$\dot{x} = J(x)\frac{\partial H}{\partial x} + S(x)\frac{\partial H}{\partial x} + F(x), \qquad x \; \varepsilon \; R^n \tag{8.2}$$

If one considers the system with destabilizing vector and one linear output, one gets (8.3), where y is a vector denoting the output of the system. In addition, if ξ is the estimated state vector of x, and η the estimated output in terms of ξ, then an observer to (8.2) can be designed as given in (8.4), where K is a vector of constant gains.

$$\begin{aligned} \dot{x} &= J(y)\frac{\partial H}{\partial x} + S(y)\frac{\partial H}{\partial x} + F(y), \qquad x \; \varepsilon \; R^n \\ y &= C\frac{\partial H}{\partial x}, \qquad y \; \varepsilon \; R^m \end{aligned} \tag{8.3}$$

$$\begin{aligned} \dot{\xi} &= J(y)\frac{\partial H}{\partial \xi} + S(y)\frac{\partial H}{\partial \xi} + F(y) + K(y-\eta) \\ \eta &= C\frac{\partial H}{\partial \xi} \end{aligned} \tag{8.4}$$

The master-slave synchronization by Hamiltonian forms is achieved after accomplishing 2 Theorems [184]:

Theorem 1: The state x of the nonlinear system given in (8.3) can be global, exponential and asymptotically estimated by the state of an observer of the form (8.4), if the pair of matrices (C,S) are observables.

Theorem 2: The state x of the nonlinear system (8.3) can be global, exponential and asymptotically estimated by the state of an observer of the form (8.4), if and only if there exists a constant matrix K such that the symmetric matrix in (8.5) be negative definite.

$$\begin{aligned}[W-KC]+[W-KC]^T &= [S-KC]+[S-KC]^T \\ &= 2[S-\frac{1}{2}(KC+C^TK^T)]\end{aligned} \tag{8.5}$$

8.1.1 Master-slave synchronization of two fractional-order Chen chaotic oscillators

Lets us consider the fractional-order Chen chaotic oscillator given in (8.6), where the variables $x_{m1}=x, x_{m2}=y$ and $x_{m3}=z$, with the subindex m denoting that it is associated with the master system, and it is similar to the original and the associated energy function is given in (8.7). In this manner, the Hamiltonian system is given in (8.8), and it becomes the master. The slave system is proposed by adding a gain vector K that is multiplied by the difference of the state variables of the master x_m and slave x_s systems and is associated with the synchronization error. The gain vector is obtained by verifying that it contains the pair of matrices (C,S), and can be obtained by applying the Sylvester criterion for negative definite matrices. The description of the FOCO in the form of an observer system is described by (8.9).

$$\begin{aligned}D_t^{q_1}x_{m1} &= a(x_{m2}-x_{m1}) \\ D_t^{q_2}x_{m2} &= (c-a)x_{m1}-x_{m1}x_{m3}+cx_{m2} \\ D_t^{q_3}x_{m3} &= x_{m1}x_{m2}-bx_{m3}\end{aligned} \tag{8.6}$$

$$H(x)=\frac{1}{2}[x_{m1}^2+x_{m2}^2+x_{m3}^2] \tag{8.7}$$

$$\begin{bmatrix}D_t^{q_1}x_{m1}\\ D_t^{q_2}x_{m2}\\ D_t^{q_3}x_{m3}\end{bmatrix}=\begin{bmatrix}0 & a-c/2 & 0\\ c/2-a & 0 & 0\\ 0 & 0 & 0\end{bmatrix}\frac{\partial H}{\partial x}+\begin{bmatrix}-a & c/2 & 0\\ c/2 & c & 0\\ 0 & 0 & -b\end{bmatrix}\frac{\partial H}{\partial x}+\begin{bmatrix}0\\ -x_{m1}x_{m3}\\ x_{m1}x_{m2}\end{bmatrix} \tag{8.8}$$

$$\begin{bmatrix} D_t^{q_1} x_{s1} \\ D_t^{q_2} x_{s2} \\ D_t^{q_3} x_{s3} \end{bmatrix} = \begin{bmatrix} 0 & a-c/2 & 0 \\ c/2-a & 0 & 0 \\ 0 & 0 & 0 \end{bmatrix} \frac{\partial H}{\partial x} + \begin{bmatrix} -a & c/2 & 0 \\ c/2 & c & 0 \\ 0 & 0 & -b \end{bmatrix} \frac{\partial H}{\partial x} + \\ + \begin{bmatrix} 0 \\ -x_{s1}x_{s3} \\ x_{s1}x_{s2} \end{bmatrix} + \begin{bmatrix} k_1 \\ k_2 \\ k_3 \end{bmatrix} (y - \eta) \tag{8.9}$$

With the observer in Hamiltonian form, one can derive the slave system as given in (8.10), where the oberver's gains are set to $k_1 = 10, k_2 = 10, k3 = 10$.

$$\begin{aligned} D_t^{q_1} x_{s1} &= a(x_{s2} - x_{s1}) + 10(x_{m1} - x_{s1}) \\ D_t^{q_2} x_{s2} &= (c-a)x_{s1} - x_{s1}x_{s3} + cx_{s2} + 10(x_{m2} - x_{s2}) \\ D_t^{q_3} x_{s3} &= x_{s1}x_{s2} - bx_{s3} + 10(x_{m3} - x_{s3}) \end{aligned} \tag{8.10}$$

The synchronization in a master-slave topology of the FOCO given in (8.6) is performed by using the coefficient values from the first two rows of the optimized design parameters (a, b and c) given in Table 7.1. The optimized parameters are: $a = 34.217571, b = 2.4901145$ and $c = 26.873330$, and the synchronization among the state variables of the master and slave systems is shown in Fig. 8.1. The case when the synchronization does not occur is shown in Fig. 8.2(a), and in Fig. 8.2(b) when it is successful by applying the Hamiltonian technique. The time series of the master and slave systems are shown in Fig. 8.3, and the synchronization errors are shown in Fig. 8.4. It can be seen that the synchronization is accomplished around iteration 800.

By using another set of design parameters from the optimized results given in Table 7.1, we choose $a = 33.633243, b = 2.0457396$ and $c = 26.955938$. The synchronization among the state variables of the master and slave FOCOs by applying Hamiltonian forms and observer approach are also accomplished. The behaviors are also quite similar to the ones shown above for another set of values of the coefficients. The corresponding synchronization errors are shown in Fig. 8.5, where it can be seen that now the synchronization is accomplished around iteration 700.

8.1.2 Master-slave synchronization of two fractional-order Zhang hyper-chaotic oscillators

The second case study is the synchronization of the hidden fractional-order hyper-chaotic oscillator based on the 4D Zhang equations given in (8.11), where the four state variables have subindex m to identify the master system and subindex s to the salve system. The associated energy function is given in (8.12), and the Hamiltonian form can be described by (8.13)

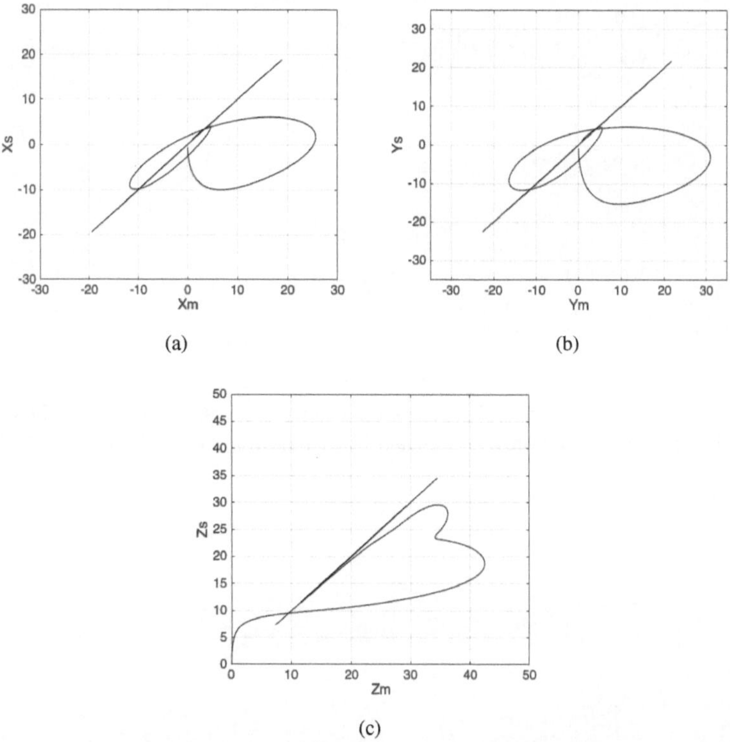

Figure 8.1: Synchronization for the master and slave state variables: (a) x, (b) y and (c) z the the fractional-order Chen chaotic oscillator by applying Hamiltonian forms with $a = 34.217571, b = 2.4901145$ and $c = 26.873330$.

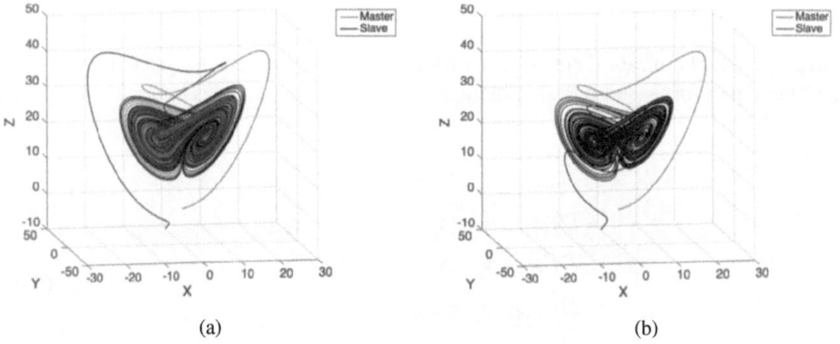

Figure 8.2: Synchronization of the fractional-order Chen chaotic oscillators in 3D by using the optimized values of $a = 34.217571, b = 2.4901145$ and $c = 26.873330$.

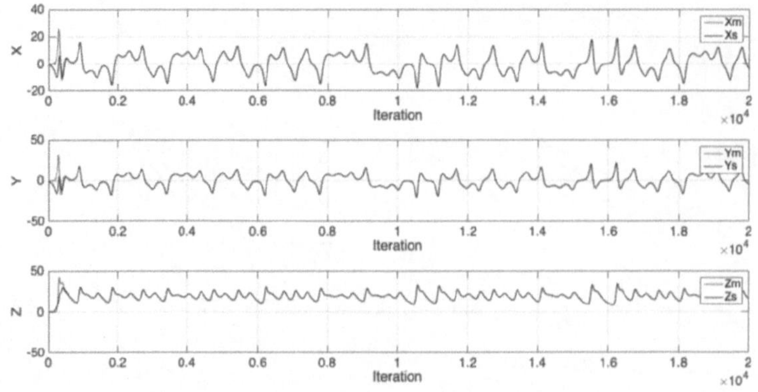

Figure 8.3: Chaotic time series of the master and slave state variables from the FOCO given in (8.6) by applying Hamiltonian forms and by setting $a = 34.217571, b = 2.4901145$ and $c = 26.873330$.

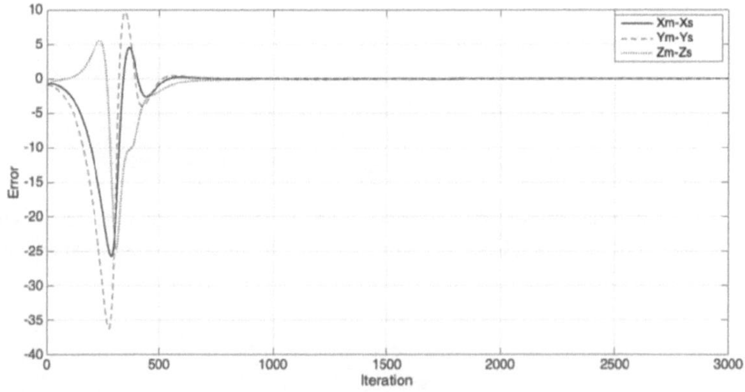

Figure 8.4: Synchronization errors of the fractional-order Chen chaotic oscillator by applying Hamiltonian forms and by using the optimal values: $a = 34.217571, b = 2.4901145$ and $c = 26.873330$.

$$\begin{aligned} D_t^{q_1} x_{m1} &= a(x_{m2} - x_{m1}) + x_{m4} \\ D_t^{q_2} x_{m2} &= -x_{m1} x_{m3} + x_{m4} \\ D_t^{q_3} x_{m3} &= x_{m1} x_{m2} - b \\ D_t^{q_4} x_{m4} &= -c x_{m1} \end{aligned} \tag{8.11}$$

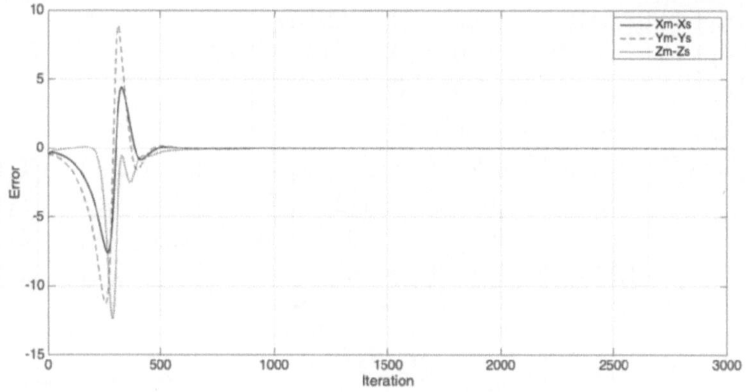

Figure 8.5: Synchronization errors of the fractional-order Chen chaotic oscillator by applying Hamiltonian forms and by using the optimal values: $a = 33.633243, b = 2.0457396$ and $c = 26.955938$.

$$H(x) = \frac{1}{2}[x_{m1}^2 + x_{m2}^2 + x_{m3}^2 + x_{m4}^2] \tag{8.12}$$

$$\begin{bmatrix} D_t^{q_1} x_{m1} \\ D_t^{q_2} x_{m2} \\ D_t^{q_3} x_{m3} \\ D_t^{q_3} x_{m4} \end{bmatrix} = \begin{bmatrix} 0 & a/2 & 0 & c/2+1/2 \\ -a/2 & 0 & 0 & 1/2 \\ 0 & 0 & 0 & 0 \\ -c/2-1/2 & -1/2 & 0 & 0 \end{bmatrix} \frac{\partial H}{\partial x} + \\ + \begin{bmatrix} -a & a/2 & 0 & 1/2-c/2 \\ a/2 & 0 & 0 & 1/2 \\ 0 & 0 & 0 & 0 \\ 1/2-c/2 & 1/2 & 0 & 0 \end{bmatrix} \frac{\partial H}{\partial x} + \begin{bmatrix} 0 \\ -x_{m1}x_{m3} \\ x_{m1}x_{m2} - b \\ 0 \end{bmatrix} \tag{8.13}$$

The slave system is proposed by adding the gain vector K that multiplies the difference between the states variables of the master and slave, and is associated with the synchronization error. The gain vector is obtained by verifying that it contains the pair of matrices (C, S) and by applying the Sylvester criterion for negative definite matrices, to derive (8.14). With the observer defined, the slave system is given in (8.15), where the gain vector is set to $k_1 = 5, k_2 = 5, k_3 = 2, k_4 = 1$.

$$\begin{bmatrix} D_t^{q_1}x_{s1} \\ D_t^{q_2}x_{s2} \\ D_t^{q_3}x_{s3} \\ D_t^{q_3}x_{s4} \end{bmatrix} = \begin{bmatrix} 0 & a/2 & 0 & c/2+1/2 \\ -a/2 & 0 & 0 & 1/2 \\ 0 & 0 & 0 & 0 \\ -c/2-1/2 & -1/2 & 0 & 0 \end{bmatrix} \frac{\partial H}{\partial x} + \\ + \begin{bmatrix} -a & a/2 & 0 & 1/2-c/2 \\ a/2 & 0 & 0 & 1/2 \\ 0 & 0 & 0 & 0 \\ 1/2-c/2 & 1/2 & 0 & 0 \end{bmatrix} \frac{\partial H}{\partial x} + \begin{bmatrix} 0 \\ -x_{s1}x_{s3} \\ x_{s1}x_{s2}-b \\ 0 \end{bmatrix} + \begin{bmatrix} k_1 \\ k_2 \\ k_3 \\ k_4 \end{bmatrix} (y-\eta) \tag{8.14}$$

$$\begin{aligned} D_t^{q_1}x_{s1} &= a(x_{s2}-x_{s1})+x_{s4}+5(x_{m1}-x_{s1}) \\ D_t^{q_2}x_{s2} &= -x_{s1}x_{s3}+x_{s4}+5(x_{m2}-x_{s2}) \\ D_t^{q_3}x_{s3} &= x_{s1}x_{s2}-b+2(x_{m3}-x_{s3}) \\ D_t^{q_4}x_{s4} &= -cx_{s1}+(x_{m4}-x_{s4}) \end{aligned} \tag{8.15}$$

The optimized system based on (8.11) is synchronized by using the design parameters given in the first two rows in Table 7.3. The first set of values is $a = 5.0336401, b = 33.155891$ and $c = 1.0355292$. The synchronization of the four state variables of the master and slave systems is shown in Fig. 8.6. The chaotic time series of the master and slave systems are shown in Fig. 8.7, whose corresponding synchronization errors can be appreciated in Fig. 8.8, where it can be seen that the synchronization is accomplished around iteration 3500.

The synchronization of the hidden fractional-order hyper-chaotic oscillator based on the 4D Zhang equations given in (8.11), by using the second set of optimal design parameters, corresponds to the values: $a = 5.4763167, b = 36.575757$ and $c = 1.0999028$. In this case, the synchronization errors between the master and slave systems are shown in Fig. 8.9, where it can be seen that the synchronization is accomplished around iteration 3200.

8.2 Synchronization of FOCOs by applying the OPCL technique

The Open-Plus-Closed-Loop (OPCL) synchronization technique is based on the control systems combination. It is a heterogeneous synchronization technique that allows us to obtain the master and slave design parameters. From a dynamical system described by $\dot{x} = f(x)$, the system of equations associated to the master chaotic oscillator is given by (8.16), where $x_{1m}(t), x_{2m}(t)$ and $x_{3m}(t)$ denote the state variables. In a similar way, by using the subindex s instead of m,

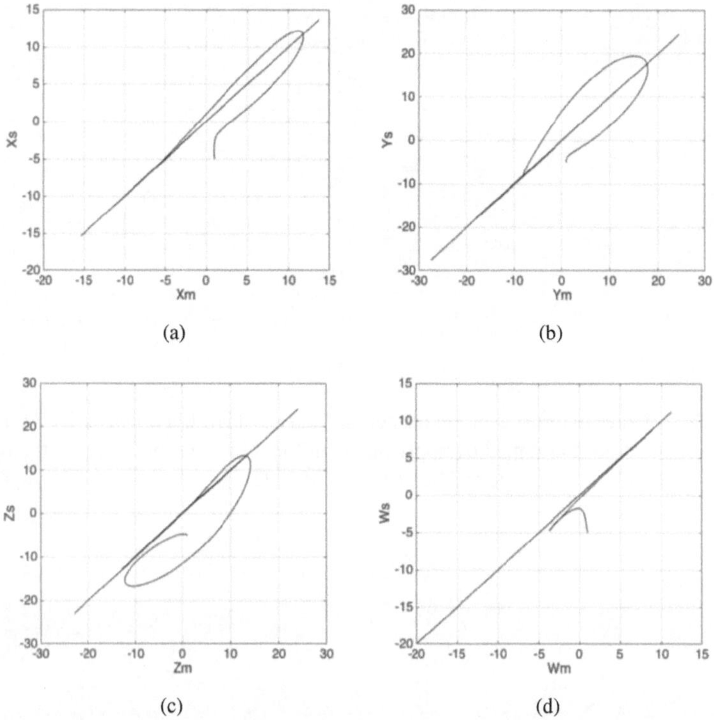

Figure 8.6: Synchronization of the master and slave state variables of the hidden fractional-order Zhang hyper-chaotic oscillator by applying Hamiltonian forms and using the optimized values: $a = 5.0336401, b = 33.155891$ and $c = 1.0355292$. State variables: (a) x, (b) y, (c) z, and (d) w.

the slave chaotic oscillator has the state variables $x_{1s}(t), x_{2s}(t)$ and $x_{3s}(t)$, and it described by (8.17). $D(v(t), u(t))$ is given in (8.18), with D_1 and D_2 as open loop and closed loop parts, respectively, and given by (8.19) and (8.20).

$$\frac{d}{dt}u(t) = F(u(t)) = F(x_{1m}(t), x_{2m}(t), x_{3m}(t)); \quad u \in R^3 \tag{8.16}$$

$$\frac{d}{dt}v(t) = F(v(t)) + D(v(t), u(t)); \qquad v \in R^3 \tag{8.17}$$

$$D(v(t), u(t)) = D_1(u(t)) + D_2(v(t), u(t)); \tag{8.18}$$

$$D_1(u(t)) = \frac{du(t)}{dt} - F(u(t)); \tag{8.19}$$

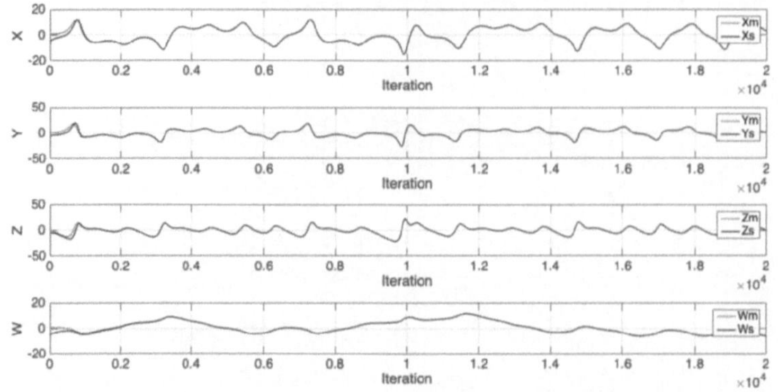

Figure 8.7: Chaotic time series of the master and slave hidden fractional-order Zhang hyper-chaotic oscillators, synchronized by applying Hamiltonian forms and by using the optimized values $a = 5.0336401, b = 33.155891$ and $c = 1.0355292$.

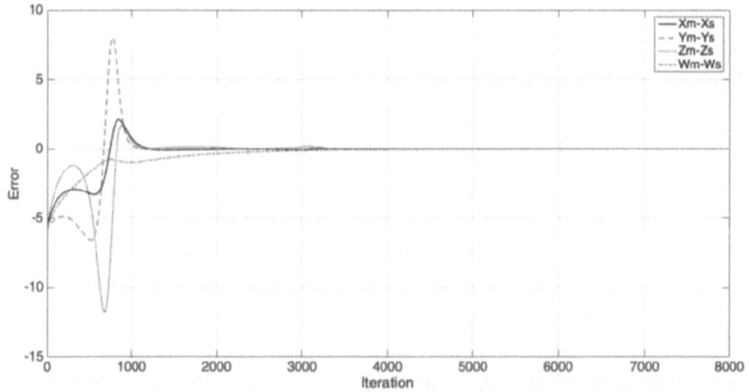

Figure 8.8: Synchronization errors of the hidden fractional-order Zhang hyper-chaotic oscillator by applying Hamiltonian forms and by using the optimized values: $a = 5.0336401$, $b = 33.155891$ and $c = 1.0355292$.

$$D_2(v(t), u(t)) = \left(H - \frac{\delta}{\delta t} F(u(t))\right) e(t) \tag{8.20}$$

H is an arbitrary constant Hurwitz matrix, so that the simplicity of the slave system depends on how this matrix is chosen, and some constants can be added to perform the gain system function. The synchronization error is evaluated as $e(t) = v(t) - u(t)$. When the OPCL synchronization is achieved, the error tends to zero and it can be verified by Taylor's series [199]. One important condition is that if the real parts of the eigenvalues from H are negative, the synchronization

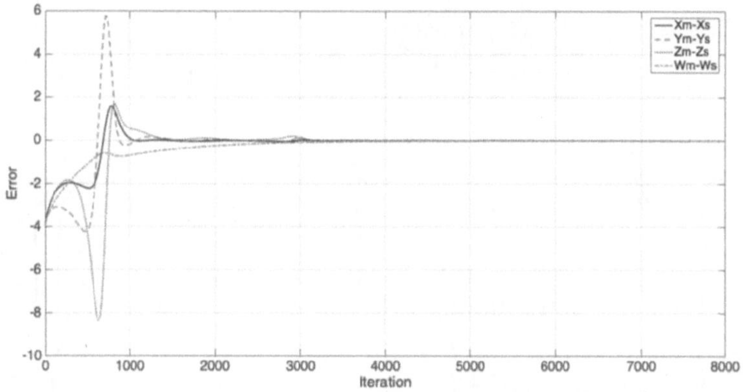

Figure 8.9: Synchronization errors of the hidden fractional-order Zhang hyper-chaotic oscillator by applying Hamiltonian forms and by using the optimized values: $a = 5.4763167$, $b = 36.575757$ and $c = 1.0999028$.

will be successful. This is a necessary condition but not enough since there may be an H with eigenvalues equal to zero and the synchronization still can occur [185].

8.2.1 Master-slave synchronization of two hidden fractional-order Zhang hyper-chaotic oscillators

Lets us consider again the 4D hidden FOCO described by (8.11). One can propose that the open loop part in the slave system be null ($D_1(u(t)) = 0$), this simplifies the analysis. For the closed part, the master system partial derivative can be described by (8.21), and H is proposed in (8.22), where $P1$, $P2$, $P3$ and P_4 are constant values. The eigenvalues of H determine that P_1, P_2, P_3 and P_4 must be negatives. For example: by setting $p_1 = -2$, $p_2 = -2$, $P_3 = -4$ and $P_4 = -4$, the eigenvalues are equal to $\lambda_1 = -4, \lambda_2 = -7.03, \lambda_3 = -3.0 - j0.1884$ and $\lambda_4 = -3.0 + j0.1884$, all of them with negative real part, and thereby the condition described above is accomplished. Therefore, the closed loop contribution is given in (8.23). Finally, with the open-closed loop contribution, the slave FOCO for the hidden Zhang system is given in (8.24).

$$\frac{\delta}{\delta t} F(u(t)) = \begin{pmatrix} -a & a & 0 & 1 \\ -x_{m3} & 0 & -x_{m1} & 1 \\ x_{m2} & x_{m1} & 0 & 0 \\ -c & 0 & 0 & 0 \end{pmatrix} \tag{8.21}$$

$$H = \begin{pmatrix} P_1 - a & a & 0 & 1 \\ 0 & P_2 & 0 & 1 \\ 0 & 0 & P_3 & 0 \\ -c & 0 & 0 & P_4 \end{pmatrix} \tag{8.22}$$

$$\begin{aligned} D_2 &= \left(\begin{pmatrix} P_1 - a & a & 0 & 1 \\ 0 & P_2 & 0 & 1 \\ 0 & 0 & P_3 & 0 \\ -c & 0 & 0 & P_4 \end{pmatrix} - \begin{pmatrix} -a & a & 0 & 1 \\ -x_{m3} & 0 & -x_{m1} & 1 \\ x_{m2} & x_{m1} & 0 & 0 \\ -c & 0 & 0 & 0 \end{pmatrix} \right) (v_t - u_t) \\ &= \begin{pmatrix} P_1 * (x_{s1} - x_{m1}) \\ x_{m3} * (x_{s1} - x_{m1}) + P_2 * (x_{s2} - x_{m2}) + x_{m1} * (x_{s3} - x_{m3}) \\ -x_{m2} * (x_{s1} - x_{m1}) - x_{m1} * (x_{s2} - x_{m2}) + P_3 * (x_{s3} - x_{m3}) \\ P_4 * (x_{s4} - x_{m4}) \end{pmatrix} \end{aligned} \tag{8.23}$$

$$\begin{aligned} D_t^{q_1} x_{m1} &= a(x_{m2} - x_{m1}) + x_{m4} + P_1 * (x_{s1} - x_{m1}) \\ D_t^{q_2} x_{m2} &= -x_{m1}x_{m3} + x_{m4} + x_{m3} * (x_{s1} - x_{m1}) + P_2 * (x_{s2} - x_{m2}) + x_{m1} * (x_{s3} - x_{m3}) \\ D_t^{q_3} x_{m3} &= x_{m1}x_{m2} - b - x_{m2} * (x_{s1} - x_{m1}) - x_{m1} * (x_{s2} - x_{m2}) + P_3 * (x_{s3} - x_{m3}) \\ D_t^{q_4} x_{m4} &= -cx_{m1} + P_4 * (x_{s4} - x_{m4}) \end{aligned} \tag{8.24}$$

By using the optimized parameters $a = 5.0336401, b = 33.155891$ and $c = 1.0355292$, and by setting $P_1 = -2$, $P_2 = -2$, $P_3 = -4$ and $P_4 = -4$, the synchronization between the master and slave FOCOs occurs as shown in Fig. 8.10. Figure 8.11 shows the attractor among three state variables when the synchronization does not occur and when it is successful by applying the OPCL technique. The chaotic time series of the master and slave FOCOs are shown in Fig. 8.12. The synchronization errors are shown in Fig. 8.13, where it can be observed that the minimum error occurs around iteration 1600.

By using another set of optimized design parameters from Table 7.3, e.g., by setting $a = 5.4763167, b = 36.575757$ and $c = 1.0999028$, and with $P_1 = -2$, $P_2 = -2$, $P_3 = -4$ and $P_4 = -4$, the synchronization between the master and slave hidden fractional-order Zhang hyper-chaotic oscillators generates the errors shown in Fig. 8.14, where one can observe that the minimum error occurs around iteration 1400.

8.3 Image encryption by using optimized FOCOs

The synchronization techniques described above can be used to implement an encryption system to process digital data that can be associated with different

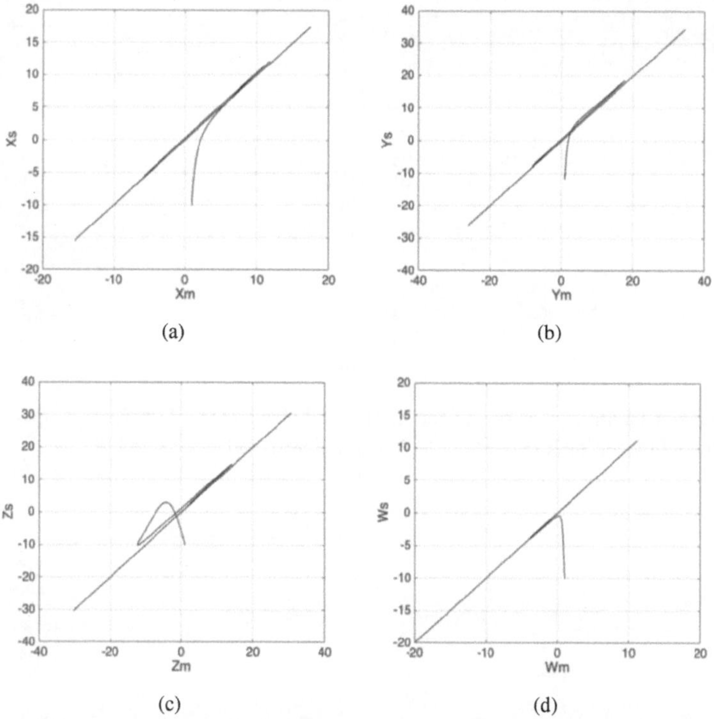

Figure 8.10: Master-slave synchronization of two hidden fractional-order Zhang hyper-chaotic oscillators by applying OPCL and by setting: $a = 5.0336401, b = 33.155891$ and $c = 1.0355292$. State variables: (a) x, (b) y, (c) z and (d) w.

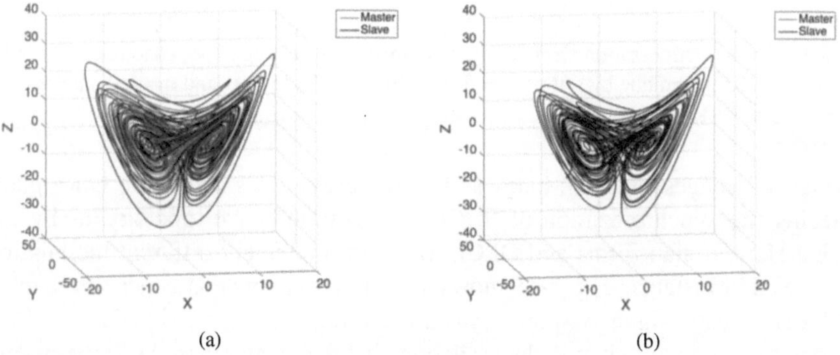

Figure 8.11: Portraits of the master-slave synchronization of the fractional-order Zhang hyper-chaotic oscillator by applying OPCL technique by setting $a = 5.0336401, b = 33.155891$ and $c = 1.0355292$. The synchronization: (a) does not occur, and (b) it is successful.

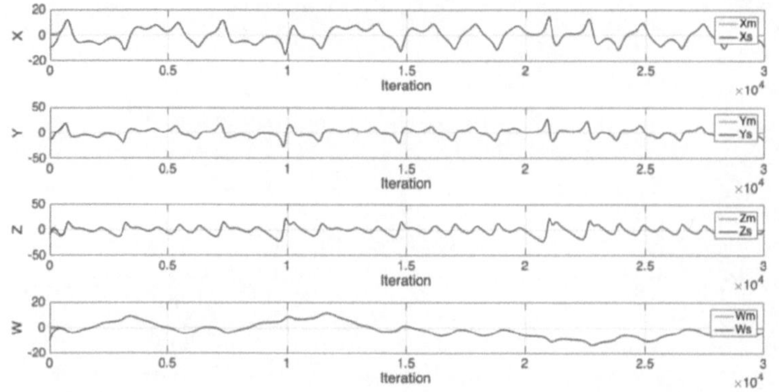

Figure 8.12: Chaotic time series of the master and slave Zhang FOCOs by applying OPCL technique and by setting $a = 5.0336401, b = 33.155891$ and $c = 1.0355292$.

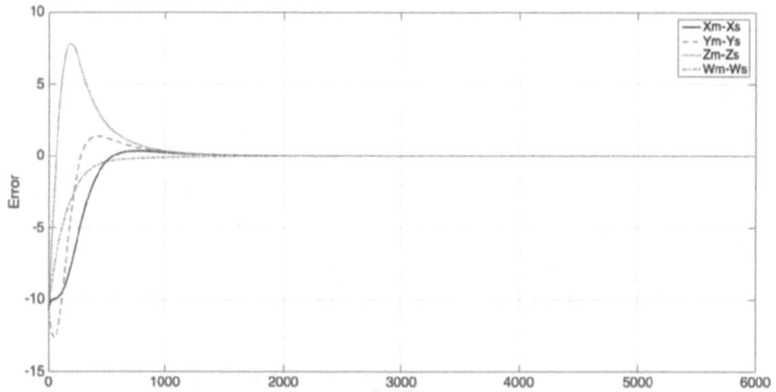

Figure 8.13: Synchronization errors of the fractional-order Zhang hyper-chaotic oscillator by applying OPCL technique by setting $a = 5.0336401, b = 33.155891$ and $c = 1.0355292$.

formats of images, such as: black and white, grey scales and color. Once guaranteeing the synchronization of FOCOs in a master-slave topology, as by applying Hamiltonian forms and OPCL, one can implement different modulation schemes. For instance, Fig. 8.15 shows a masking scheme that consists of a master block (master signal S_M) that transmits an image (original signal S_O), which is contaminated with chaotic data (chaotic signal S_C), and then the slave system (slave signal S_S) recovers the original data. As one can infer, to recover the 100% of data, both the master and the slave FOCOs must be perfectly synchronized, which remains a challenge to guarantee a perfect recovery of data. The channel

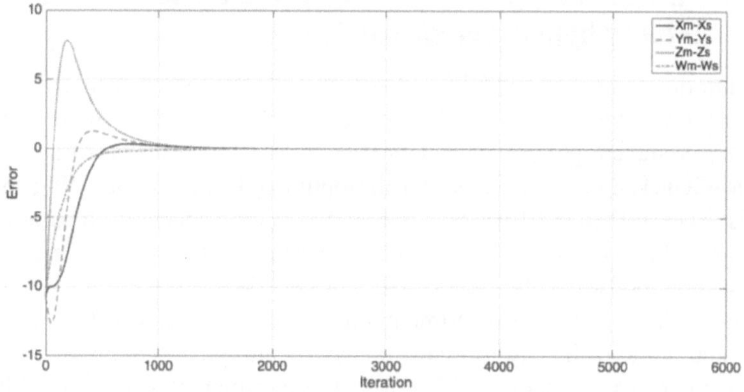

Figure 8.14: Synchronization errors of the fractional-order Zhang hyper-chaotic oscillator by applying OPCL technique by setting $a = 5.4763167, b = 36.575757$ and $c = 1.0999028$.

connects both blocks, the master and the slave and it can be wireless but the synchronization requires a detailed modeling of the protocol like wi-fi and the device where the FOCOs are implemented, which can be embedded devices or analog electronics [4]. The FOCOs can generate the same data when they are perfectly synchronized, otherwise an error can occur and then some data can be lost. The masking operations are quite simple: the transmitter adds chaotic data to the original image, and the receiver subtracts the chaotic data to recover the original image. In this manner, the challenge is the implementation of a synchronized system that generates or mitigates the very small synchronization error.

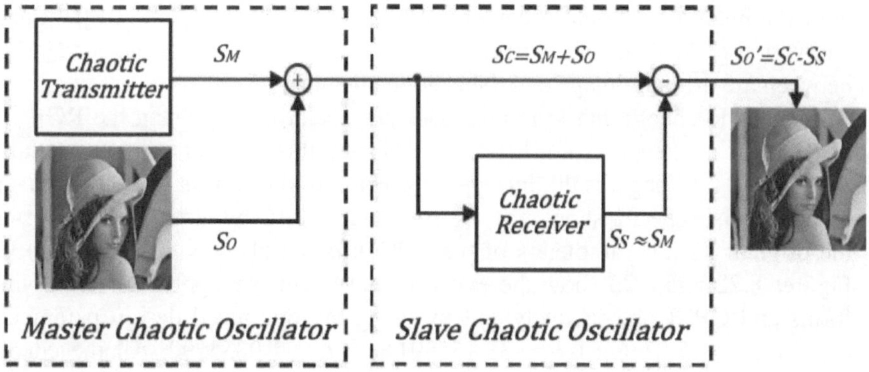

Figure 8.15: Secure communication system based on the synchronization of two fractional-order chaotic oscillator in a master-slave topology.

8.3.1 Image encryption by using hidden fractional-order Zhang hyper-chaotic oscillators

The Hamiltonian forms and OPCL synchronization techniques are applied herein along with two hidden fractional-order Zhang hypechaotic oscillators connected in a master-slave topology. The encryption of grey images is performed by using the original design parameters and the optimized ones by applying the metaheuristics described in Chapters 4, 5, 6 and 7. The original design parameters are set to: $a = 12.0$, $b = 40.0$ and $c = 3.6$. The optimized design parameters are set to: $a = 5.0336401, b = 33.155891$ and $c = 1.0355292$.

Figure 8.16 shows the experimental results of the masking of the cameraman black and white of 256×256-pixels image, by applying Hamiltonian forms and observer approach, and Fig. 8.17 shows the experimental results by applying the OPCL technique, by setting: $a = 12.0$, $b = 40.0$ and $c = 3.6$. Figures 8.18 and 8.19 show the experimental results for the same image, by applying Hamiltonian forms and OPCL techniques, respectively, but now by using the optimized design parameters that are set to: $a = 5.0336401, b = 33.155891$ and $c = 1.0355292$. As one sees, the chaotic masking is performed between the image and each of the four state variables x, y, z and w, to appreciate which is the most suitable to guarantee the best encryption. One can perform different tests as already done in chaotic image encryption schemes [3]. In this chapter only the correlation analysis is performed to appreciate that the state variable x is more adequate to generate the highest randomness. In the figures shown below, the column on the left shows the original grey image, the centered column shows the chaotic channel data, and the column in the right column shows the recovered image. The four rows are associated to the encryption of the original grey image by using the four state variables. The first row is the result of the image encryption by using the state variable x, the second by y, the third by z, and the fourth by w of the hidden fractional-order Zhang hyper-chaotic oscillator. All the image encryption results are summarized in Table 8.1, which includes the correlations between the original image and the chaotic channel.

The same chaotic masking processes are performed by using the RGB image known as Peppers of 512×512-pixels. In this manner, Fig. 8.20 shows the chaotic masking results by applying Hamiltonian forms and observer approach, and Fig. 8.21 shows the experimental results by applying OPCL with the original design parameters of the FOCO: $a = 12.0$, $b = 40.0$ and $c = 3.6$. Figures 8.22 and 8.23 show the experimental results by applying Hamiltonian forms and OPCL, respectively, but by using the optimized design parameters equal to: $a = 5.0336401, b = 33.155891$ and $c = 1.0355292$. All the chaotic masking results are summarized in Table 8.2, which shows the correlations between the original RGB image and the chaotic channel for the four state variables x, y, z and w.

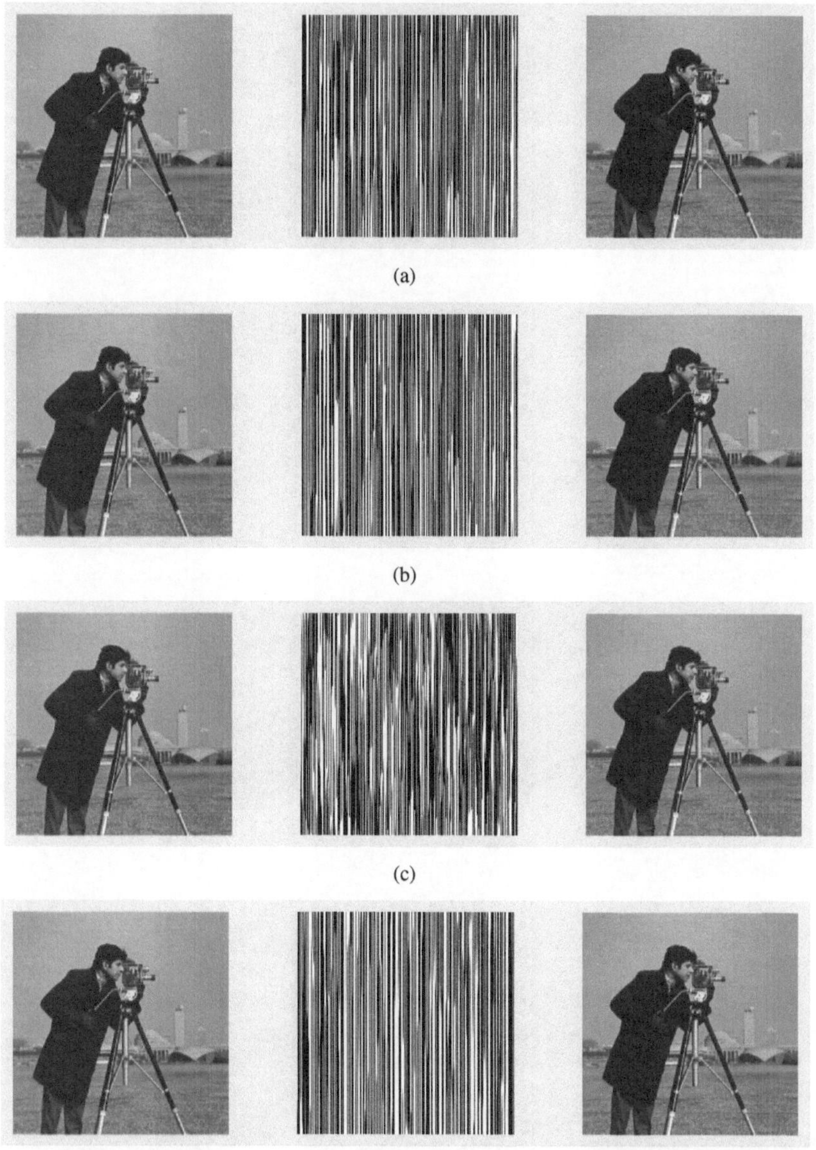

(a)

(b)

(c)

(d)

Figure 8.16: Original (first column), chaotic masking (second column), and recovered (third column) cameraman (256×256-pixels) images when applying Hamiltonian forms for the synchronization of the fractional-order Zhang hyper-chaotic oscillators with $a = 12.0$, $b = 40.0$ and $c = 3.6$, and by using the chaotic time series of the state variable: (a) x, (b) y, (c) z, and (d) w.

(a)

(b)

(c)

(d)

Figure 8.17: Original (first column), chaotic masking (second column), and recovered (third column) cameraman (256×256-pixels) images when applying OPCL for the synchronization of the fractional-order Zhang hyper-chaotic oscillators with $a = 12.0$, $b = 40.0$ and $c = 3.6$, and by using the chaotic time series of the state variable: (a) x, (b) y, (c) z, and (d) w.

Another well-known image that is used to test encryption methods is the Lena RGB of 512×512-pixels, which is masked with chaotic data by applying both

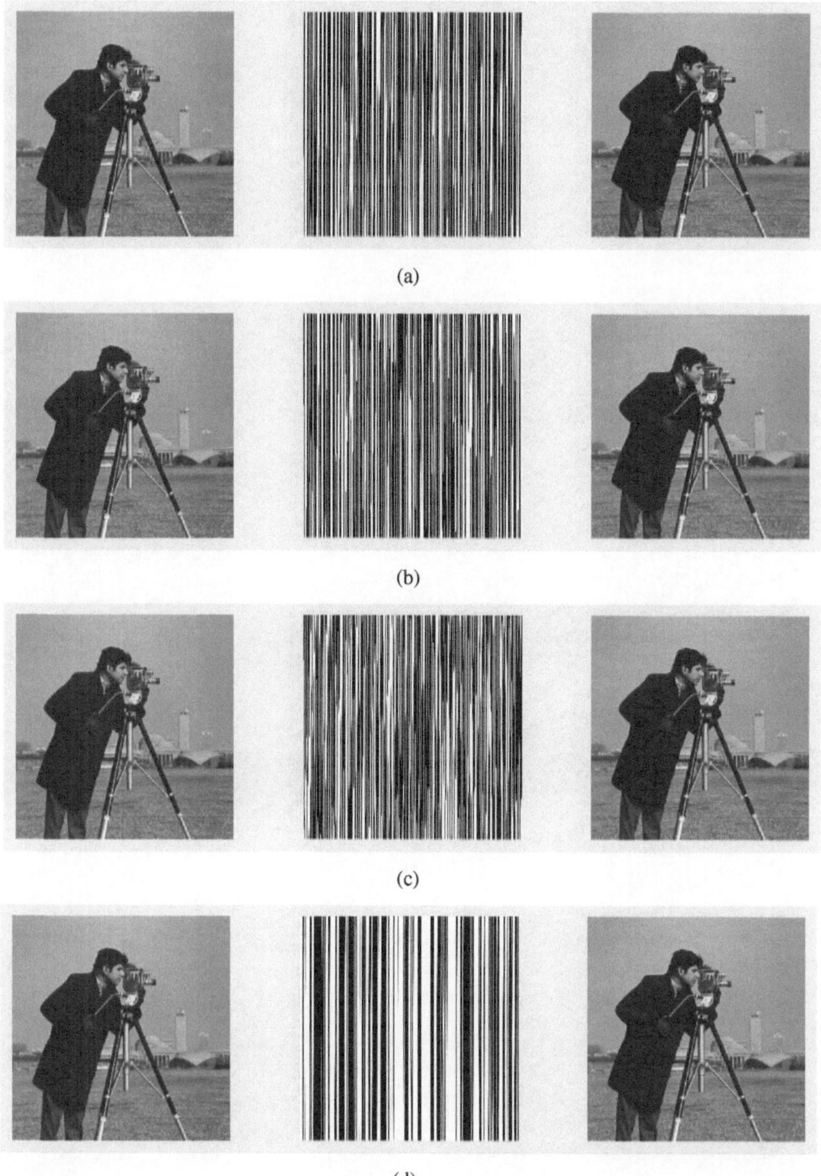

Figure 8.18: Original (first column), chaotic masking (second column), and recovered (third column) cameraman (256×256-pixels) images when applying Hamiltonian forms for the synchronization of the fractional-order Zhang hyper-chaotic oscillators with $a = 5.0336401, b = 33.155891$ and $c = 1.0355292$, and by using the chaotic time series of the state variable: (a) x, (b) y, (c) z, and (d) w.

(a)

(b)

(c)

(d)

Figure 8.19: Original (first column), chaotic masking (second column), and recovered (third column) cameraman (256×256-pixels) images when applying OPCL for the synchronization of the fractional-order Zhang hyper-chaotic oscillators with $a = 5.0336401, b = 33.155891$ and $c = 1.0355292$, and by using the chaotic time series of the state variable: (a) x, (b) y, (c) z, and (d) w.

(a)

(b)

(c)

(d)

Figure 8.20: Original (first column), chaotic masking (second column), and recovered (third column) Peppers (512×512-pixels) RGB images when applying Hamiltonian forms for the synchronization of the fractional-order Zhang hyper-chaotic oscillators with $a = 12.0$, $b = 40.0$ and $c = 3.6$, and by using the chaotic time series of the state variable: (a) x, (b) y, (c) z, and (d) w.

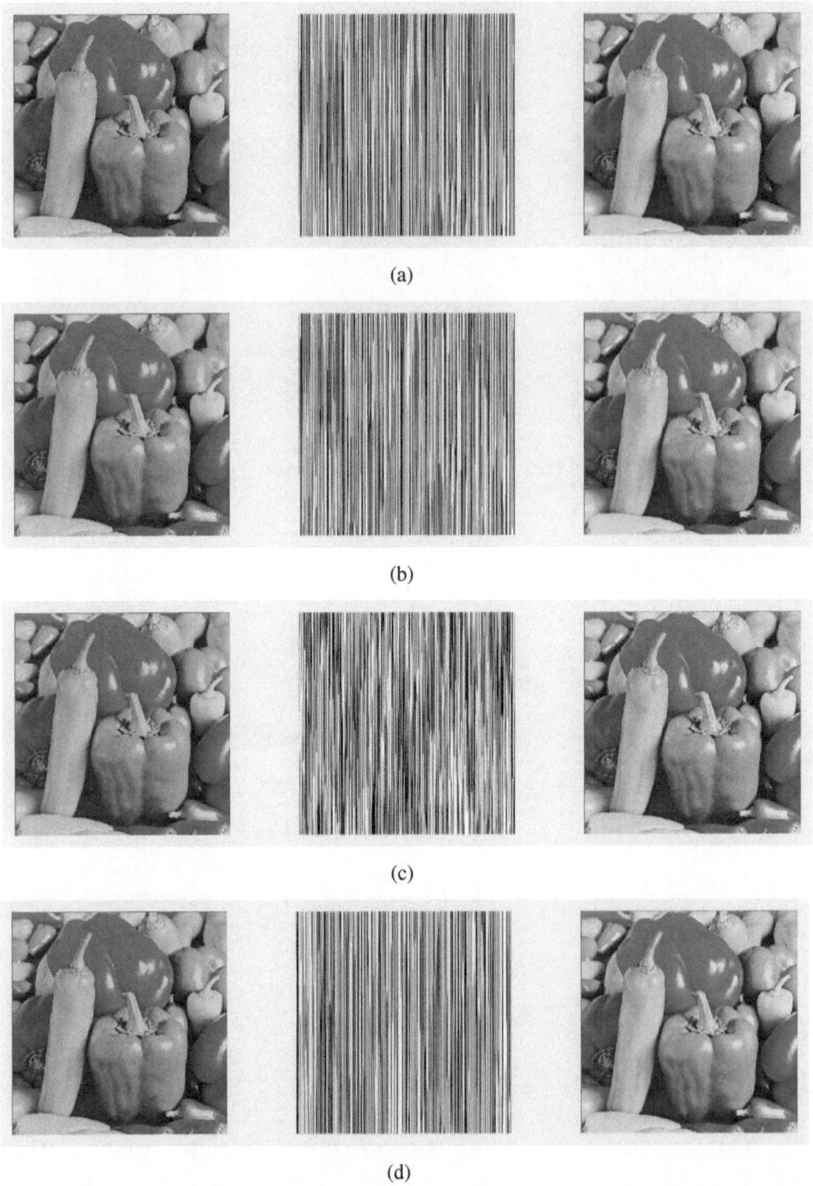

Figure 8.21: Original (first column), chaotic masking (second column), and recovered (third column) Peppers (512×512-pixels) RGB images when applying OPCL for the synchronization of the fractional-order Zhang hyper-chaotic oscillators with $a = 12.0$, $b = 40.0$ and $c = 3.6$, and by using the chaotic time series of the state variable: (a) x, (b) y, (c) z, and (d) w.

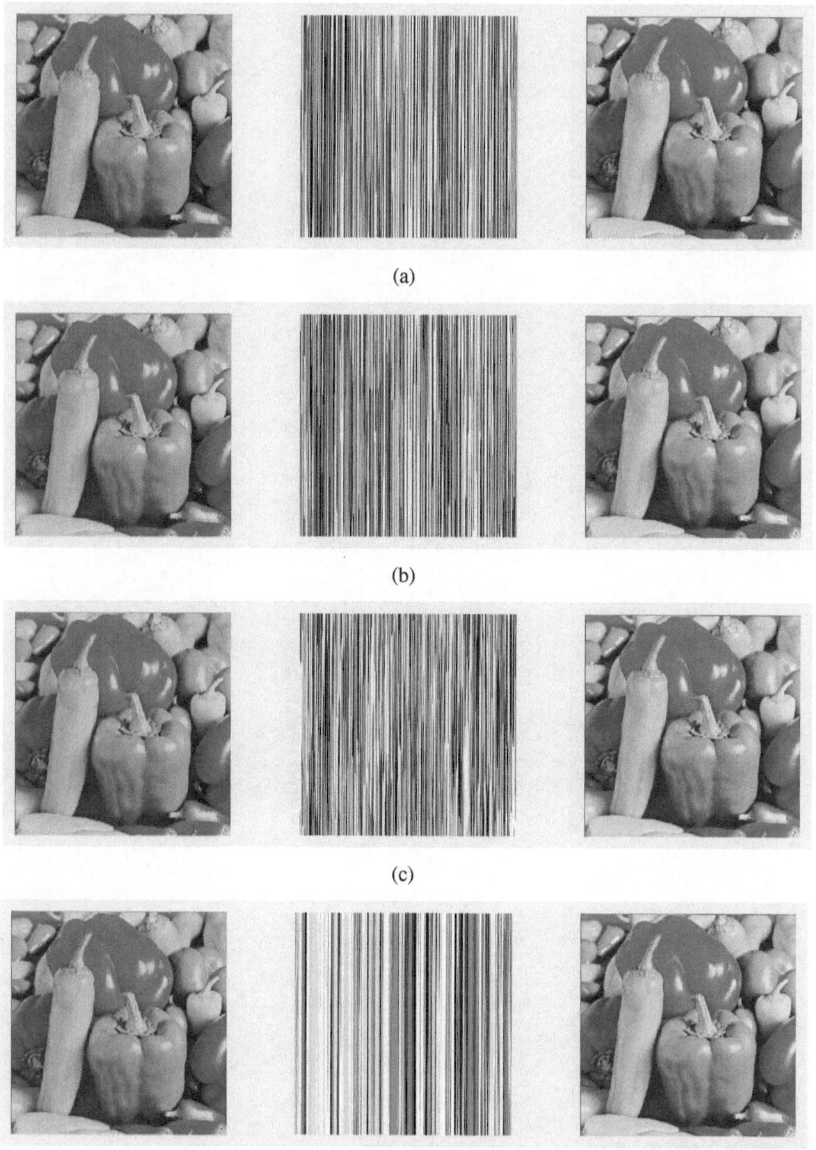

(a)

(b)

(c)

(d)

Figure 8.22: Original (first column), chaotic masking (second column), and recovered (third column) Peppers (512×512-pixels) RGB images when applying Hamiltonian forms for the synchronization of the fractional-order Zhang hyper-chaotic oscillators with $a = 5.0336401, b = 33.155891$ and $c = 1.0355292$, and by using the chaotic time series of the state variable: (a) x, (b) y, (c) z, and (d) w.

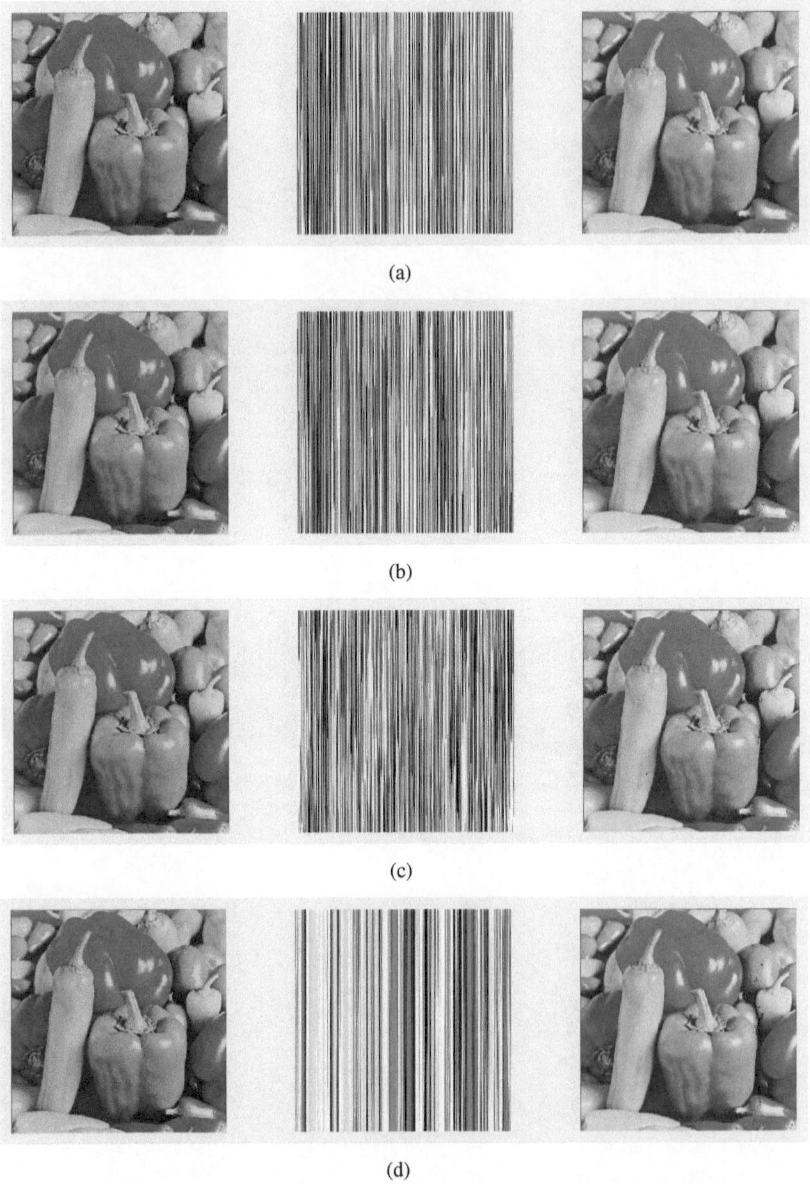

Figure 8.23: Original (first column), chaotic masking (second column), and recovered (third column) Peppers (512×512-pixels) RGB images when applying OPCL for the synchronization of the fractional-order Zhang hyper-chaotic oscillators with $a = 5.0336401$, $b = 33.155891$ and $c = 1.0355292$, and by using the chaotic time series of the state variable: (a) x, (b) y, (c) z, and (d) w.

Table 8.1: Correlations between the original and chaotic masking of Cameraman's image by applying Hamiltonian forms and OPCL techniques for the fractional-order Zhang hyper-chaotic oscillator.

Synchronization Technique	Transmission Variable	Correlation Original Parameters $(S_O+S_C)/S_O$	Correlation Optimized Parameters $(S_O+S_C)/S_O$
Hamiltonian	x	0.0030	0.0001
	y	0.0105	0.0014
	z	0.0045	0.0017
	w	0.0118	0.0018
OPCL	x	0.0015	0.0009
	y	0.0102	0.0048
	z	0.0048	0.0038
	w	0.0140	0.0072

Table 8.2: Correlations between the original and chaotic masking of Peppers' image by applying Hamiltonian forms and OPCL techniques for the fractional-order Zhang hyper-chaotic oscillator.

Synchronization Technique	Transmission Variable	Correlation Original Parameters $(S_O+S_C)/S_O$	Correlation Optimized Parameters $(S_O+S_C)/S_O$
Hamiltonian	x	0.0033	0.0001
	y	0.0201	0.0053
	z	0.0027	0.0014
	w	0.0156	0.0131
OPCL	x	0.0030	0.0002
	y	0.0050	0.0005
	z	0.0203	0.0011
	w	0.0159	0.0127

Hamiltonian forms and OPCL techniques for the master-slave synchronization of the fractional-order Zhang hyper-chaotic oscillators. Figure 8.24 shows the experimental results of the chaotic masking of the RGB Lena image by applying Hamiltonian forms, and Fig. 8.25 by the OPCL technique by setting: $a = 12.0$, $b = 40.0$ and $c = 3.6$. Figure 8.26 and 8.27 shows the experimental results with the Hamiltonian forms and OPCL techniques, respectively, but with the optimized parameters $a = 5.0336401, b = 33.155891$ and $c = 1.0355292$. All the simulation results are summarized in Table 8.3, which shows the correlations between the original image and the chaotic channel.

Table 8.3: Correlations between the original and chaotic masking of Lena's image by applying Hamiltonian forms and OPCL techniques for the fractional-order Zhang hyper-chaotic oscillator.

Synchronization Technique	Transmission Variable	Correlation Original Parameters $(S_O + S_C)/S_O$	Correlation Optimized Parameters $(S_O + S_C)/S_O$
Hamiltonian	x	0.0031	0.0011
	y	0.0049	0.0109
	z	0.0037	0.0017
	w	0.0173	0.0091
OPCL	x	0.0029	0.0005
	y	0.0047	0.0035
	z	0.0035	0.0019
	w	0.0175	0.0107

8.3.2 FPGA-based design of a chaotic secure communication system using FOCOs

According to Chapters 6 and 7, the fractional-order chaotic oscillators can be implemented on reconfigurable and reprogrammable devices such as the FPAA and FPGA. In both cases, the FOCOs can be used to generate binary random sequences and then one can develop applications between two systems, equivalent to the transmission-reception systems. In this manner, one can implement a chaotic secure communication system, as the one shown in Fig. 8.15. The channel can be wired or wireless, and for the case of using serial communication, one can take advantage of the serial protocol interface known as RS-232. If the optimized FOCOs given in Chapters 6 and 7 are implemented on FPGAs, they can be connected as shown in Fig. 8.28 to design a chaotic secure communication system based on FOCOs. It is worthy mentioning that one can also use single-board-computers as the Raspberry-Pi to synthesize the FOCOs and then one can communicate with the transmitter and receiver by bluetooth or wi-fi protocols. Both FOCOs that are in the transmitter and receiver must be synchronized as described in the previous section, or by applying a different technique, and their communication can be extended to network topologies, as the systems described in [200].

(a)

(b)

(c)

(d)

Figure 8.24: Original (first column), chaotic masking (second column), and recovered (third column) Lena (512×512-pixels) RGB images when applying Hamiltonian forms for the synchronization of the fractional-order Zhang hyper-chaotic oscillators with $a = 12.0$, $b = 40.0$ and $c = 3.6$, and by using the chaotic time series of the state variable: (a) x, (b) y, (c) z, and (d) w.

(a)

(b)

(c)

(d)

Figure 8.25: Original (first column), chaotic masking (second column), and recovered (third column) Lena (512×512-pixels) RGB images when applying OPCL for the synchronization of the fractional-order Zhang hyper-chaotic oscillators with $a = 12.0$, $b = 40.0$ and $c = 3.6$, and by using the chaotic time series of the state variable: (a) x, (b) y, (c) z, and (d) w.

(a)

(b)

(c)

(d)

Figure 8.26: Original (first column), chaotic masking (second column), and recovered (third column) Lena (512×512-pixels) RGB images when applying Hamiltonian forms for the synchronization of the fractional-order Zhang hyper-chaotic oscillators with $a = 5.0336401$, $b = 33.155891$ and $c = 1.0355292$, and by using the chaotic time series of the state variable: (a) x, (b) y, (c) z, and (d) w.

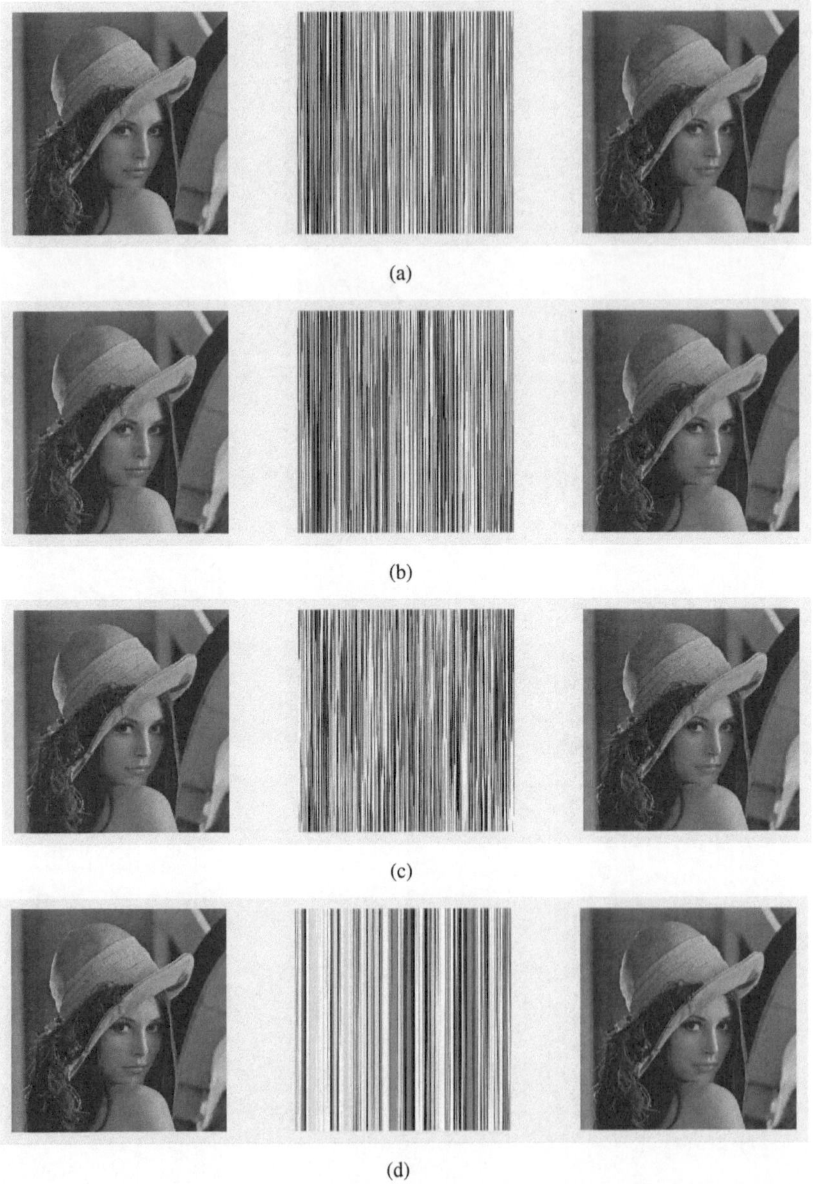

Figure 8.27: Original (first column), chaotic masking (second column), and recovered (third column) Lena (512×512-pixels) RGB images when applying OPCL for the synchronization of the fractional-order Zhang hyper-chaotic oscillators with $a = 5.0336401, b = 33.155891$ and $c = 1.0355292$, and by using the chaotic time series of the state variable: (a) x, (b) y, (c) z, and (d) w.

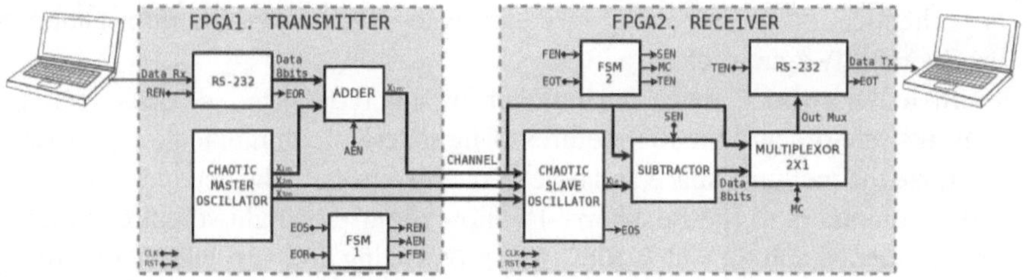

Figure 8.28: Chaotic secure communication system based on the synchronization of two fractional-order chaotic oscillators connected in a master-slave topology.

The chaotic secure communication systems shown in Fig. 8.28 associates two FPGAs that are used to implement the master and slave FOCOs. The transmitter embeds the master FOCO synthesized in the FPGA1, whose finite state machine (FSM1 block) controls the transmission protocol. The signal EOS activates the secure communication system when the master-slave synchronization is successful and the error among the state variables is the lowest one. Then signal REN enables the serial communication through RS-232 protocol in the receiver block. At this time, the transmitter can send information that is ready to be received by the block receiver, and the signal EOR is activated. The FSM enables the adder block in the transmitter to mask the image with the chaotic data generated by the FOCO, the AEN signal is activated to send the encrypted data (original image + chaotic data) to the receiver block. The FOCO in the slave block is embedded in the FPGA2, and is synchronized with the master FOCO in the FPGA1. The slave FOCO is controlled by the finite state machine (FSM2 block), that is enabled with the FEN signal. The encrypted data that is coming to the receiver block in the PFGA2, is recovered by subtracting the chaos when the signal SEN is activated. Finally, the multiplexer block is enabled with the signal MC to transmit the encrypted and recovered data through the RS-Transmitter block controlled by the TEN signal to the personal computer. When the transmission finishes, the signal EOT is activated and the secure communication system is ready to process new data.

The FPGAs are quite useful for fast prototyping and verification, but in the last years, the single-board-computer based on the embedded system well-known as Raspberry-Pi, has shown its usefulness to develop application in the internet of Things (IoT). The single-board-computer is a low-cost system and suitable for lightweight applications, and it can be programmed in different language programs, such as: Python and C/C++. The application of the Raspberry-Pi in the IoT includes different collection of objects, services, humans and devices connections that can communicate among them, share data and information to accomplish a desired task. Some forecasting issues on the IoT applications can be found in [201, 202], and the new applications require transparent services of authentication, guaranteeing of data privacy, security of the information and invulnerability against attacks [3].

In IoT networks, when the number of interconnected devices increases, requires security and the IoT requires of lightweight communication protocols to optimize the medium, space and time of the resources in the network. The single-board-computers as the Raspberry-Pi allow the use of lightweight communication protocols, and they also allow fast prototyping and verification of a design. A very useful and free communication protocol is known as MQTT (Message Queue Telemetry Transport), which includes machine-to-machine (M2M) connections, is of low power consumption and designed as a message transport system recognizing publishers and subscribers, with efficient distribution to one or more receivers. This lightweight communication protocol requires a broker or agent to manage the topics from the publisher/subscriber and to coordinate the communication among the receivers.

Figure 8.29, shows a communication system following the scheme sketched in Fig. 8.15, but now by using the MQTT communication protocol that is easily coupled to the IoT applications. In this protocol one can send and receive different data, such as: text, images, audio and video. Such kind of data can be encrypted by using FOCOs, as described in this chapter, and by synchronizing FOCOs to guarantee security during the transmission of information from the transmitter (publisher) to the receiver (subscriber).

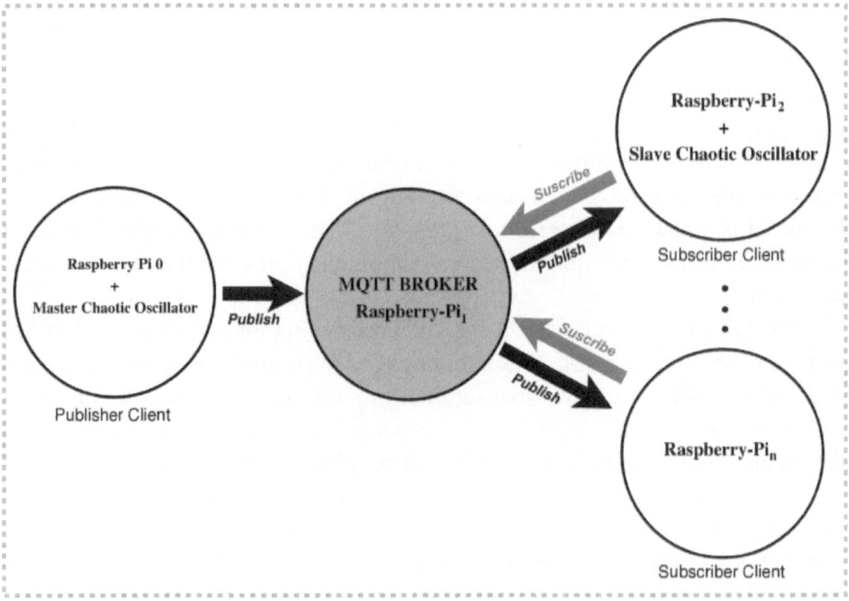

Figure 8.29: MQTT protocol for data transmission and application to the IoT.

8.4 Design of random number generators from the optimized FOCOs

The generation of random numbers or binary random sequences can be performed by using different entropy sources. In the case of chaotic systems, one can explore the selection of the state variables having the highest Lyapunov exponent $LE+$ and Kaplan-Yorke dimension D_{KY}, which can be optimized by applying the metaheuristics described and adapted in Chapters 4, 5, 6 and 7. The randomness of the generated bit strings can be tested by different statistical programs, such as NIST [203], and TESTU01 [204]. If the chaotic signal changes slowly, one can apply a transformation of the bit strings, such as the mod255 [205], otherwise the bit strings will have many logic "1" and many logic "0" so that the data will not be as random as desired.

Thinking on logic "1" and "0", one of the statistical tests to evaluate the randomness is the monobit test, which evaluates if the number of "1s" and "0s" is the same, i.e., fifty-fifty. This can be accomplished when the chaotic signal is passed through a comparator that fixes a threshold to classify if the digital word is associated to a logic "1" (if it is above the umbral) or a logic "0" (if it is equal or below the umbral). Moving the umbral can ensure better results in the monobit test. See for example the case shown in Fig. 8.30, where one sees the chaotic time series associated to the state variable x of the fractional-order Zhang hyper-chaotic oscillator given in (8.11) and generated by using the design parameters $a = 5.0336401, b = 33.155891$ and $c = 1.0355292$. If the threshold is established to 0.0 the monobit test for the generation of 100,000 bit, provides a percentage of 50.51% classified as logic "0", and 49.48% classified as "1". Fixing the threshold to -0.2, provides 49.80% of logic "0" and 50.19% of "1".

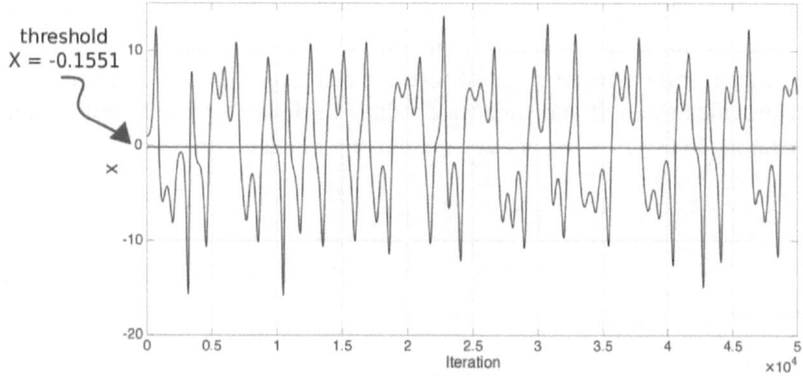

Figure 8.30: Chaotic time series of state variable x of the fractional-order Zhang hyper-chaotic oscillator and a threshold set to -0.1551, to classify the digital word as logic "1" (above this value) or logic "0" (equal or below this value).

Another experiment setting the threshold to -0.1551, provides almost fifty-fifty, so that finding the optimal threshold value is another challenge when using a chaotic system.

Figure 8.31 shows the bit string of 200 bit generated from the state variable x of the FOCO given in (8.11) and by setting the threshold to -0.1551.

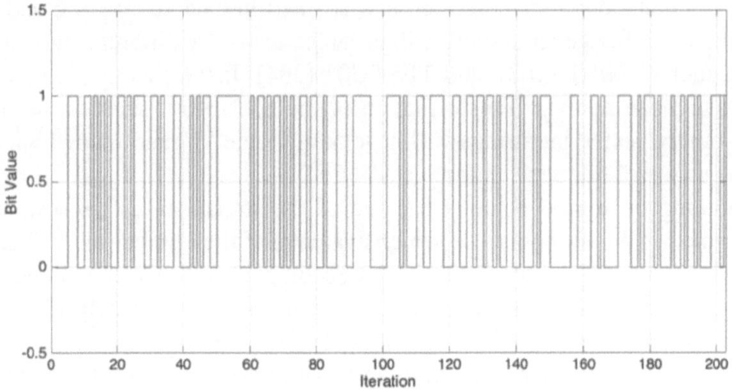

Figure 8.31: Bit string generated from (8.11) by using the optimized design parameters $a = 5.0336401, b = 33.155891$ and $c = 1.0355292$, and by setting the threshold to -0.1551 in the chaotic time series in Fig. 8.30.

The randomness is evaluated by applying different tests as NIST [203] and TESTU01 [204]. They include a set of libraries of statistical tests programmed in ANSI C, and the test are empirical but reasonably good to verify the randomness of random number generators. Some examples of NIST tests by using 100 bit strings of 1,000,000 bit each one, and for different chaotic systems is given in [3]. Recall that nowadays one can find different FOCOs that can be optimized in their dynamical characteristics, so that one can search for the best state variable that provides the best statistical test values to design optimal random number generators. In this manner, one can guarantee a much better encryption of images.

References

[1] Guillén-Fernández, O., A. Meléndez-Cano, E. Tlelo-Cuautle, J. Cruz Núñez-Pérez and J. de Jesus Rangel-Magdaleno. 2019. On the synchronization techniques of chaotic oscillators and their FPGA-based implementation for secure image transmission. PloS One 14(2): e0209618.

[2] Hamiche, H., O. Megherbi, R. Kara, S. Djennoune, R. Saddaoui and M. Laghrouche. 2017. A new implementation of an impulsive synchronisation of two discrete-time hyperchaotic systems using Arduino-Uno boards. International Journal of Modelling, Identification and Control 28(2): 177–186.

[3] Tlelo-Cuautle, E., J.D. Díaz-Muñoz, A.M. González-Zapata, R. Li, W.D. León-Salas, F.V. Fernández, O. Guillén-Fernández and I. Cruz-Vega. 2020. Chaotic image encryption using Hopfield and Hindmarsh–Rose neurons implemented on FPGA. Sensors 20(5): 1326.

[4] Tlelo-Cuautle, E., A. Dalia Pano-Azucena, O. Guillén-Fernández and A. Silva-Juárez. 2020. Analog/Digital Implementation of Fractional Order Chaotic Circuits and Applications. Springer.

[5] Singh, J.P., J. Koley, A. Akgul, B. Gurevin and B. Krishna Roy. 2019. A new chaotic oscillator containing generalised memristor, single op-amp and RLC with chaos suppression and an application for the random number generation. The European Physical Journal Special Topics 228(10): 2233–2245.

[6] Rezk, A.A., A.H. Madian, A.G. Radwan and A.M. Soliman. 2020. Multiplierless chaotic pseudo random number generators. AEU-International Journal of Electronics and Communications 113: 152947.

[7] Vaidyanathan, S., I. Pehlivan, L.G. Dolvis, K. Jacques, M. Alcin, M. Tuna and I. Koyuncu. 2020. A novel ANN-based four-dimensional two-disk

hyperchaotic dynamical system, bifurcation analysis, circuit realisation and FPGA-based TRNG implementation. International Journal of Computer Applications in Technology 62(1): 20–35.

[8] Yildiz, M.Z., O.F. Boyraz, E. Guleryuz, A. Akgul and I. Hussain. 2019. A novel encryption method for dorsal hand vein images on a microcomputer. IEEE Access 7: 60850–60867.

[9] Al-Hazaimeh, O.M., M.F. Al-Jamal, N. Alhindawi and A. Omari. 2017. Image encryption algorithm based on lorenz chaotic map with dynamic secret keys. Neural Computing and Applications, pp. 1–11.

[10] Azzaz, M.S., C. Tanougast, A. Maali and M. Benssalah. 2019. An efficient and lightweight multi-scroll chaos-based hardware solution for protecting fingerprint biometric templates. International Journal of Communication Systems, pp. e4211.

[11] Jin, Q., F. Min and C. Li. 2019. Infinitely many coexisting attractors of a dual memristive Shinriki oscillator and its FPGA digital implementation. Chinese Journal of Physics 62: 342–357.

[12] Sambas, A., S. Vaidyanathan, E. Tlelo-Cuautle, S. Zhang, O. Guillen-Fernandez, Y. Hidayat and G. Gundara 2019. A novel chaotic system with two circles of equilibrium points: Multistability, electronic circuit and FPGA realization. Electronics 8(11): 1211.

[13] Wang, F., R. Wang, H.H.C. Iu, C. Liu and T. Fernando. 2019. A novel multi-shape chaotic attractor and its FPGA implementation. IEEE Transactions on Circuits and Systems II: Express Briefs 66(12): 2062–2066.

[14] Varsakelis, C. and P. Anagnostidis. 2016. On the susceptibility of numerical methods to computational chaos and superstability. Communications in Nonlinear Science and Numerical Simulation 33: 118–132.

[15] Parker, T.S. and L. Chua. 2012. Practical Numerical Algorithms for Chaotic Systems. Springer New York.

[16] Lambert, J.D. 1973. Computational Methods in Ordinary Differential Equations. Wiley, 1973.

[17] Pano-Azucena, A.D., E. Tlelo-Cuautle, G. Rodriguez-Gomez and L.G. De la Fraga. 2018. FPGA-based implementation of chaotic oscillators by applying the numerical method based on trigonometric polynomials. AIP Advances 8(7): 075217.

[18] Runge, C. 1895. über die numerische Auflösung von differentialgleichungen. Mathematische Annalen 46(2): 167–178.

[19] Kutta, W. 1901. Beitrag zur naherungsweisen integration totaler differentialgleichungen. Z. Math. Phys. 46: 435–453.

[20] Petrávs, I. 2011. Fractional-order Nonlinear Systems: Modeling, Analysis and Simulation. Springer Science & Business Media.

[21] Oustaloup, A. 1981. Fractional order sinusoidal oscillators: Optimization and their use in highly linear fm modulation. IEEE Transactions on Circuits and Systems 28(10): 1007–1009, October 1981.

[22] Arenta, A., R. Caponetto, L. Fortuna and D. Porto. 2002. Nonlinear non-integer order circuits and systems. World Scientific Series on Nonlinear Science, Series A, 38.

[23] Ahmad, W.M. and J.C. Sprott. 2003. Chaos in fractional-order autonomous nonlinear systems. Chaos, Solitons & Fractals 16(2): 339–351.

[24] Azar, A.T., A.G. Radwan and S. Vaidyanathan. 2018. Fractional Order Systems: Optimization, Control, Circuit Realizations and Applications. Academic Press.

[25] Rajagopal, K., S. Çiçek, A.J.M. Khalaf, V.-T. Pham, S. Jafari, A. Karthikeyan and P. Duraisamy. 2018. A novel class of chaotic flows with infinite equilibriums and their application in chaos-based communication design using DCSK. Zeitschrift für Naturforschung A 73(7): 609–617.

[26] Volos, C.K., S. Jafari, J. Kengne, J.M. Munoz-Pacheco and K. Rajagopal. 2019. Nonlinear dynamics and entropy of complex systems with hidden and self-excited attractors. Entropy 21(4): 370; DOI: 10.3390/e21040370.

[27] Baleanu, D., J.A.T. Machado and A.C.J. Luo. 2011. Fractional Dynamics and Control. Springer Science & Business Media.

[28] Li, C., X. Liao and J. Yu. 2003. Synchronization of fractional order chaotic systems. Physical Review E 68(6): 067203.

[29] Martínez-Guerra, R. and C.A. Pérez-Pinacho. 2018. Advances in Synchronization of Coupled Fractional Order Systems: Fundamentals and Methods. Springer.

[30] Azar, A.T., S. Vaidyanathan and A. Ouannas. 2017. Fractional Order Control and Synchronization of Chaotic Systems, volume 688. Springer.

[31] Tepljakov, A. 2017. Fractional-order Modeling and Control of Dynamic Systems. Springer.

[32] Pano-Azucena, A.D., E. Tlelo-Cuautle, J.M. Muñoz-Pacheco and L.G. de la Fraga. 2019. FPGA-based implementation of different families of fractional-order chaotic oscillators applying Grünwald–Letnikov method.

Communications in Nonlinear Science and Numerical Simulation 72: 516–527.

[33] Rajagopal, K., S. Jafari, S. Kacar, A. Karthikeyan and A. Akgül. 2019. Fractional order simple chaotic oscillator with saturable reactors and its engineering applications. Information Technology and Control 48(1): 115–128.

[34] Silva-Juarez, A., E. Tlelo-Cuautle, L.G. de la Fraga and R. Li. 2020. FPAA-based implementation of fractional-order chaotic oscillators using first-order active filter blocks. Journal of Advanced Research 1: 1–16.

[35] Oldham, K. and J. Spanier. 1974. The Fractional Calculus Theory and Applications of Differentiation and Integration to Arbitrary Order. Elsevier.

[36] Miller, K.S. and B. Ross. 1993. An Introduction to the Fractional Calculus and Fractional Differential Equations. Wiley-Interscience.

[37] Podlubny, I. 1999. Fractional differential equations: an introduction to fractional derivatives, fractional differential equations, to methods of their solution and some of their applications. Mathematics in Science and Engineering, 198.

[38] Caputo, M. 1967. Linear models of dissipation whose Q is almost frequency independent II. Geophysical Journal International 13(5): 529–539.

[39] Dorcak, L., J. Prokop and I. Kostial. 1994. Investigation of the properties of fractional-order dynamical systems. In Proc. 11th Int. Conf. Process Control, pp. 19–20.

[40] Pan, I. and S. Das. 2012. Intelligent Fractional Order Systems and Control: An Introduction, volume 438. Springer.

[41] Deng, W. and J. Lü. 2007. Generating multi-directional multi-scroll chaotic attractors via a fractional differential hysteresis system. Physics Letters A 369(5-6): 438–443.

[42] Ford, N.J. and A.C. Simpson. 2001. The numerical solution of fractional differential equations: speed versus accuracy. Numerical Algorithms 26(4): 333–346.

[43] Chen, Y., I. Petras and D. Xue. 2009. Fractional order control—A tutorial. In 2009 American Control Conference, pp. 1397–1411, June 2009.

[44] Podlubny, I. 1998. Fractional Differential Equations: An Introduction to Fractional Derivatives, Fractional Differential Equations, to Methods of their Solution and some of their Applications, volume 198. Elsevier.

[45] Cafagna, D. and G. Grassi. 2012. On the simplest fractional-order memristor-based chaotic system. Nonlinear Dynamics 70(2): 1185–1197, October 2012.

[46] Garrappa, R. 2014. Short tutorial: Solving fractional differential equations by Matlab codes. Department of Mathematics University of Bari, Italy.

[47] Danca, M.F. and N. Kuznetsov. 2018. Matlab code for Lyapunov exponents of fractional-order systems. International Journal of Bifurcation and Chaos 28(05): 1850067.

[48] Diethelm, K., N.J. Ford and A.D. Freed. 2002. A predictor-corrector approach for the numerical solution of fractional differential equations. Nonlinear Dynamics 29(1-4): 3–22.

[49] Muñoz-Pacheco, J.M., E. Zambrano-Serrano, O. Félix-Beltrán, L.C. Gómez-Pavón and A. Luis-Ramos. 2012. Synchronization of PWL function-based 2D and 3D multi-scroll chaotic systems. Nonlinear Dynamics 70(2): 1633–1643.

[50] Diethelm, K. and A.D. Freed. 1998. The FracPECE subroutine for the numerical solution of differential equations of fractional order. Forschung und wissenschaftliches Rechnen 1999: 57–71.

[51] Garrappa, R. 2010. On linear stability of predictor–corrector algorithms for fractional differential equations. International Journal of Computer Mathematics 87(10): 2281–2290.

[52] Lorenz, E.N. 1963. Deterministic nonperiodic flow. Journal of the Atmospheric Sciences 20(2): 130–141.

[53] Tavazoei, M.S. and M. Haeri. 2007. A necessary condition for double scroll attractor existence in fractional-order systems. Physics Letters A 367(1-2): 102–113.

[54] Rössler, O.E. 1976. An equation for continuous chaos. Physics Letters A 57(5): 397–398.

[55] Chen, G. and T. Ueta. 1999. Yet another chaotic attractor. International Journal of Bifurcation and Chaos 9(07): 1465–1466.

[56] Lü, J., G. Chen and S. Zhang. 202. Dynamical analysis of a new chaotic attractor. International Journal of Bifurcation and Chaos 12(05): 1001–1015.

[57] Liu, C., T. Liu, L. Liu and K. Liu. 2004. A new chaotic attractor. Chaos, Solitons & Fractals 22(5): 1031–1038.

[58] Sprott, J.C. 1994. Some simple chaotic flows. Physical Review E 50(2): R647.

[59] Wu, R. and C. Wang. 2016. A new simple chaotic circuit based on memristor. International Journal of Bifurcation and Chaos 26(09): 1650145.

[60] Matsumoto, T. 1984. A chaotic attractor from Chua's circuit. IEEE Transactions on Circuits and Systems 31(12): 1055–1058.

[61] Lü, J. and G. Chen. 2006. Generating multiscroll chaotic attractors: Theories, methods and applications. International Journal of Bifurcation and Chaos 16(04): 775–858.

[62] Li, C., J.C. Sprott, W. Thio and H. Zhu. 2014. A new piecewise linear hyperchaotic circuit. IEEE Transactions on Circuits and Systems II: Express Briefs 61(12): 977–981.

[63] Volos, C., J.-O. Maaita, S. Vaidyanathan, V.-T. Pham, I. Stouboulos and I. Kyprianidis. 2016. A novel four-dimensional hyperchaotic four-wing system with a saddle–focus equilibrium. IEEE Transactions on Circuits and Systems II: Express Briefs 64(3): 339–343.

[64] Karakaya, B., A. Gülten and M. Frasca. 2019. A true random bit generator based on a memristive chaotic circuit: Analysis, design and FPGA implementation. Chaos, Solitons & Fractals 119: 143–149.

[65] Prakash, P., K. Rajagopal, I. Koyuncu, J.P. Singh, M. Alcin, B.K. Roy and M. Tuna. 2020. A novel simple 4-D hyperchaotic system with a saddle-point index-2 equilibrium point and multistability: Design and FPGA-based applications. Circuits, Systems, and Signal Processing, pp. 1–22.

[66] Yu, F., L. Liu, B. He, Y. Huang, C. Shi, S. Cai, Y. Song, S. Du and Q. Wan. 2019. Analysis and FPGA realization of a novel 5D hyperchaotic four-wing memristive system, active control synchronization, and secure communication application. Complexity.

[67] Leonov, G.A., N.V. Kuznetsov, O.A. Kuznetsova, S.M. Seledzhi and V.I. Vagaitsev. 2011. Hidden oscillations in dynamical systems. Trans. Syst. Contr. 6: 54–67.

[68] Carbajal-Gómez, V.H., E. Tlelo-Cuautle and F.V. Fernández. 2013. Optimizing the positive Lyapunov exponent in multi-scroll chaotic oscillators with differential evolution algorithm. Applied Mathematics and Computation 219(15): 8163–8168.

[69] Trejo-Guerra, R., E. Tlelo-Cuautle, V.H. Carbajal-Gómez and G. Rodriguez-Gomez. 2013. A survey on the integrated design of chaotic oscillators. Applied Mathematics and Computation 219(10): 5113–5122.

[70] Carbajal-Gomez, V.H., E. Tlelo-Cuautle, J.M. Muñoz-Pacheco, L.G. de la Fraga, C. Sanchez-Lopez and F.V. Fernandez-Fernandez. 2019. Optimization and CMOS design of chaotic oscillators robust to PVT variations. Integration 65: 32–42.

[71] Garcia-Ortega, J.-M., E. Tlelo-Cuautle and C. Sanchez-Lopez. 2007. Design of current-mode gm-c filters from the transformation of opamp-rc filters. Journal of Applied Sciences 7(9): 1321–1326.

[72] Muñiz-Montero, C., L.V. García-Jiménez, L.A. Sánchez-Gaspariano, C. Sánchez-López, V.R. González-Díaz and E. Tlelo-Cuautle. 2017. New alternatives for analog implementation of fractional-order integrators, differentiators and pid controllers based on integer-order integrators. Nonlinear Dynamics 90(1): 241–256.

[73] Tahir, F.R., R. Ali and L. Fortuna. 2014. Analog programmable electronic circuit-based chaotic Lorenz system. Basrah Journal for Engineering Science 14(1): 39–47.

[74] Anadigm. 2014. Dynamically Reconfigurable dpASP, 3rd Generation, AN231E04 Datasheet Rev 1.2 www.anadigm.com.

[75] Anadigm designer 2 user manual. 2014. www.anadigm.com.

[76] Charef, A., H.H. Sun, Y.Y. Tsao and B. Onaral. 1992. Fractal system as represented by singularity function. IEEE Transactions on Automatic Control 37(9): 1465–1470.

[77] Chen, L., W. Pan, R. Wu, K. Wang and Y. He. 2016. Generation and circuit implementation of fractional-order multi-scroll attractors. Chaos, Solitons & Fractals 85: 22–31.

[78] Wang, S.-P., S.-K. Lao, H.-K. Chen, J.-H. Chen and S.-Y. Chen. 2013. Implementation of the fractional-order Chen-Lee system by electronic circuit. International Journal of Bifurcation and Chaos 23(02): 1350030.

[79] Li, C. and G. Chen. 2004. Chaos and hyperchaos in the fractional-order Rössler equations. Physica A: Statistical Mechanics and its Applications 341: 55–61.

[80] Yu, Y. and H.-X. Li. 2008. The synchronization of fractional-order Rössler hyperchaotic systems. Physica A: Statistical Mechanics and its Applications 387(5-6): 1393–1403.

[81] Grigorenko, I. and E. Grigorenko. 2003. Chaotic dynamics of the fractional Lorenz system. Physical Review Letters 91(3): 034101.

[82] Djouambi, A., A. Charef and A. BesançOn. 2007. Optimal approximation, simulation and analog realization of the fundamental fractional order

transfer function. International Journal of Applied Mathematics and Computer Science 17(4): 455–462.

[83] Xiang-Rong, C., L. Chong-Xin and W. Fa-Qiang. 2008. Circuit realization of the fractional-order unified chaotic system. Chinese Physics B 17(5): 1664.

[84] Jia, H.Y., Q. Tao and Z.Q. Chen. 2014. Analysis and circuit design of a fractional-order Lorenz system with different fractional orders. Systems Science & Control Engineering: An Open Access Journal 2(1): 745–750.

[85] Li, C. and J. Yan. 2007. The synchronization of three fractional differential systems. Chaos, Solitons & Fractals 32(2): 751–757.

[86] Jia, H.Y., Q. Tao and Z.Q. Chen. 2014. Analysis and circuit design of a fractional-order Lorenz system with different fractional orders. Systems Science & Control Engineering: An Open Access Journal 2(1): 745–750.

[87] Vinagre, B.M., I. Podlubny, A. Hernandez and V. Feliu. 2000. Some approximations of fractional order operators used in control theory and applications. Fractional Calculus and Applied Analysis 3(3): 231–248.

[88] Oustaloup, A., F. Levron, B. Mathieu and F.M. Nanot. 2000. Frequency-band complex noninteger differentiator: characterization and synthesis. IEEE Transactions on Circuits and Systems I: Fundamental Theory and Applications 47(1): 25–39.

[89] Carlson, G. and C. Halijak. 1964. Approximation of fractional capacitors (1/s)^(1/n) by a regular newton process. IEEE Transactions on Circuit Theory 11(2): 210–213.

[90] Matsuda, K. and H. Fujii. 1993. H (infinity) optimized wave-absorbing control-analytical and experimental results. Journal of Guidance, Control, and Dynamics 16(6): 1146–1153.

[91] Krishna, B.T. 2011. Studies on fractional order differentiators and integrators: A survey. Signal Processing 91(3): 386–426.

[92] Krishna, B.T. and K.V.V.S. Reddy. 2008. Active and passive realization of fractance device of order 1/2. Active and Passive Electronic Components, 2008.

[93] Jia, H.Y., Q. Tao and Z.Q. Chen. 2014. Analysis and circuit design of a fractional-order Lorenz system with different fractional orders. Systems Science & Control Engineering: An Open Access Journal 2(1): 745–750.

[94] Tacha, O.I., J.M. Munoz-Pacheco, E. Zambrano-Serrano, I.N. Stouboulos and V.-T. Pham. 2018. Determining the chaotic behavior in a fractional-

order finance system with negative parameters. Nonlinear Dynamics 94(2): 1303–1317.

[95] Liu, L. and C. Liu. 2014. Theoretical analysis and circuit verification for fractional-order chaotic behavior in a new hyperchaotic system. Mathematical Problems in Engineering, 2014.

[96] Dorvcák, L., J. Terpák, I. Petrávs and F. Dorvcáková. 2007. Electronic realization of the fractional-order systems. Acta Montanistica Slovaca 12(3): 231–237.

[97] Ruo-Xun, Z. and Y. Shi-Ping. 2009. Chaos in fractional-order generalized Lorenz system and its synchronization circuit simulation. Chinese Physics B 18(8): 3295.

[98] Chen, D., C. Liu, C. Wu, Y. Liu, X. Ma and Y. You. 2012. A new fractional-order chaotic system and its synchronization with circuit simulation. Circuits, Systems, and Signal Processing 31(5): 1599–1613.

[99] Van Valkenburg, M.E. 1982. Analog Filter Design. Holt, Rinehart and Winston.

[100] Lu, J., G. Chen, X. Yu and H. Leung. 2004. Design and analysis of multiscroll chaotic attractors from saturated function series. IEEE Transactions on Circuits and Systems I: Regular Papers 51(12): 2476–2490.

[101] Lu, J.G. and G. Chen. 2006. A note on the fractional-order Chen system. Chaos, Solitons & Fractals 27(3): 685–688.

[102] Pandey, A., R.K. Baghel and R.P. Singh. 2012. Analysis and circuit realization of a new autonomous chaotic system. International Journal of Electronics and Communication Engineering 5(4): 487–495.

[103] Rajagopal, K., S. Cicek, A. Akgul, S. Jafari and A. Karthikeyan. 2019. Chaotic cuttlesh: king of camouage with self-excited and hidden flows, its fractional-order form and communication designs with fractional form. Discrete & Continuous Dynamical Systems-B 22(11): 0.

[104] Ruan, J., K. Sun, J. Mou, S. He and L. Zhang. 2018. Fractional-order simplest memristor-based chaotic circuit with new derivative. The European Physical Journal Plus 133(1): 3.

[105] Qi, G., G. Chen, S. Du, Z. Chen and Z. Yuan. 2005. Analysis of a new chaotic system. Physica A: Statistical Mechanics and its Applications 352(2-4): 295–308.

[106] Niu, Y., X. Sun, C. Zhang and H. Liu. 2020. Anticontrol of a fractional-order chaotic system and its application in color image encryption. Mathematical Problems in Engineering, 2020.

[107] Zhou, C., Z. Li and F. Xie. 2019. Coexisting attractors, crisis route to chaos in a novel 4D fractional-order system and variable-order circuit implementation. The European Physical Journal Plus 134(2): 73.

[108] Dadras, S., H.R. Momeni, G. Qi and Z.-L. Wang. 2012. Four-wing hyperchaotic attractor generated from a new 4D system with one equilibrium and its fractional-order form. Nonlinear Dynamics 67(2): 1161–1173.

[109] Pham, V.-T., S. Takougang Kingni, C. Volos, S. Jafari and T. Kapitaniak. 2017. A simple three-dimensional fractional-order chaotic system without equilibrium: Dynamics, circuitry implementation, chaos control and synchronization. AEU-international Journal of Electronics and Communications 78: 220–227.

[110] Li, X. and Z. Li. 2019. Hidden extreme multistability generated from a fractional-order chaotic system. Indian Journal of Physics 93(12): 1601–1610.

[111] Zhang, S., Y. Zeng, Z. Li and C. Zhou. 2018. Hidden extreme multistability, antimonotonicity and offset boosting control in a novel fractional-order hyperchaotic system without equilibrium. International Journal of Bifurcation and Chaos 28(13): 1850167.

[112] Barzilai, J. and J.M. Borwein. 1988. Two-point step size gradient methods. IMA Journal of Numerical Analysis 8: 141–148.

[113] Golberg, D.E. and K. Deb. 1991. Foundations of Genetic Algorithms. Chapter A comparison of selection schemes used in genetic algorithms. pp. 69–93. Morgan Kaufmann, San Mateo, California.

[114] De Jong, A.K. 1975. An Analysis of the Behavior of a Class of Genetic Adaptive Systems. PhD thesis, University of Michigan, Ann Arbor, Michigan, USA.

[115] Booker, L.B. 1982. Intelligent Behavior as an Adaptation to the Task Environment. PhD thesis, University of Michigan, AnArbor, Michigan, USA. Logic of Computers Group.

[116] Brindle, A. 1981. Genetic Algorithms for Function Optimization. PhD thesis, University of Alberta, Edmonton, Alberta, Canada. Department of Computer Science.

[117] Baker, J.E. 1987. Reducing bias and inefficiency in the selection algorithm. pp. 14–22. *In*: Grefenstette, J.J. (ed.). Proceedings of the Second International Conference on Genetic Algorithms, July 1987.

[118] Grefenstette, J.J. and J.E. Baker. 1989. How genetic algorithms work: A critical look at implicit parallelism. pp. 20–27. *In*: Schaffer, J.D. (ed.).

Proceedings of the Third International Conference on Genetic Algorithms, pages 20–27, June 1989.

[119] Baker, J.E. 1985. Adaptive selection methods for genetic algorithms. pp. 101–111. *In*: Grefenstette, J.J. (ed.). Proceedings of the First International Conference on Genetic Algorithms.

[120] Syswerda. G. 1989. Uniform crossover in genetic algorithms. pp. 2–9. *In*: Schaffer, J.D. (ed.). Proceedings of the Third International Conference on Genetic Algorithms.

[121] Mitchell, M. 1996. An Introduction to Genetic Algorithms. The MIT Press, Cambridge, Massachusetts.

[122] Michalewicz, Z. 1996. Genetic Algorithms + Data Structures = Evolution Programs. Springer-Verlag, New York, USA, 3rd edition.

[123] Storn, R. and K. Price. 1997. Differential evolution-a simple and efficient heuristic for global optimization over continuous spaces. Journal of Global Optimization 11(4): 341–359, December 1997.

[124] Mezura, E., J. Velázquez, and C.A. Coello. 2006. A comparative study of differential evolution variants for global optimization. In GECCO '06, pp. 485–492, New York, NY, USA, 2006. ACM Press.

[125] Chakraborty, U.K. 2008. Advances in Differential Evolution. Studies in Computational Intelligence. Springer.

[126] Zielinski, K. and R. Laur. 2008. Stopping criteria for differential evolution in constrained single-objective optimization. In Advances in Differential Evolution. Springer.

[127] Hansen, N. 2006. Comparisons results among the accepted papers to the special session on real-parameter optimization at CEC-05, 2006. http://www.ntu.edu.sg/home/epnsugan/index_files/CEC-05/compareresults.pdf.

[128] Kennedy, J. and R.C. Eberhart. 1995. Particle swarm optimization. In Proceedings of the IEEE international conference on neural networks, pp. 1942–1948.

[129] Wang, D., D. Tan and L. Liu. 2018. Particle swarm optimization algorithm: an overview. Soft Computing 2(22): 387–408.

[130] Pederson, M.E.H. and A.J. Chipperfield. 2010. Simplifying particle swarm optimization. Applied Soft Computing 10(2): 618–628.

[131] Tanabe, R. and A. Fukunaga. 2013. Success-history based parameter adaptation for differential evolution. In 2013 IEEE Congress on Evolutionary Computation, pp. 71–78.

[132] Zhang, J. and A.C. Sanderson. 2009. JADE: Adaptive differential evolution with optional external archive. IEEE Transactions on Evolutionary Computation 13(5): 945–958.

[133] Tanabe, R. and A.S. Fukunaga. 2014. Improving the search performance of SHADE using linear population size reduction. In 2014 IEEE Congress on Evolutionary Computation (CEC), pp. 1658–1665.

[134] Tanabe, R. 2020. LSHADE source code. https://ryojitanabe.github.io/publication.

[135] Silva-Juárez, A., C.J. Morales-Pérez, L.G. de la Fraga, E. Tlelo-Cuautle and J. de Jesús Rangel-Magdaleno. 2019. On maximizing the positive Lyapunov exponent of chaotic oscillators applying DE and PSO. International Journal of Dynamics and Control 7(4): 1157–1172.

[136] Silva-Juarez, A., G. Rodriguez-Gomez, L.G. de la Fraga, O. Guillen-Fernandez and E. Tlelo-Cuautle. 2019. Optimizing the Kaplan-Yorke dimension of chaotic oscillators applying DE and PSO. Technologies 7(2): 38.

[137] Carbajal-Gómez, V.H., E. Tlelo-Cuautle, F.V. Fernández, L.G. de la Fraga and C. Sánchez-López. 2014. Maximizing Lyapunov exponents in a chaotic oscillator by applying differential evolution. International Journal of Nonlinear Sciences and Numerical Simulation 15(1): 11–17.

[138] de la Fraga, L.G., E. Tlelo-Cuautle, V.H. Carbajal-Gómez and J.M. Munoz-Pacheco. 2012. On maximizing positive Lyapunov exponents in a chaotic oscillator with heuristics. Revista mexicana de física 58(3): 274–281.

[139] Tlelo-Cuautle, E., L.G. de la Fraga, V.-T. Pham, C. Volos, S. Jafari and A. de Jesus Quintas-Valles. 2017. Dynamics, FPGA realization and application of a chaotic system with an infinite number of equilibrium points. Nonlinear Dynamics 89(2): 1129–1139, July 2017.

[140] Wolf, A., J.B. Swift, H.L. Swinney and J.A. Vastano. 1985. Determining Lyapunov exponents from a time series. Physica D: Nonlinear Phenomena 16(3): 285–317.

[141] Hernández-Gómez, R. and C.A. Coello-Coello. 2013. MOMBI: A new metaheuristic for many-objective optimization based on the R2 indicator. In 2013 IEEE Congress on Evolutionary Computation, pp. 2488–2495.

[142] Hernández-Gómez, R. 2013. A new multi-objective evolutionary algorithm based on the R2 indicator. Master's thesis, Cinvestav, Computer Science Deparment. https://www.cs.cinvestav.mx/TesisGraduados/2013/TesisRaquelHernandez.pdf.

[143] Deb, K., A. Pratap, S. Agarwal and T. Meyarivan. 2002. A fast and elitist multiobjective genetic algorithm: NSGA-II. IEEE Transactions on Evolutionary Computation 6(2): 182–197.

[144] Zhang, Q. and H. Li. 2007. MOEA/D: A multiobjective evolutionary algorithm based on decomposition. IEEE Trans. on Evol. Comp. 11(6): 712–731.

[145] Deb, K. and R.B. Agrawal. 1995. Simulated binary crossover for continuous search space. Complex Syst. 9: 115–148, April 1995.

[146] Deb, K. 2020. NSGA-II source code. http://www.iitk.ac.in/kangal/codes.shtml.

[147] de la Fraga, L.G. 2020. Book code. http://cs.cinvestav.mx/~fraga/OptCode.tar.gz.

[148] Zhang, Q., W. Liu and H. Li. 2009. The performance of a new version of MOEA/D on CEC09 unconstrained MOP test instances. In 2009 IEEE Congress on Evolutionary Computation, pp. 203–208.

[149] Li, H. and Q. Zhang. 2009. Multiobjective optimization problems with complicated pareto sets, MOEA/D and NSGA-II. IEEE Transactions on Evolutionary Computation 13(2): 284–302, April 2009.

[150] Li, K., S. Kwong, J. Cao, M. Li, J. Zheng and R. Shen. 2012. Achieving balance between proximity and diversity in multi-objective evolutionary algorithm. Information Sciences 182(1): 220–242, January 2012.

[151] Chasalow, S.D. and R.J. Brand. 1995. Algorithm AS 299: Generation of simplex lattice points. Journal of the Royal Statistical Society. Series C (Applied Statistics) 44(4): 534–545.

[152] Brockhoff, D., T. Wagner and H. Trautmann. 2012. On the properties of the R2 indicator. In Proceedings of the 14th Annual Conference on Genetic and Evolutionary Computation, GECCO 12, pp. 465–472, New York, NY, USA. Association for Computing Machinery.

[153] de la Fraga, L.G. and E. Tlelo-Cuautle. 2015. Optimizing an amplifier by a many objective algorithm based on R2 indicator. In 2015 IEEE International Symposium on Circuits and Systems (ISCAS), pp. 265–268.

[154] Hernández Gómez, R. 2018. Parallel Hyper-Heuristics for Multi-Objective Optimization. PhD thesis, Cinvestav, Computer Science Deparment. https://www.cs.cinvestav.mx/TesisGraduados/2018/TesisRaquelHernandez.pdf.

[155] Hernández Gómez, R. EMO project. http://computacion.cs.cinvestav.mx/~rhernandez/.

[156] Ishibuchi, H., H. Masuda, Y. Tanigaki and Y. Nojima. 2015. Modified distance calculation in generational distance and inverted generational

distance. *In*: Gaspar-Cunha, A., C. Henggeler Antunes and C. Coello (eds.). Evolutionary Multi-Criterion Optimization. EMO 2015. Lecture Notes in Computer Science, vol 9019. Springer, Cham.

[157] Osyczka, A. and S. Kundu. 1995. A new method to solve generalized multicriteria optimization problems using the simple genetic algorithm. Structural Optimization 10(2): 94–99.

[158] Samko, S.G., A.A. Kilbas, O.I. Marichev. 1993. Fractional integrals and derivatives, volume 1. Gordon and Breach Science Publishers, Yverdon Yverdon-les-Bains, Switzerland.

[159] Magin, R.L. 2006. Fractional calculus in bioengineering, volume 2. Begell House Redding.

[160] Grigorenko, I. and E. Grigorenko. 2006. Erratum: Chaotic dynamics of the fractional Lorenz system [phys. rev. lett. 91, 034101 (2003)]. Physical Review Letters 96(19): 199902.

[161] Diethelm, K., N.J. Ford and A.D. Freed. 2002. A predictor-corrector approach for the numerical solution of fractional differential equations. Nonlinear Dynamics 29(1-4): 3–22.

[162] Diethelm, K. 2010. The Analysis of Fractional Differential Equations: An Application-oriented Exposition using Differential Operators of Caputo Type. Springer Science & Business Media.

[163] Ahmed, E., A.M.A. El-Sayed and H.A.A. El-Saka. 2006. On some Routh–Hurwitz conditions for fractional order differential equations and their applications in Lorenz, Rössler, Chua and Chen systems. Physics Letters A 358(1): 1–4.

[164] Deng, W., C. Li and J. Lü. 2007. Stability analysis of linear fractional differential system with multiple time delays. Nonlinear Dynamics 48(4): 409–416.

[165] Tavazoei, M.S. and M. Haeri. 2008. Chaotic attractors in incommensurate fractional order systems. Physica D: Nonlinear Phenomena 237(20): 2628–2637.

[166] Tavazoei, M.S. and M. Haeri. 2009. A note on the stability of fractional order systems. Mathematics and Computers in Simulation 79(5): 1566–1576.

[167] Chlouverakis, K.E. and J.C. Sprott. 2005. A comparison of correlation and Lyapunov dimensions. Physica D: Nonlinear Phenomena 200(1-2): 156–164.

[168] Omidinasab, F. and V. Goodarzimehr. 2020. A hybrid particle swarm optimization and genetic algorithm for truss structures with discrete variables. Journal of Applied and Computational Mechanics 6(3): 593–604.

[169] Song, Y., F. Zhang and C. Liu. 2020. The risk of block chain financial market based on particle swarm optimization. Journal of Computational and Applied Mathematics 370: 112667.

[170] Bonnah, E., S. Ju and W. Cai. 2020. Coverage maximization in wireless sensor networks using minimal exposure path and particle swarm optimization. Sensing and Imaging 21(1): 4.

[171] Storn, R. and K. Price. 1997. Differential evolution–a simple and efficient heuristic for global optimization over continuous spaces. Journal of Global Optimization 11(4): 341–359.

[172] Kennedy, J. and R, Eberhart. 1995. Particle swarm optimization. In Proceedings of ICNN'95-International Conference on Neural Networks, volume 4, pp. 1942–1948. IEEE.

[173] Hegger, R., H. Kantz and T. Schreiber. 1999. Practical implementation of nonlinear time series methods: The TISEAN package. Chaos: An Interdisciplinary Journal of Nonlinear Science 9(2): 413–435.

[174] Wang, Z., C. Volos, S.T. Kingni, A.T. Azar and V.-T. Pham. 2017. Four-wing attractors in a novel chaotic system with hyperbolic sine nonlinearity. Optik 131: 1071–1078.

[175] Muñoz-Pacheco, J.M., D.K. Guevara-Flores, O.G. Félix-Beltrán, E. Tlelo-Cuautle, J.E. Barradas-Guevara and C.K. Volos. 2018. Experimental verification of optimized multiscroll chaotic oscillators based on irregular saturated functions. Complexity, 2018.

[176] Lu, J., G. Chen, X. Yu and H. Leung. 2004. Design and analysis of multiscroll chaotic attractors from saturated function series. IEEE Transactions on Circuits and Systems I: Regular Papers 51(12): 2476–2490, December 2004.

[177] Muñoz-Pacheco, J.M. and E. Tlelo Cuautle. 2010. Electronic design automation of multi-scroll chaos generators. Bentham Science Publishers.

[178] Abraham, A. and L. Jain. 2005. Evolutionary multiobjective optimization. In Evolutionary Multiobjective Optimization, pp. 1–6. Springer.

[179] Danca, M.F. and N. Kuznetsov. 2018. MATLAB code for Lyapunov exponents of fractional-order systems. International Journal of Bifurcation and Chaos 28(05): 1850067.

[180] Tlelo-Cuautle, E., L.G. de la Fraga and J. Rangel-Magdaleno. 2016. Engineering Applications of FPGAs. Springer.

[181] Boccaletti, S., J. Kurths, G. Osipov, D.L. Valladares and C.S. Zhou. 2002. The synchronization of chaotic systems. Physics Reports 366(1-2): 1–101.

[182] Pecora, L.M. and T.L. Carroll. 1990. Synchronization in chaotic systems. Physical Review Letters 64(8): 821.

[183] Pecora, L.M. and T.L. Carroll. 1991. Synchronizing chaotic circuits. IEEE Trans. Circ. Sys. 38: 453–456.

[184] Sira-Ramirez, H. and C. Cruz-Hernández. 2001. Synchronization of chaotic systems: A generalized hamiltonian systems approach. International Journal of Bifurcation and Chaos 11(05): 1381–1395.

[185] Lerescu, A.I., N. Constandache, S. Oancea and I. Grosu. 2004. Collection of master–slave synchronized chaotic systems. Chaos, Solitons & Fractals 22(3): 599–604.

[186] Vaidyanathan, S., S. Sampath and A.T. Azar. 2015. Global chaos synchronisation of identical chaotic systems via novel sliding mode control method and its application to Zhu system. International Journal of Modelling, Identification and Control 23(1): 92–100.

[187] Chen, X., J.H. Park, J. Cao and J. Qiu. 2017. Sliding mode synchronization of multiple chaotic systems with uncertainties and disturbances. Applied Mathematics and Computation 308: 161–173.

[188] Rajagopal, K., G. Laarem, A. Karthikeyan and A. Srinivasan. 2017. FPGA implementation of adaptive sliding mode control and genetically optimized PID control for fractional-order induction motor system with uncertain load. Advances in Difference Equations 2017(1): 273.

[189] Nosrati, K., C. Volos and A. Azemi. 2017. Cubature Kalman filter-based chaotic synchronization and image encryption. Signal Processing: Image Communication 58: 35–48.

[190] Hussein Abd, M., F.R. Tahir, G.A. Al-Suhail and V.-T. Pham. 2017. An adaptive observer synchronization using chaotic time-delay system for secure communication. Nonlinear Dynamics 90(4): 2583–2598.

[191] Wang, Y., H.R. Karimi and H. Yan. 2018. An adaptive event-triggered synchronization approach for chaotic Lur's systems subject to aperiodic sampled data. IEEE Transactions on Circuits and Systems II: Express Briefs 66(3): 442–446.

[192] Vaidyanathan, S., C. Volos, V.-T. Pham and K. Madhavan. 2015. Analysis, adaptive control and synchronization of a novel 4-D hyperchaotic

hyperjerk system and its SPICE implementation. Archives of Control Sciences 25(1): 135–158.

[193] Li, Z. and D. Xu. 2004. A secure communication scheme using projective chaos synchronization. Chaos, Solitons & Fractals 22(2): 477–481.

[194] Boulkroune, A., A. Bouzeriba and T. Bouden. 2016. Fuzzy generalized projective synchronization of incommensurate fractional-order chaotic systems. Neurocomputing 173: 606–614.

[195] Vaidyanathan, S., A. Akgul, S. Kaçar, and U. Çavuşoğlu. 2018. A new 4-D chaotic hyperjerk system, its synchronization, circuit design and applications in rng, image encryption and chaos-based steganography. The European Physical Journal Plus 133(2): 46.

[196] Smaoui, N., M. Zribi and T. Elmokadem. 2017. A novel secure communication scheme based on the Karhunen–Loéve decomposition and the synchronization of hyperchaotic Lü systems. Nonlinear Dynamics 90(1): 271–285.

[197] Daltzis, P.A., C.K. Volos, H.E. Nistazakis, A.D. Tsigopoulos and G.S. Tombras. 2018. Analysis, synchronization and circuit design of a 4D hyperchaotic hyperjerk system. Computation 6(1): 14.

[198] Vaseghi, B., M.A. Pourmina and S. Mobayen. 2017. Secure communication in wireless sensor networks based on chaos synchronization using adaptive sliding mode control. Nonlinear Dynamics 89(3): 1689–1704.

[199] Atlee Jackson, E. and I. Grosu. 1995. An open-plus-closed-loop (OPCL) control of complex dynamic systems. Physica D: Nonlinear Phenomena 85(1-2): 1–9.

[200] López-Mancilla, D., G. López-Cahuich, C. Posadas-Castillo, C.E. Castañeda, J.H. García-López, J.L. Vázquez-Gutiérrez and E. Tlelo-Cuautle. 2019. Synchronization of complex networks of identical and nonidentical chaotic systems via model-matching control. Plos One 14(5): e0216349.

[201] Mahmoud, R., T. Yousuf, F. Aloul and I. Zualkernan. 2015. Internet of things (IoT) security: Current status, challenges and prospective measures. In 2015 10th International Conference for Internet Technology and Secured Transactions (ICITST), pp. 336–341. IEEE.

[202] Alrashdi, I., A. Alqazzaz, E. Aloufi, R. Alharthi, M. Zohdy and H. Ming. 2019. Ad-iot: Anomaly detection of iot cyberattacks in smart city using machine learning. In 2019 IEEE 9th Annual Computing and Communication Workshop and Conference (CCWC), pp. 0305–0310. IEEE.

[203] Rukhin, A., J. Soto, J. Nechvatal, M. Smid and E. Barker. 2001. A statistical test suite for random and pseudorandom number generators for cryptographic applications. Technical report, Booz-Allen and Hamilton Inc., McLean VA.

[204] L'Ecuyer, P. and R. Simard. 2007. TestU01: A library for empirical testing of random number generators. ACM Transactions on Mathematical Software (TOMS) 33(4): 1–40.

[205] García-Guerrero, E.E., E. Inzunza-González, O.R. López-Bonilla, J.R. Cárdenas-Valdez and E. Tlelo-Cuautle. 2020. Randomness improvement of chaotic maps for image encryption in a wireless communication scheme using PIC-microcontroller via Zigbee channels. Chaos, Solitons & Fractals 133: 109646.

Index

O

P

Q

R

S

Authors' Biographies

Esteban Tlelo-Cuautle received a B.Sc. degree from Instituto Tecnológico de Puebla (ITP) México in 1993. He then received both M.Sc. and Ph.D. degrees from Instituto Nacional de Astrofísica, Óptica y Electrónica (INAOE), México in 1995 and 2000, respectively. In 2001 he was appointed as Professor-Researcher at INAOE. He has authored 4 books, edited 11 books and around 300 works published in book chapters, international journals and conferences. He serves as Associate Editor in IEEE Transactions on Circuits and Systems I: Regular Papers, Engineering Applications of Artificial Intelligence, Electronics, Integration the VLSI Journal, PLOS ONE, and Frontiers of Information Technology and Electronics Engineering. His research interests include analog signal processing, synthesis and design of integrated circuits, optimization by metaheuristics, design and applications of chaotic systems, security in the internet of things, symbolic analysis and analog/RF and mixed-signal design automation tools.

Luis Gerardo de la Fraga received the BS degree in electrical engineering from the Veracruz Institute of Technology, in Veracruz, Mexico in 1992; he received the M.Sc. degree from the National Institute of Astrophysics, Optics, and Electronics (INAOE), Puebla, Mexico, in 1994; and the Ph.D. degree from the Autonomous University of Madrid, Spain, in 1998. He develop his predoctoral work in the National Center of Biotechnology (CNB) in Madrid, Spain. Since 2000 he has been with the Computer Science Department at the Center of Research and Advanced Studies (Cinvestav), in Mexico City. His research areas include computer vision, application of evolutionary algorithms, applied mathematics, and network security. He is very enthusiastic about open software and GNU/Linux systems. Dr. de la Fraga has published more than 30 articles in international journals, 6 book chapters, 1 book and more than 50 articles in international conferences. He had graduated 27 M.Sc. and 3 Ph.D. students. He is member of the Mexican Academy of Sciences, and Senior Member of the IEEE society. He is also member of ACM society.

Omar Guillén-Fernández received a Bachelor degree from Instituto Tecnológico de Veracruz in 2015, and the M.Sc. degree in Electronics from Instituto Nacional de Astrofísica, Óptica y Electrónica (INAOE) at México, in 2018. He is currently pursuing his Ph.D. degree at INAOE. He has co-authored one book, 2 book chapters, 6 journal papers and a couple of conference proceedings. His topics of interest are oriented towards chaotic systems, synchronization techniques, security, embedded systems, optimization and integrated circuit design and applications.

Alejandro Silva-Juárez received his Engineering degree on Electronics and his M.Sc. degree from Instituto Tecnológico de Puebla, México. He is pursuing his Ph.D. degree at Instituto Nacional de Astrofísica, Óptica y Electrónica (INAOE). He has co-authored one book, 3 journal papers and a couple of conference proceedings. His research interests include modeling, simulation and optimization of chaotic systems, synthesis of analog circuits and fractional-order chaotic systems and applications.